Entangled Itineraries

Entangled Itineraries

Materials, Practices, and Knowledges across Eurasia

Edited by **Pamela H. Smith**

University of Pittsburgh Press

Published by the University of Pittsburgh Press, Pittsburgh, Pa., 15260

Manufactured in the United States of America
Printed on acid-free paper
10 9 8 7 6 5 4 3 2 1

Cataloging-in-Publication data is available from the Library of Congress

ISBN 13: 978-0-8229-6577-0
ISBN 10: 0-8229-6577-1

Cover art:Rectangular world map from Fatimid treatise,
Book of Curiosities of the Sciences and Marvels for the Eyes.

Cover design: Melissa Dias-Mandoly

In memory of Ronald E. Smith (1931–2018),
whose love still journeys with me

Contents

Acknowledgments

This volume traveled a long and looping itinerary, wending its way from the Max Planck Institute for the History of Science in Berlin (MPIWG) around the globe, and back and forth between multiple scholarly hubs and New York City in a constant whirring (pinging?) of electronic files and emails, sometimes stalled, sometimes frenetic. First and foremost—and on behalf of all the participants and contributors to this volume—I offer wholehearted thanks to MPIWG, the origin point of that itinerary, and especially to Prof. Dr. Lorraine Daston who invited me to organize the working group on "Itineraries of Materials, Recipes, Techniques, and Knowledge in the Early Modern World." The initiative ran from 2013 to 2015, with two large workshops in March and July 2014 and a two-week writing session with most of the final contributors in July 2015.

At the MPIWG, Raine Daston created a node of intense activity in the history of science, which shaped all who came within its force field. Her meticulous attention to detail—human, material, and logistical—has been felt in her brilliant scholarship as well as in her ability to foster an extraordinary space of knowledge exchange and creation in Berlin. Thank you, Raine, for all your very hard work, expansive generosity, genuine wisdom, and depth of insight.

My deep gratitude also goes to Josephine Fenger, a remarkable hub of activity in her own right. Josephine took charge of organizing all the workshops and at the last minute obtained all images, rights, and permissions—an extremely complicated feat. Daria Coscodan, Regina Held, Nina Ludwig, Tanja Neuendorf, and Cathleen Paethe also assisted in myriad ways. The MPIWG is an astonishing place; long may it endure! At Columbia University, Shabnam Aslam and Rae Binstock helped with editing and formatting. Many more wonderful presentations were given and papers written than could appear in this volume, and I thank all the participants, including Sonja Brentjes, Harold Cook, Sven Dupré, Johan Elverskog, Asaf Goldschmidt, Harun Küçük, Shigehisa Kuriyama, Elaine Leong, Maria Mavroudi, Luca Mola, Carla Nappi, Hyunhee Park, Dhruv Raina, Claire Sable, Angela Schottenhammer, Tim Stanley, Mary Terrall, Francesca Trivellato, and Yulian Wu.

I have very much appreciated working with editor Abby Collier at the University of Pittsburgh Press, and I am especially grateful for her care in finding two anonymous readers who provided remarkably timely, generous, and knowledgeable readings and sage advice for every chapter. We are all very much in their debt, as well as to the excellent copyeditor, Pippa Letsky, for preventing many gaps and gaffes in this volume. Of course, we take responsibility for the remaining ones. We hope they provide not irritation but, rather, a spur to further scholarship on the itineraries of materials, objects, practices, people, and knowledge. Finally, I am enormously grateful to the contributors to this volume from whom I learned so much. Our intensive two weeks together in Berlin, discussing, eating, arguing, writing, and above all, educating each other, was an experience of deep and sustained colloquy that I hope every scholar, young and old, has the good fortune to enjoy at some point in life.

Entangled Itineraries

Part 1

Overview

Chapter 1

Nodes of Convergence, Material Complexes, and Entangled Itineraries

Pamela H. Smith

> The artifact, in short, is the crystallization of activity within a relational field.
>
> **Tim Ingold**

> Science is not just knowledge about matter; it is also knowledge that comes through matter.
>
> **Francesca Bray**

Some time ago global historians of science moved away from a model of knowledge-making in which a discretely bounded fixed corpus of knowledge from one part of the globe diffuses outward—transmitted from point to point, arriving whole or in parts, and replacing local knowledge systems. Instead, knowledge is now understood to be constituted in part by movement. In other words, the routes that materials, practices, and knowledge take can be more important than their roots or originary forms. This perspective informs the present volume, in which contributors follow the routes of materials, people, techniques, and practices (both esoteric and exoteric), ways of knowing, and codified knowledge systems across Eurasia, tracing their itineraries as they weave in and out of "nodes of convergence" or "relational fields." Through such movement, these materials and practices—and the knowledge systems that both give them meaning and in turn are formed by them—are transformed and constituted anew.

Nodes of Convergence

Tim Ingold's suggestive description (quoted above) of artefacts as the crystallization of activity within a relational field makes the same point: materials and objects are constructed by the material, social, economic, intellectual, and emotional contexts in which they find themselves—the "relational field." As these artifacts move, they do not remain stable, although perceived properties or characteristics might sometimes continue to be noted (such as the putrid smell of asafetida as detailed in Leung and Chen's chapter). Materials are not stable across distance and time—their perceived properties can be transformed, as demonstrated by asafetida, sometimes viewed as a remedy, sometimes a poison. Just as the uses to which they are put and their relative value—economic and cultural—can vary, so too can the methods by which they are investigated, and even their status as compelling objects of study and knowledge. Such meanings and significance of materials and objects transmuted—sometimes drastically—as they traveled into and through new relational fields. The chapters in this volume demonstrate how materials and things altered as they moved and were mobilized in new systems of production and in new regimes of attention and knowledge-making and as they were codified in tools, techniques of production, objects, and writing. Materials, techniques, recipes, objects, and books both carried and made knowledge as they moved. These itineraries of materials never constituted placid, smooth streams of cumulative knowledge-making; rather, they were nonlinear, sometimes swelling in one relational field, and shrinking to almost nothing and disappearing into a substratum in another, from which a technique might sometimes emerge and reenter the textual tradition.[1] In this volume, we examine several such "relational fields" into which and through which materials moved. Some of these fields formed in busy, crowded, cosmopolitan cities such as Chang'an and Quanzhou during the Tang period in China (treated by Leung and Chen in chapter 7), or the Central Asian crossroads of Kucha, a hub for the trade of materials that could only to be obtained in the Central Asian heartland along the Tarim basin and the area called Bactria by the Greeks (touched upon by Yoeli-Tlalim in chapter 3), or the nexus for European and South East Asian trade formed by the region of Malacca (chapter 5), or the important cities of Reyy and Baghdad, centers of Arabic scholarship written by Persian literati, the echoes of which resonated in Bursa in the Ottoman Empire (chapter 11), even in southern France (chapter 8), and in Assam during the British period (chapter 6). These hubs, nexuses, centers of exchange and calculation—places, or nodes, of convergence where people, materials, texts, objects, and practices came together—produced new sorts of objects (both material and intellectual), new techniques, knowledges, and epistemologies. Indeed, they produced new species of materials as well. The essays in this vol-

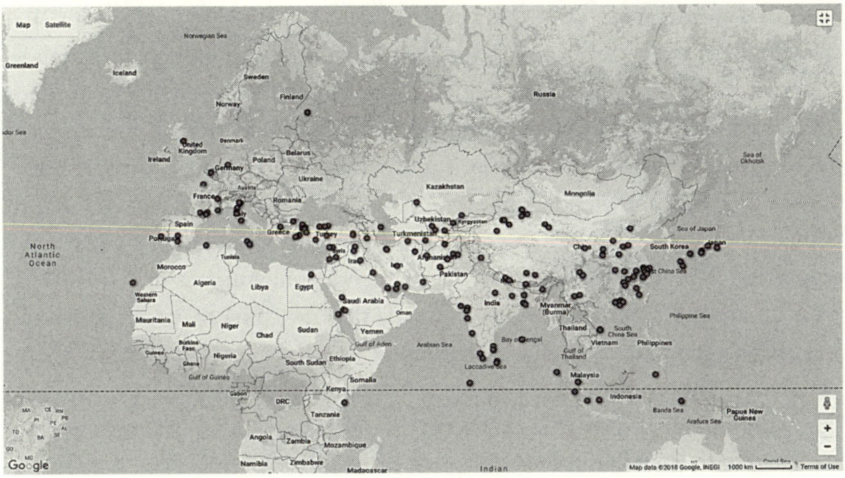

FIGURE 1.1. Google fusion map showing the hubs and nodes mentioned in this volume. All local place names have been included in the map.

ume follow materials such as hing/*awei*/asafedita (chapter 7); spiritual, medicinal, and iatrochemical practices (chapters 3, 5, 8, 9, 11, 12); objects such as a wooden skeleton (chapter 12) and inkstones (chapter 10); knowledge systems such as alchemy (chapter 11); and peoples, as well, into these relational fields and their nodes of convergence, often but not always constituted by diverse centers of trade, which, by their very nature as human crossroads, constituted complex fields of relations—economic, social, cultural, and material. Materials, practices, objects, and people enter or are drawn into these fields, where they are transformed and thrust thereby into new paths of circulation and movement (figure 1.1).

Human-Material Interactions and the Formation of Material Complexes

Francesca Bray, in the epigraph above, makes a profound point. The declarative and positive knowledge of modern science proclaimed by the new philosophers in sixteenth- and seventeenth-century Europe, which culminated in nineteenth- and twentieth-century notions of scientific modernity, consisted not just in "knowledge" in the guise of abstract theories about matter, but itself emerged *through* matter. In other words, the skilled techniques by which humans manipulated materials brought into being knowledge systems. Materials undertake complex itineraries as they are worked by humans. Humans have always interacted with their environment in a variety of ways for survival, and human communities have always manipulated natural materials—minerals, metals, plants, and animals, to name just a few—in order to produce effects and products, whether useful, spirit-filled, ornamental, or as models of the workings of nature. Out of the experience gained through these interactions, skills and knowledge emerge.[2] Through the reciprocal engagement of humans and materials that this working entails, humans come to know the properties of materials, which in turn engenders techniques and skills that grow up around these properties. Humans thereby come to assign value and meaning to the materials and their properties—they come to "know" them. "Knowing" always involves assigning value and meaning, but the principles of "knowing" are not always the same everywhere, as the history of epistemology has taught us. Social and cultural preoccupations vary, regimes of attention change, economic values fluctuate, which all transform the questions people ask and the methods they use to answer them. Thus, knowledge and epistemologies, too, are shaped by the itineraries that materials follow and the relational fields in which they are formed and employed.

We can thus trace the emergence of ideas, theories, and knowledge systems starting in the interaction of the human hand with the material world. This interaction with natural materials and the subsequent production of goods and knowledge emerge from very long durée systematic observation of patterns in nature, as well as long-term and sustained experimentation with natural substances. These engagements with the environment are often codified—sometimes in myths, songs, rituals, or mnemonic devices, and, while the manipulation of natural materials could be codified in texts, it was more often—and perhaps more effectively—transmitted, not in texts, but in persons, from master to apprentice, or via rituals and the spoken word. The process of "coming to know" through the manipulation and trading of materials is an always complex and often partially obscured process through which people, materials, techniques, theories, and ideas move across geographic and epistemic space to form "amalgams" or "assemblages"—the systems of knowledge that include materials, people, practices, and ideas—that in this volume we call "material complexes."[3] Conceiving of the movement of these material complexes in a staged way—as an "itinerary" through which the material complex forms—helps us trace the interaction among materials, human making, and the formation of knowledge systems over long spans of distance and time.

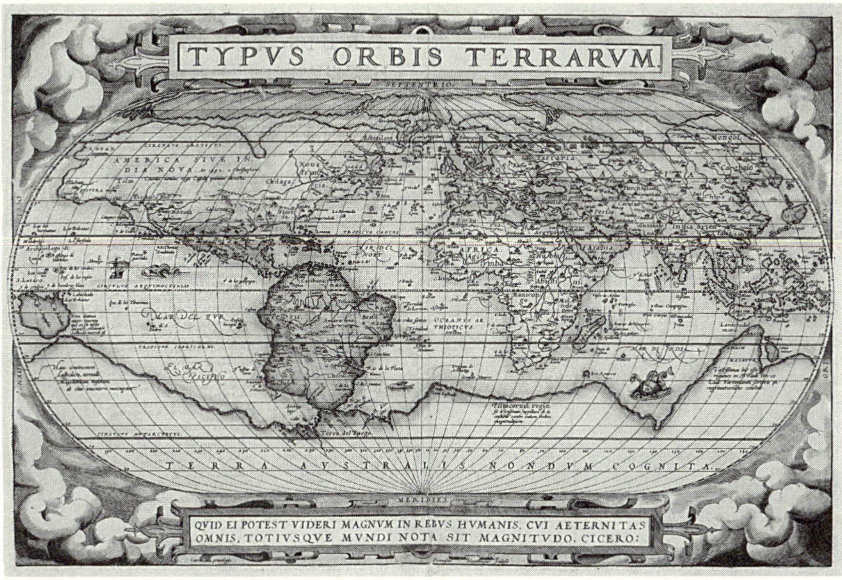

FIGURE 1.2. Abraham Ortelius, *Theatrum Orbis Terrarum* (Antwerp, 1570). Public domain.

Entangled Itineraries

"Itinerary" can connote linear movement, but the itineraries traced in this volume are looping, circuitous, and sometimes circulatory, returning again to the same relational nodes, to be transformed once more. It would be misleading to imagine these routes as threading across planar space, in the way we have been taught to see it since the fifteenth- and sixteenth-century flat projections constructed by European scholars and artisans (figure 1.2) and to imagine the movement of materials and ideas following linear, flattened itineraries, without boundaries, hindrances, filters, or blockages. We could also think of these staged routes as like those mapped out by the coastal place names in portolans (figure 1.3), and by many other diverse itinerary maps, such as an eleventh-century Egyptian map from the *Book of Curiosities* (figure 1.4); or the 1628 version of the *Wubei Zhi* chart of Zheng He showing his fifteenth-century voyages (figure 1.5), or the thirteen-meter-long "Routen-rolle" recording Duke August of Saxony's journey in 1565, measured with an odometer and compass (held by the Sächsische Landes-Bibliothek), and even on our modern Google "Route finders."

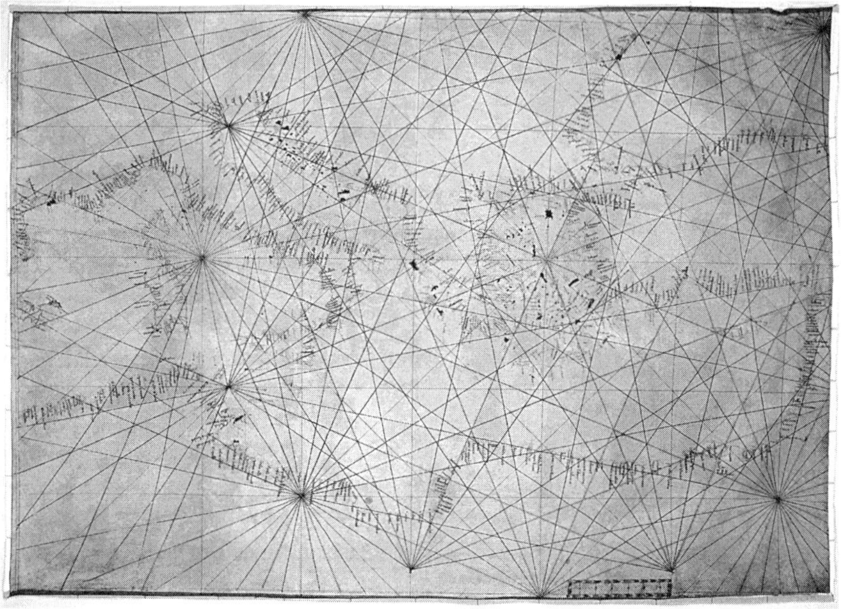

FIGURE 1.3. Nautical portolan of the Mediterranean (fourteenth century). 43 x 59 cm. Names of coastal towns form the itinerary-style mapping, all set within points of the compass and directions of the winds. Geography and Map Division, Library of Congress. Public domain.

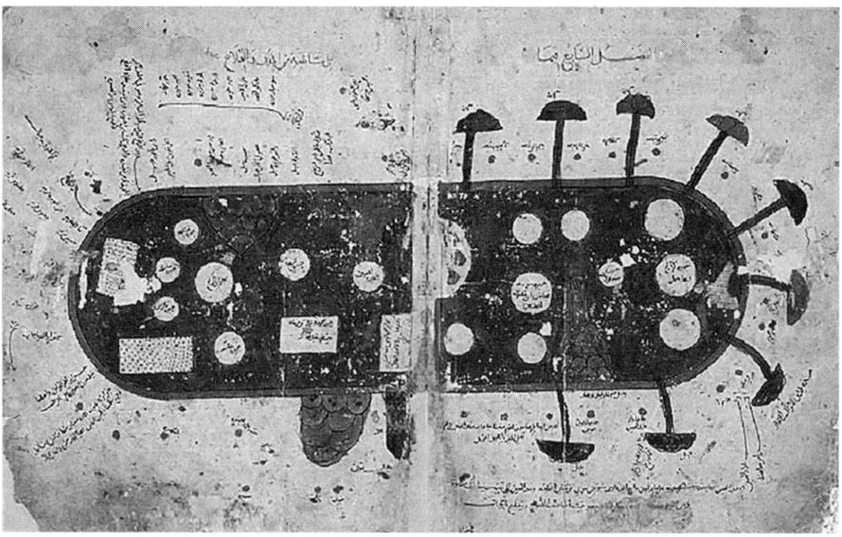

FIGURE 1.4. *Kitāb Gharā'ib al-funūn wa-mulah al-'uyūn* (The book of curiosities of the sciences and marvels for the eyes; Egypt, eleventh century), Map of the Indian Ocean: Book 2, Chapter 7: "On the cities and forts along the shore [of the Indian Ocean]" (MS. Arab. c. 90, fols. 29b-30a). © Bodleian Library.

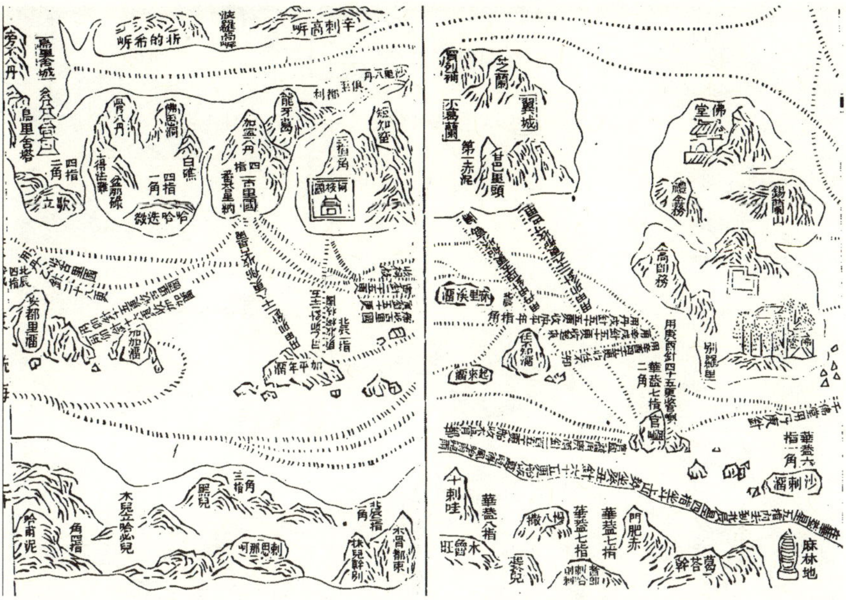

FIGURE 1.5. *Wubei Zhi* map, an ocean travel chart (1628) believed to be based on Zheng He's expeditions (1405–1433). Wikimedia, public domain.

But even such itinerary maps are flat, linear, and inert, and instead we must imagine their space alive with human beings acting in multiple communities and urban hubs, along sometimes intersecting, sometimes diverging routes—always replete with contingencies—with the traces of other systems of knowing, pieces of manuscripts and texts carried along to alight on new shores in fragments to be understood partially and translated into new languages and systems of knowing. These convoluted, collapsing, and crisscrossing routes perhaps resemble more a woven textile, or a textual palimpsest, in their intercalated, interwoven character. But textile and paper may also be too flat to make clear that the itineraries in this volume have peaks, hanging valleys, even cul-de-sacs. In the discussions that gave rise to this volume, the contributors came to see the necessity of devising a sort of three- or four-dimensional itinerary map, rather like Matthew of Paris's map (figure 1.6) when he gets close to the Holy Land (figure 1.7) and seems to need to amplify the place names with their spiritual significance, or the 1364 Japanese map of the Five Regions of India (*Go-Tenjiku zu*), which includes place names but also attempts to map the spiritual hubs and nodes of India based on the travel records of the Chinese monk Xuanzang (figure 1.8a–b). Our own attempt to plot the hubs and nodes of each chapter's itineraries on a Google Fusion

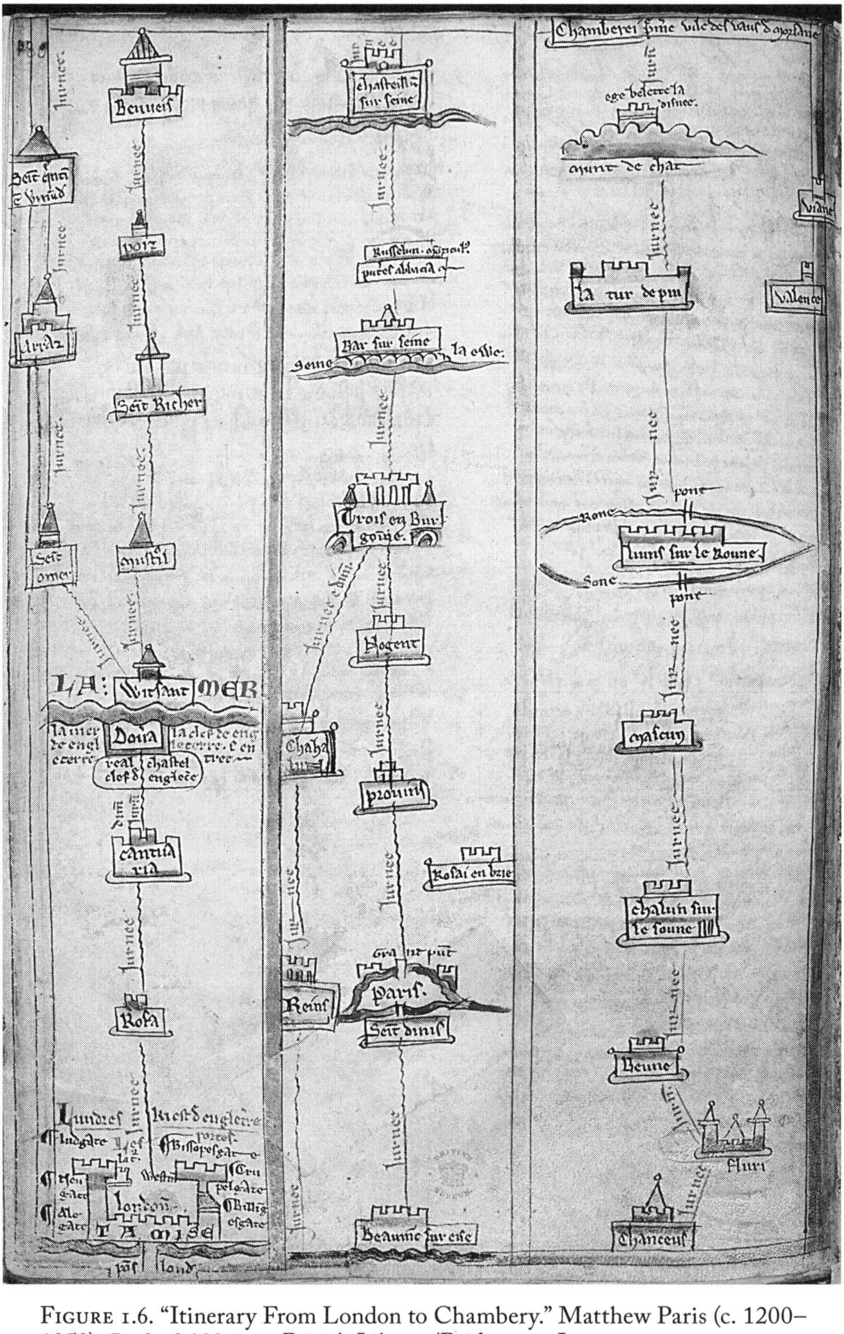

Figure 1.6. "Itinerary From London to Chambery." Matthew Paris (c. 1200–1259), *Book of Additions*. British Library/Bridgeman Images.

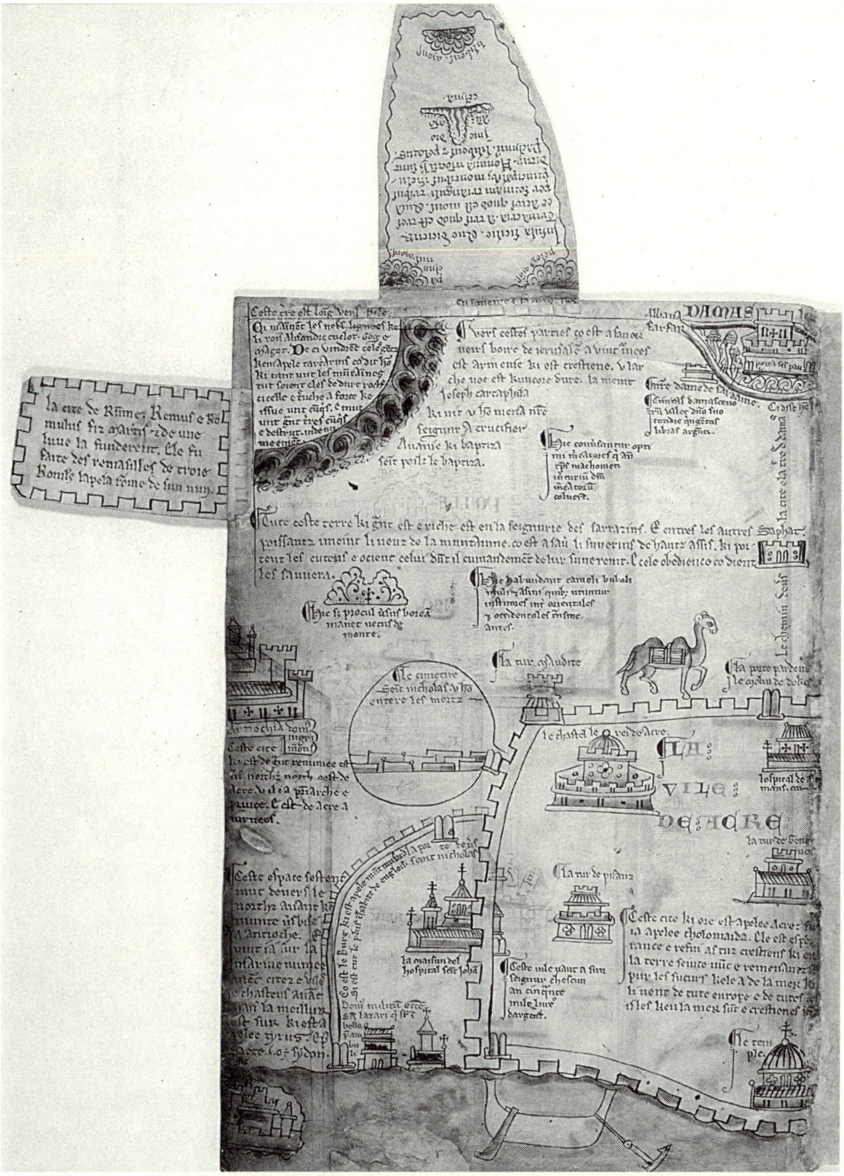

FIGURE 1.7. "Acre and the Holy Land." Matthew Paris (c. 1200–1259), *Book of Additions*. British Library/Bridgeman Images.

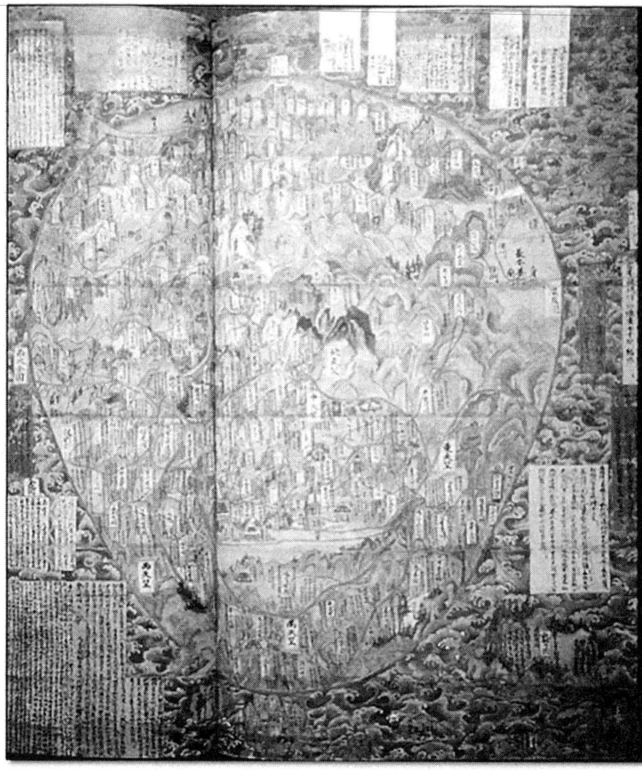

FIGURE 1.8A. *Go-Tenjiku zu* (Map of the Five Regions of India), 177 x 166.5 cm (1364). Lake Anotatta and the four great rivers are shown toward the top of the center. The entries in the boxes in the ocean portion of the map are extracts from the *Da Tang xiyu ji*. Horyuji Temple, Nara, Japan. Fair use.

table was of course far too flat (figure 1.1), while the visual complexity of the *Go-Tenjiku zu* map seemed illegible; thus, in the end, we settled for the power of written historical narrative to describe the hubs and nodes of each chapter's relational fields.

The essays in this volume contain both case studies of entangled itineraries of knowledge-making about specific materials and practices and methodological considerations of the mechanics of such knowledge-making—mechanics that involved the accretion of "material complexes" over long spans of distance and time. The authors are specialists of quite different fields of scholarship, and as we have learned from other collaborations, this has proved immensely productive, allowing the group to cross the boundaries of their individual fields to see in stark relief the "centrisms" of their own fields,

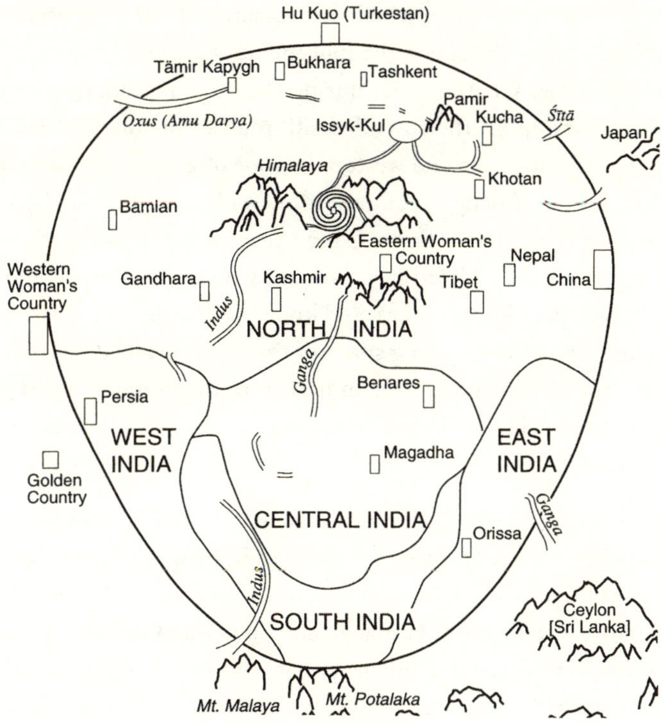

Figure 1.8b. Schematic rendering of the Horyuji *Go-Tenjiku zu* by Unno Kazutaka. "Cartography in Japan," 373.

which include taken-for-granted vocabularies, categories, and chronologies of periodization.

Place

Much recent work in the history of science has focused on the circulation of knowledge within Europe, across the Atlantic World, and between the two poles of East Asia and western Europe. Recent research has also begun to uncover the knowledge networks of East Asia, and, in the new "thalassology" with the focus on oceans, historians have begun to examine the circulation of knowledge around the Mediterranean and the Indian Ocean.[4] This scholarship has resulted in much new information about circulation, exchange, and the transformations and translations of knowledge as well as new conceptual and methodological perspectives. For historians of science, technology, and medicine, this work has changed conceptualizations about circulation, about the interrelationship of local and global, and about the formation and constitution of scientific knowledge. Much less researched—in part because of the nature of the sources—has been the movement of knowledge across Eurasia, and especially across Central Asia, during the same period, despite the long history of scholarship on the so-called silk routes.[5] Scholarship on the relationship between East Asia and western Europe often takes no account of the space in between, as if people, commodities, books, and information traveled seamlessly across this space. This volume began in a series of workshops that sought to explore this neglected space, focusing on the movement and circulation of materials, peoples, practices, and knowledge across the Eurasian continent, in some cases over very long spans of time and distance. Our inspiration to give attention to this space came in part from the vision mapped out some time ago by Andre Gunder Frank in his short work, *The Centrality of Central Asia* in which he noted:

> Central Asia is also central to the civilizations of the outlying peoples. . . . It is not clear where civilized peoples and spaces end, and where they interpenetrate with those of Central Asia. None of the civilizations are pristine. All of them were formed and even defined through interaction with Central Asia. Moreover, Central Asia is where all the outlying peoples and their civilizations connected and interacted with each other. Indeed, for millennia the Pulse of Asia (Huntington 1907) probably came from its Central Asian heartbeat. Central Asia is truly the "missing link" in Eurasian and world history. . . . Yet Central Asia is perhaps both the most important and the most neglected part of the world and its history.[6]

It could be argued that seagoing itineraries linking the Mediterranean Sea to the coastal regions of the Indian Ocean world are similarly neglected despite the fact that textual sources such as *Periplus of the Erythraean Sea*, a Greco-Roman work from the first century CE, and archaeological evidence from Arikamedu in southern India and even from the coastal regions of China indicate the movement of objects over vast distances ferried by boats and ships. Like traversing the varied and often difficult terrain of Central Asia, crossing this maritime space required distinct types of knowledge ranging from shipbuilding technologies to understanding patterns of monsoon winds.

Our aim to trace itineraries across the lands and waters of Eurasia raised many challenges: We found that the span of the itineraries expanded as we followed them. The period most interesting to us when we began our work together encompassed the period from about 1350 to about 1750 ("late medieval/early modern," as it is called by some) because of its importance in global history, as population and commerce began to burgeon and economic integration of the globe, especially with regard to the trade in precious metals and luxury commodities across long-distance commercial networks began its long development. From 1000 CE the lands across Eurasia had witnessed many shared developments: increased conflicts between nomads and settled urban dwellers as populations grew and cultivated land use expanded. With increased domestic animal husbandry and settled urban living, pandemics spread across the whole of Eurasia, the "Black Death" only one of them. During the early modern period, the lands of Eurasia also shared the bloody advent of gunpowder warfare, just as they had shared the new kinds of wounds and killing that bronze weaponry had first occasioned millennia before. This new warfare contributed to a greater consolidation of ruling power in the hands of fewer centers, whose growing numbers of bureaucratic functionaries began to harbor ideas of universal rule. At the same time, a remarkable vernacularization of canonical textual corpora took place, and millennial expectations flourished in the major religions around 1500. Of course, trade had flowed across Eurasia, around the Indian Ocean, and over the Mediterranean for thousands of years, but in this late medieval/early modern period, larger parts of the globe became connected by an increasing variety of sea and land routes. Commodities and tribute bounced and jostled over these routes, and knowledge was accreted along them. Embodied knowledge moved in individuals as they migrated or were resettled in new territories, and it moved along with sailors, soldiers, and merchants as they pursued trade and war. Knowledge traveled in objects, instruments, manuscripts, and printed books as trade routes opened up and collectors avidly sought rare and beautiful things, and it moved as factors sent back specimens and information to the metropolis. It moved as new institutions of rule, such as the remarkable civil bureaucracy of the Song dynasty, the Pax Mongolica, the "Mandalic" polities of mainland Southeast Asia and their coastal sultanates who dominated the trade routes of the maritime "silk road," as well as Italian merchants, and then the European trading companies and colonial administrations all created new configurations of social and intellectual relational fields—and in some cases even sponsored information-gathering projects. Materials, objects, and knowledge moved within and among these fields not just geographically but also epistemically, as knowledge systems of different social and cultural groups intersected. Knowledge systems formed and transformed, were filtered and sometimes disappeared, as goods and people traveled from local settings and vernacular modes of expression—such as the ships of many realms, the ceramics manufactories in Jingdezhen and Puebla, shipbuilding arsenals in Calcutta, Venice, Ragusa, and Istanbul, artisanal workshops, collectors' cabinets and gardens in all parts of Eurasia, to name only a few examples at random—to the knowledge and written forms of *Bencao* (materia medica) and *pulu* texts (catalogues of various things) or into the codifications of astronomy and astrology produced, for example, by the Mongols as they absorbed Chinese scholarship or by Rashīd al-Dīn (1247–1318) as he surveyed the regions and practices surrounding him from his center at the Ilkanid court. These itineraries helped to shape at opposite ends of Eurasia the new epistemologies of *kaozheng* (evidential studies) and the new experimental philosophy in Europe. At the same time that this motion brought about the making of new knowledge, it also formed new hierarchies of intellectual authority.

While this late medieval/early modern period remains at the heart of this volume, we came to see that such a bounded period was inadequate, for, when we set out to follow itineraries of moving materials across Eurasia, we discovered that these materials possessed much longer histories, and it seemed artificial to contain the span of the volume by arbitrary chronological points (especially as those points often derived from Eurocentric historiography). This does not mean that we think of this long history of continuous movement around Eurasia as one placidly flowing continuous stream without major ruptures and discontinuities. On the contrary, the formation of new social and political groupings as people and empires moved often caused the meaning, indeed even the substances, of materials to be transformed, knowledge to be accreted or lost, regimes of attention to change, and what constituted "knowing" a material or thing (i.e., epistemology) to be reoriented. We think that our approach of following materials where they lead may point the way to new modes of periodizing global historical narratives. For example, the rise and decline of vibrant Central Asian urban centers from prehistoric times up to the present—crucial for the history of flows of materials across Eurasia—simply does not allow the telling of a linear narrative, much less a progressivist one (and includes sharp ruptures). This comes across clearly when one considers the ups and downs of, for example, the cities of Merv, Baghdad, and Reyy. The largest city in the world in the twelfth century, Merv depended on trade in high value materials, such as lapis lazuli. At its height, Merv had twelve thousand hydraulic workers laboring on its sophisticated system of waterworks.[7] A narrative that follows the object of historical inquiry—whether

FIGURE 1.9. Orthographic projection of Eurasia, showing the landmass. Creative Commons Attribution-Share Alike 3.0 Unported license. Public domain.

it is a material, an object, a knowledge system, or an idea—in and out of nodes of convergence can foster a new mode of periodization. This mode would not strive so much to mark out clearly delineated blocs of time as to focus on transformative nodes at particular moments, or over long temporal spans, to show how their relational fields—material, social, economic, infrastructural, intellectual—brought about transformation.[8] In the same vein, a focus on the itineraries of material complexes through nodes and hubs rather than along linear and cumulative trajectories can show how attention (scholarly or economic), activity (intellectual or utilitarian), and knowledge (scientific or spiritual) swelled and receded around these complexes in different relational fields.[9] The chapters in this volume focus on a diverse set of such material complexes, but they all aim to provide such narratives.

Distance

Another challenge confronting the contributors to this volume is the expansive spatial dimension of the area of study. Although we are fully aware that the concept of Eurasia itself has a history, we have chosen to approach this space simply as the largest continuous landmass and continent—covering around 55,000,000 square kilometers (21,000,000 square miles), or just over one-third of the Earth's total land area (figure 1.9). The division of Europe and Asia into two different continents is a historical and cultural construct. No clear natural or physical boundary separates them, of course, thus the entire span is of interest to the contributors in this volume because it formed the space for regular paths of human exchange—both by sea and by land—over the very longue durée.

When the contributors to this volume came together in Berlin at the Max Planck Institute for the History of Science (MPIWG), we had initially set out to investigate the space of Eurasia strictly defined, but it soon became clear that this continental approach was simultaneously too vast and too restrictive—too vast in its temporal and geographical distances, as well as in the confusion of the various periodizing schemas imposed upon it, and too restrictive in the sense that the itineraries of materials and ideas flowing across "Eurasia" often moved through South Asia and Africa as well. This led us to seek out a methodology by which we could take on the vastness of time and (nearly global) space and thus gave rise both to the contributors' focus on individual cases and to the volume's framework of "entangled itineraries," "material complexes," and "nodes of convergence." As a result, our focus shifted from tracing itineraries across a space strictly defined as "Eurasia" to a kind of road test of the method of tracing the routes of material complexes over long temporal and geographical spaces that inevitably encompassed more than Central Asia and Eurasia. While most essays in this volume simultaneously make methodological points and form case studies of material complexes in motion, those in Part 2, "Entangled Itineraries: Modes of Approach," are explicitly methodological. The essays in Part 3, "Material Complexes in Motion," are, by and large, case studies of the itineraries of materials and objects, and those in Part 4, "Convergences and the Emergence of New Objects of Knowledge," trace the formation and transformation of new objects of knowledge within changing relational fields.

Another challenge for the volume's authors is the dynamism of itineraries—our materials, recipes, practices, and people are moving, and not just over space. For, to take just one of these, materials are not necessarily stable things but, rather, are nexuses in knowledge systems around which meaning and practice cohere and agglomerate. When historical sources talk about rhubarb, for example, what are they talking about—a plant, a medicine, a food, or a spiritual practice (or all of these simultaneously)? Relational fields transformed knowledge systems and their material complexes. A central example of this can be found in materia medica and medicinal substances—particularly interesting materials to follow because they moved over long distances and across distant expanses of social stratifications, and they functioned simultaneously as desirable items of medicine, religion, and trade (chapters 3, 5, 7, 8). They thus formed one important component of the complex of materials, practices, and knowledge that formed and reformed around the vexed relationship of body and spirit—a relationship that was a locus of intensive conceptual and practical activity across the entire expanse of Eurasia.

Historical Sources

A final challenge confronted by the contributors to the volume is sources. Because they investigate the movement of materials, peoples, and practices (all carriers of components and types of knowledge that do not necessarily take a written form), the methodological and evidentiary challenges are multiple. When following nontextual materials and processes, a first question must be that of how materials, recipes, and techniques function as knowledge? It is one of the efforts of the contributors to this volume to trace the passage of matter and materials through practices of making and using into the realm of ideas and scientific theories, and to understand how the epistemic role of materials change and are sometimes stabilized as objects of intellectual and practical focus en route.

All historians rely upon material evidence that is already preselected for them both by archival intention and also, of course, very much by chance. The contributors in this volume had to come to terms with how much we historians, like archaeologists and art historians, rely upon the evidence of the elite, such as those things that have ended up as museum objects, as well as high-value portable things that traveled especially easily or well—textiles and spices, as well as the things that by their nature (some medicinals, for example) transcended local settings and temporalities. Like all historians, we saw how much we must always rely on words and texts to tell us about materials and things.

An additional problem of working on Eurasia is the dearth of source material that is easily accessible to scholars today in centers and institutions of scholarship. This was compounded for the contributors of this volume by the fact that the movement of our objects of study often left few traces of how their movement from one place to another was negotiated, mediated, or translated; instead, they suddenly appeared in a new place in a written record or as an artefact. Only rarely do we get a glimpse of the process of translation, such as that provided by the Khotanese translation of the Sanskrit *Siddhasāra* recounted in Yoeli-Tlalim's chapter 3. Indeed, most of the itineraries of this volume include silences and lacunae, where once a babble of different languages and a hum of human production and activity must have resounded. A vital part of recovering Eurasian transmissions of knowledge lies in a multitude of forgotten languages: languages either forgotten by mainstream history of science, such as Syriac or Uighur, or dead languages, only recovered and deciphered as a result of the archaeological digs in Central Asia of the early twentieth century, such as Khotanese or the two Tocharian languages (Agnean, also known as Tocharian A, and Kuchean, or Tocharian B). Sources in these languages—along with sources in Arabic, Persian, Turkish, Hebrew, Sanskrit, Tibetan, Manchu, Tangut, and Chinese—hold the promise of allowing us to write multicentric "Eurasian" histories of science. Such histories will always require multiple linguistic and historiographical expertises, and, although the challenges are by no means overcome by the contributors to this volume (especially as the MPIWG meetings shifted from a sole focus on "Eurasia" to concentrate instead on the methodology of tracing "entangled itineraries"), we hope the volume nevertheless forms a stepping stone along the way to realizing such histories.

The essays in this volume give insight into both the mechanics of knowledge-making and the content of the knowledge that gets made or "stabilized" in the process. One principal dimension of the mechanics of knowledge-making identified by the contributors was in the "flow" of materials, objects, techniques, and knowledge—as pointed out by Ko in chapter 10, asking what are the constraints and affordances of this flow? What are the causes of "fluid" or "sticky" flow? Are images less "sticky" than concepts, for example? Do they move more easily? Do materials and techniques flow more easily as "cultural stowaways"—as part of larger systems of meaning, such as religion? How might this transform them? As Marcia Norton points out in her examination of the introduction of chocolate into Europe, the place of this substance within the complex of religious meaning made chocolate both stickier and more slippery, as European churchmen struggled with and appropriated its meaning.[10]

Another recurring question concerning the mechanics of movement was why some objects and materials were stripped of their religious or cultural value or their context as they moved from one place to another, as treated by Alberts in chapter 5.[11] Did this stripping of cultural value turn these into "thin things" that can be contrasted to "thick things," as Harold Cook argues in *Matters of Exchange*. Things might not be stripped so much as filtered, as Thomas Allsen shows in the case of the Mongols.[12] Finally, the contributors took into account the perhaps obvious relational fields formed by large-scale structuring factors, such as states and empires, religious institutions, as well as manufacturing and production sequences, and even families and lineages, in their examinations of the relational fields of the hubs and nodes along the itineraries (chapters 5, 6, 9, 12).

The products and contents of the knowledge made by this movement includes materials of all kinds, such as commodities, medicines, products of industry, as well as objects of art, religious ritual, and household use; but also personae, professions, and specialists emerge and "are made" as a result of these material itineraries, as do these specialists' modes of proof, authority, and "knowledge that works." Regimes of attention and new epistemologies form around these materials and the specialists they bring into being. This specialist knowledge and specialist epistemologies are codified in written texts but also in ritual and workshop practices, in objects and tools, in oral and bodily systems of operation and knowledge, and in oral modes of expression—poetry, song, epics, and tales.

Our collective investigations in this volume illuminate how knowledge systems, epistemologies, objects, texts, materials, and practices that might appear to be stable or "natural" are, in fact, agglomerations accreted and transformed over time and space. They thus reveal the importance of tracing the routes over which they traveled. When we see how far in space and time the components of these agglomerations have traveled (across Eurasia and around the globe), it is puzzling to reflect on how historians could ever have imagined the fortunes and formation of knowledge systems as a story of uncovering the origins of recognizably stable bodies of knowledge. It reminds us once again just how powerful a tool historiography—and origin myths—have been in the formation of identity, whether ethnic, national, imperial, or epistemic.[13] Historiography, too, plays a role in the dynamics of relational fields, our own included. The "Eurasian histories" of this volume help reveal the multiple sources and the diverse and entangled itineraries from which materials and knowledges emerge.

Chapter 2

Trans-Eurasian Routes of Exchange

A Brief Historical Overview

Tansen Sen and Pamela H. Smith

Long-term and long-distance flows of goods, knowledge, texts, practices, and peoples (material complexes) moved back and forth across trans-Eurasian paths of exchange—the "Afro-Eurasian ecumene," as Ptolemy conceived of this expanse, a conception followed in the visualizations of Eurasian mapmakers such as Muhammed al-Idrisi (c. 1100–1165), in his map made in service of Roger II of Sicily in 1154 (figure 2.1), and fifteenth-century German cartographers (figure 2.2).[1] From at least the second millennium BCE, the Eurasian world was integrated through the movement of peoples and the diffusion of ideas and goods. This integration took place through the routes taken by materials, practices, and knowledge, and such routes can be more important than their roots. In the same vein, Arjun Appadurai has argued that "bumps and blocks, disjunctures and differences are produced by the variety of circuits, scales and speeds which characterize the circulation of cultural elements."[2]

Appadurai explains this phenomenon through the frameworks of "the circulation of forms and the forms of circulation." By "forms," Appadurai indicates "a family of phenomena, including styles, techniques, or genres, which can be inhabited by specific voices, contents, messages, and materials." This "circulation of forms," he contends, "produces new and distinct genre experiments, many of which are forced to coexist in uneven and uneasy combinations." Associated with this circulation of forms is what Appadurai calls the "work of the imagination," what "humans do as they strive to extend their chances of survival, improve their horizons of possibility, and increase their wealth and security." Moreover, these activities also produce "locality." The local "was not just an inverted canvas on which the global was written, but . . .

FIGURE 2.1. Al-Idrîsî, Nuzhat al-mushtaq fî ikhtirâq al-âfâq, map of the world, from a copy made in 1456 by 'Alî ibn Hasan al-Hûfî al-Qâsimî in Cairo and now preserved at the Bodleian Library, Oxford, as MS. Pococke 375, fols. 3v–4. Also called the Kitab Rujar (Book of Roger of Sicily; the Tabula Rogeriana; 1154). Public domain.

itself was a product of incessant effort."[3] With regard to the relationship between the global and the local, Appadurai suggests we "accept that the global is not merely the accidental site of the fusion or confusion of circulating global elements. It is the site of the mutual transformation of circulating forms." These transformations, he explains, take place through the "work of imagination," which in turn produces locality.[4] Although Appadurai's study relates to issues of contemporary globalization and the resulting connections and circulations, the frameworks of the "circulation of forms" and the "forms

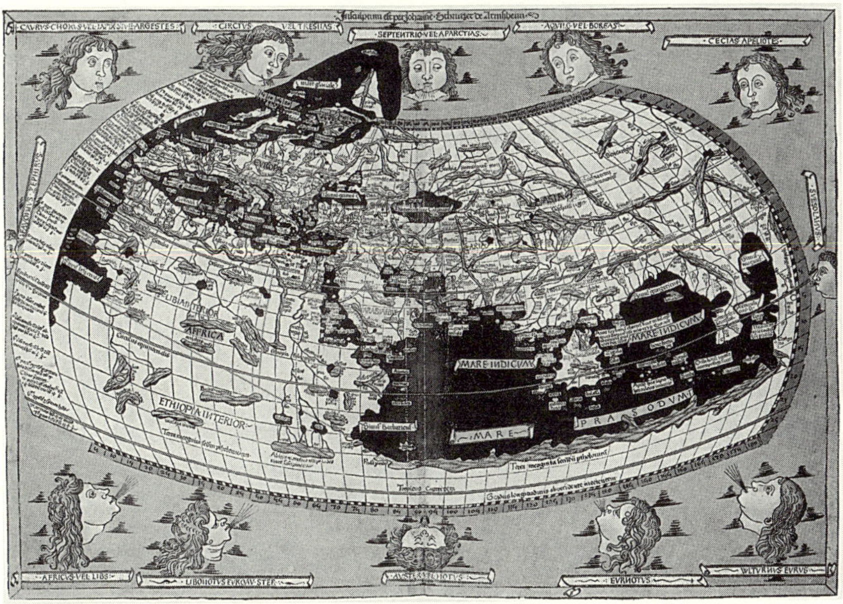

FIGURE 2.2. Claudius Ptolemy, *Oecumene,* Johannes Schnitzer, engraver (1482). Public domain.

of circulation" could, as he acknowledges, be applied to the earlier phases of cross-cultural connections.

Two other conceptual frameworks are useful in understanding the movement of materials across Eurasia and the integration this achieved. Like Appadurai, Homi Bhaba underscores the significance of disjunctures and differences during circulations, but he also stresses the importance of convergence: "Circulation takes a measure of mobility—the movement of languages, ideas, meanings, cultural forms, social systems—as it converges in specific and singular spaces of representation negotiated through a dialogue of difference. Incommensurable customs, disjunctive symbolic structures, itineraries that are diverse and yet proximate, continuities that become contingent over time—these dis-proportionate convergences generate an energy of interdisciplinary circulation." Convergence, for Bhaba, "is not about a practice or project as an end in itself, even if that end is an entangled encounter of diverse thoughtways and institutional intersections. The aim of convergence as critique is to track the spatial and temporal territories that open up within, and through, the act of circulation. The iterative dynamics of circulation and convergence reveal lateral meanings and interstitial spaces produced *in transit.*"[5] Bhaba's emphasis on the "act of circulation" emphasizes again the importance of following itineraries. In this he agrees with the notion of circulatory history put forth by Prasenjit Duara. Circulatory history, according to Duara, is a mode "in which ideas, practices and texts enter society or locale as one kind of thing and emerge from it considerably transformed to travel elsewhere even as it refers back, often narratively, to the initiating moment."[6]

As these scholars make clear, itineraries should be conceptualized as neither random nor unidirectional. Rather, already at an early date, several regions of the Eurasian world were connected through the networks of exchange that entailed deliberate, repeated, and multidirectional movements. Below are some examples of people, objects, and ideas that traveled across Eurasia and contributed to the integration of the geographical spaces and terrains, triggered various forms of circulations, and led to instances of transformation and convergence.

Early Eurasian Connections

Several millennia before the Common Era, the seafaring community of Austronesians and the nomadic tribes who inhabited the Eurasian steppe zone connected far-flung areas (including parts of Africa) through their networks of commodity exchange and seasonal migrations.[7] While the former group, originating from the South China Sea region, used their canoes to travel across the Indian Ocean and introduced new styles of pottery-making into places such as southern India, the Indo-European nomads are credited with the circulation of expertise associated with bronze-making, the use of composite bows, horse riding, and iron stirrups.[8] The movements and contributions of Austronesians and Indo-Europeans clearly indicate the shortcomings of conceptualizing or tracking the itineraries of peoples, goods, and ideas based on either continental designations or political boundaries. For example, the itinerary of Maldivian cowries demonstrates the lack of separation between Europe and Asia: from the second millennium BCE through the second millennium CE, a considerable volume of cowries from the Maldives Islands in the Indian Ocean traveled in several directions, to the west into the African mainland and Europe and to the east across the Bay of Bengal into eastern India, Myanmar, and all the way to the Yellow River valley of China as early as the second millennium BCE.[9]

Although there are no textual records of the early trade, it is apparent from tracing the itineraries of these objects that the relaying of the commodity involved diverse modes of transportation over a varied terrain—mountainous tracks, river crossings, and overland pathways—and several different groups of traders. The itinerary of cowries from the Maldives Islands to the central plains of China may have been relatively fixed, as archaeological findings in Assam, Burma, and Yunnan help identify some of the relay sites that formed part of the networks of cowrie trade.[10] From the Maldives Islands, the cowries reached ports in present-day Bengal and then were transported by overland routes through Assam and Burma into the Yunnan region where they were used as monetary units. Some may have also reached the Burmese ports first and then moved along land routes into Yunnan. From Yunnan, cowries were sent on to the Yellow River valley and to the Shang capital, Anyang.

These earlier networks of exchange served as the foundations for a deeper integration of the Eurasian world during the last five centuries of the first millennium BCE. The formation of empires, the emergence of urban centers, and a better understanding of monsoon winds in the Indian Ocean were some of the other key factors stimulating intra-Eurasian interactions during this period. The Achaemenid Empire (550–330 BCE) and the Hellenistic polities that spanned from the Indus to the Aegean Sea, the Mauryan Empire (322–185 BCE) in South Asia, and the Han Empire (206 BCE–220 CE) in China, all contributed to the building of roads and bridges that facilitated the movement of peoples and goods. The growth of local economies and the increase in foreign trade led these empires to experiment with standardized monetary and weighing systems, which encouraged further expansion of trade. The increase in population and the appearance of cities created demand for local and foreign staple goods and rich luxuries. It was during this period that the corpora of knowledge were codified, first orally and subsequently in texts, in transregional languages—Greek, Chinese, Sanskrit, and early Persian—that came to underpin the construction of remarkably durable fictive identities, ostensibly knitting together diverse peoples into cross-regional groupings, imagined at various times and places as homogenous; unified by politics, bureaucracy, culture, race, or ethnicity.

Parthian traders, Greek and Roman sailors, and South and Southeast Asian merchants connected the ports and markets of the Eurasian world. Roman glass, Chinese silk, precious and semiprecious stones and beads from Central and South Asia, and Southeast Asian gold and tin became some of the key commodities in the increasingly lucrative cross-regional commerce. Western Han officials based in the Hepu region (in present-day Guangxi province of China) owned and were buried with their Roman glassware, jewelry made of imported beads and precious and semiprecious stones (such as carnelian, agate, and lapis lazuli), and pearls originating from various regions of South Asia.[11] Glass objects from the Roman Empire moved through ports in Southeast and South Asia, entering South Asia through the commercial networks that connected the Coromandel coast in southwest India to the ports in the Red Sea. From the Coromandel coast, the imported glassware passed through Southeast Asia and reached Jiaozhi, Hepu, and Guangzhou on the southern frontiers of the Qin and Han dynasties of China.[12]

The main consumers of these foreign luxuries in China were members of the elite who had settled in the frontier regions after the southward expansions by the two empires. The beads and semiprecious stones excavated from the Hepu tombs mostly originated in South Asia. However, as some scholars have suggested, it is possible that foreign artisans based in Southeast Asia were responsible for reprocessing and even manufacturing these items according to South Asian style for local consumption and for export to China.[13] This practice of customizing commodities for foreign and local consumption was also common in long-distance commercial interactions. The oasis regions of Central Asia are known to have tailored and redesigned silk fabrics originating in China according to the requirements of markets elsewhere. Such customized Chinese silk products were transported to foreign ports and markets through South Asia. The first-century Greco-Roman work that describes navigation and trade around the Roman world, entitled *The Periplus of the Erythraean Sea*, mentions Chinese silk yarns and silk fabrics as some of the commodities exported from ports known as Barbarikon (present-day Karachi, Pakistan) and Barygaza (present-day Bharuch in Gujarat, India).[14]

The procurement of foreign goods for domestic consumption and transit trade gradually increased over time and apparently dominated long-distance commerce after the tenth century. Horses, porcelain, and silver were some of the commodities that circulated widely within the Afro-Eurasian world and contributed significantly to the integration of far-flung ports and towns through networks of commercial exchanges. Horses suitable for warfare were in high demand by Chinese courts from the Han dynasty. Central Asian and Persian horses, and later Tibetan ponies and mounts, were imported in large numbers throughout most of China's imperial history. South Asia was an important part of this supply chain. Arabian horses were first exported to the Malabar coast, and from there they were transshipped to China. During the twelfth and thirteenth centuries merchants from Kish in the Persian Gulf played an important role in supplying these Arabian horses to South Asia.[15] The lucrative nature of the trade in horses is underscored in the works of the Persian historian Waṣṣāf and the Italian traveler Marco Polo, who mentions that "some, indeed most of them, fetch fully two hundred pounds of Touraine apiece." Waṣṣāf reports that fourteen hundred horses were sold to the Pandyan ruler named Sundara on the Malabar coast by Arab traders at the price of 220 dinars of red gold each, including for those that might be lost at sea. Waṣṣāf also notes that ten thousand horses were exported annually from the Persian Gulf during the reign of Atabeg Abu Bakr (r. 1226–1260) to ports in South Asia. The revenue from such exports, according to him, amounted to 2,200,000 dinars.[16] The continuation of this transit route two centuries later is confirmed by Fei Xin, who traveled to the Malabar coast (and beyond) with the Ming admiral Zheng He (1371–1433) in the early fifteenth century. According to Fei Xin, who accompanied Zheng He on some of the voyages, horses were brought to the Malabar coast from West Asia, bred locally, and "transferred" for hundreds or thousands of coins for onward sale to Ming China.[17]

Another trading circuit through which horses were supplied to markets in China is reported by the thirteenth-century Persian-language record from South Asia known as *Tabaqāt-I Nāsirī*. According to the work, the trade in horses was carried out by the Tibetans in Bengal and Kamarupa (present-day Assam state of India). In the market town of Lankhnauti in northern Bengal, according to this work, about fifteen hundred horses were sold every morning.[18] Some of these horses, it seems, were exported to China through the maritime routes that connected the Bay of Bengal ports to the coastal regions of China. Ming sources mention that representatives from Bengal often presented horses as tribute to the court. These representatives, in addition to presenting tribute, must have also sold some of the horses in local markets or at least informed the Ming court about their availability in Bengal. Since Bengal did not produce horses, the gifts brought to Ming China were most likely procured in the markets of Lankhnauti. This circuit also featured the export of silver from the mines in China to Bengal during the precolonial period. Then, with Spanish colonization and the discovery of silver mines in the Americas, the webs of commerce expanded significantly, connecting regions such as Yunnan, Sichuan, Tibet, Bengal, Assam, and Myanmar to the Indian Ocean on one end and to the Pacific on the other.

The trade in this circuit continued into the twentieth century, with significant ramifications: since at least the Song period, the Tibetans had procured Chinese tea in exchange for horses sold on the borderlands of Sichuan and Kham. Because of political instability in the Sichuan-Kham region in the early twentieth century, the trade in tea expanded geographically to include Myanmar and British India. Tea from the Yunnan region reached Lhasa through Myanmar and north Bengal and Sikkim. The merchants engaged in this trade often went to Calcutta to remit their earnings to China, using the remittance services of the overseas Chinese and governmental financial institutions, such as the Indian branches of the Bank of China. As a result, urban centers such as Calcutta and Kunming became important parts of the web of commerce, as providers of financial services rather than simply as markets for commodity exchange.[19]

During the Han period, the envoy Zhang Qian (d. 133 BCE) gave the first written account of the trade through this Yunnan-Bengal region. "When I was in Daxia [Bactria]," he writes, "I saw bamboo canes from Qiong and cloth made in the province of Shu [Sichuan]. When I asked the people how they had gotten such articles, they replied, 'Our merchants go to buy them in the markets of Shendu [India].'"[20] This is just one of numerous records of political and commercial contact between China and western Asian lands to be found in the *Hou Han Shu*, the official history of the Chinese Han dynasty.[21] Indeed, as Andre Gunder Frank notes, "when we lack such written official testimony by contemporaries or archeological evidence, we may suppose that trade was still older and more widespread than recorded. . . . That is, trade relations preceded and exceeded diplomatic ones."[22] Key nodes for such trading activity, including Babylonia in Persia and Arikamedu in southern India, formed throughout the Eurasian world. The commercial linkages also triggered the movement of artisans and the diffusion of technical and geographical knowledge.[23] Itinerant traders also helped spread of religious ideas, images, and artifacts, including those associated with Buddhism (see Sen, this volume).

In the first century CE the founding of the Kuṣāṇa Empire, which extended from Central Asia to the Gangetic plains of India, facilitated the interweaving of commercial activity and religious transmissions across the overland and maritime routes. The Sogdian traders used the Kuṣāṇa territories to establish a vast network that, for example, supplied horses to the Chinese markets and procured silk in return. The Sogdians were also some of the earliest transmitters and translators of Buddhist doctrines entering

China.[24] Persian, Sri Lankan, and Southeast Asian sailors also contributed to the spread of Buddhism by ferrying missionaries and pilgrims around within an emerging Buddhist cosmopolis. The spread of Buddhism to foreign lands resulted in the demand for related ritual objects and other religious paraphernalia that also benefited merchant communities. This intimate association between merchant communities and the spread of Buddhist doctrines persisted throughout most of the first millennium CE.[25]

Like Buddhism, the spread of Islam from the seventh century CE triggered the circulation of religious paraphernalia, knowledge and information, art and architecture. Also comparable to Buddhism, but with more intensity, Islam prompted the movement of missionaries and pilgrims. The emphasis on Hajj as one of the pillars of the faith resulted not only in frequent travels of people from different parts of Eurasia to Mecca but also in the development of transportation facilities and an intensification of economic transactions associated with pilgrimage activities.[26] The post-seventh-century period also witnessed an unprecedented growth in maritime commerce. The networks of Muslim traders belonging to several different ethnic groups and transformations in navigational and shipbuilding technologies were some of the contributing factors in the upsurge of trade across the Indian Ocean. The ninth-century Belitung shipwreck is a prime example of this growth in maritime connections. This Arab dhow, which sailed from the Chinese coast and sank in the Indonesian waters near the island of Belitung, carried over sixty-four thousand pieces of Chinese porcelain ware, gold and silver objects, and many other objects. These items seem to have been commercial goods as well as gifts for foreign rulers.[27] Archaeological finds have demonstrated that Chinese ceramics traveled as far as East Africa during this period.[28] In fact, it was during almost the same period that a Chinese envoy from the Tang court is reported to have traveled to the Abbasid Empire.[29] It is evident that maritime routes had developed into important conduits for both commercial and diplomatic exchanges during the second half of the first millennium. The Srivijayan polity in Southeast Asia, for example, established strategic relations with the powerful Chola court in south India and at the same time maintained strong commercial connections with economically vibrant Song China. The Srivijayan ruler based in Sumatra exerted significant control over maritime commerce taking place through the Straits of Malacca. Srivijaya not only became the main transit center for commodities destined for or originating from the markets in China but also regulated ships passing through the

Straits of Malacca. Srivijaya's monopoly over maritime commerce with China was challenged in 1025 by the Chola court, which reportedly launched a massive naval attack on the Srivijayan ports. This attack signified the beginning of the militarization of the Indian Ocean world, something that was not possible without the development of navigational and shipbuilding technologies.[30]

New oceangoing ships and methods of way finding also had a significant impact on the nature and volume of commodities traded across the Eurasian world. Bulk goods including incense, aromatics, and spices became more commonly traded items as tonnage increased in the large ships. Frankincense from the Persian Gulf, pepper from southern India, and sandalwood from Southeast Asia were frequently traded across the Indian Ocean.[31] Similarly, porcelain from China was exported in larger quantities as ships, rather than camels, provided a safer mode of transportation over long distances.[32] Maritime exchanges continued to expand between the thirteenth and fifteenth centuries, at first overlapping with intense overland interactions that occurred as Mongol empires formed across the continent, a time when a Tuscan commercial center such as Siena existed on "the periphery of Mongol Eurasia."[33] Indeed, during the so-called pax Mongolica, maritime and overland routes became more intimately intertwined. This is evident from the travels of Marco Polo and Ibn Battuta, the extensive knowledge about the "world" collected by the Persian historian Rashīd al-Dīn, and the expansive dispersion of Chinese porcelain ware. As individuals and whole groups of peoples moved across Eurasia, their practices and ideas traveled with them. These long-distance movements of peoples, including the forced migrations of craftspeople under the Mongols, and informal exchanges such as the capture of prisoners in war caused sociotechnical systems—from weaving to weapons—to move along with them.

During the Mongol period ships from Yuan China began to dominate the eastern Indian Ocean networks. These Chinese ships made frequent calls at the Malabar coast of India, and Chinese diasporic communities formed during this period in Southeast Asian ports.[34] These communities became more active in the early fifteenth century, when the Ming court instituted strict restrictions on foreign trade. The Ming court also launched seven massive naval expeditions led by the commander Zheng He between 1405 and 1433. These expeditions resulted in the formation of new hubs of maritime commerce including Malacca in Southeast Asia, Cochin in southern India, and Malindi in eastern Africa. The voyages, which included several hundred ships, over twenty thousand soldiers, and advanced gunpowder weapons, also had a significant impact on the movement of people and animals, at the same time as they militarized the oceanic space and consolidated state participation in maritime trade. Only a half century later, a new group entered into these already well-established connective pathways, and these early European trading companies, especially the Portuguese, imitated many points of strategy that had informed Zheng He's voyages, including control of major choke points and attempts to monopolize private trading activity.[35]

The Global Integration of Eurasia

The integration of Eurasia through commerce and religious transmission peaked during the colonial period as commodities such as tea, spices, and textiles moved in large quantities within the region. Active missionary work by Jesuits and members of other Christian churches connected regions as distant as Rome, Goa, Macao, Malacca, and Kagoshima. Similar to Buddhist and Muslim proselytizers, these missionaries also brought with them unique ideas of observing the cosmos, curing diseases, and technical expertise. In the *longue durée* of connections and interactions across Eurasia, the European traders and missionaries were just the latest contributors to integrate a vast space that had transcended imagined continental divides far into prehistory.[36] More importantly, in the sixteenth and seventeenth centuries, the Eurasian world became integrated into the circulations across the Pacific and Atlantic oceans, resulting in new forms and more intensive circulations of people, objects, and ideas.

The web of exchange, highlighted by the foregoing survey, that knit together Eurasia has often been confusingly called the "Silk Road." However, it could be named the Jade or Horse Road, and, from other regional perspectives, the rhubarb and paper routes.[37] The concept of a singular Silk Road has also masked an equally ancient "maritime silk route," and an even earlier eastward flow of furs along a northerly "fur route," as well as a "musk route" connecting Tibet and the Islamic world from the eighth century. But, most significantly, as this list of alternative names for the route implies, it has masked the very nature of human interaction across the continent (figure 2.3).[38] Exchange across regional and political boundaries was the norm. Indeed, such flows were simply extended further east and west around the globe as the Americas were forcibly incorporated within the Afro-Eurasian ecumene from the sixteenth century on.

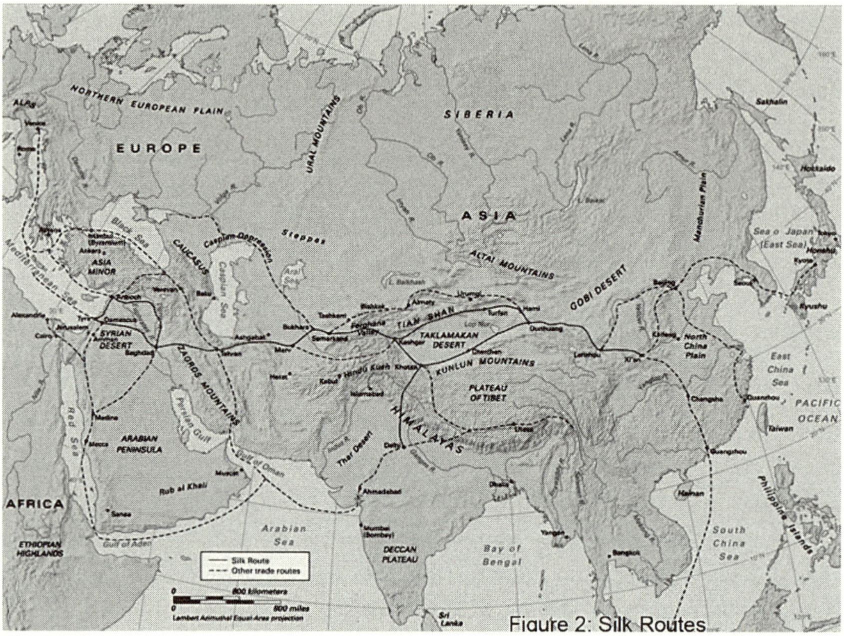

Figure 2.3. Silk routes. Trade routes in the Middle Ages.

By following these flows of materials, we can discern some general patterns in the movement of materials, practices, and knowledge. Some moved with particular speed, especially those related to the techniques of survival, both of sustenance and of war (copper alloys, the stirrup, gunpowder, to name just three). Although our primary documentary evidence for the movement of foodstuffs comes from the period when New World foods entered into already well-established trade routes at the eastern end of Eurasia, a few examples indicate how fast such crops could travel. The peanut cultivated in the Chaco region of South America, for example, could already be found growing as a food crop near Shanghai by the 1560s or 1570s at the latest; sweet potatoes had arrived in China at least by the 1560s; New World maize was established in China in 1555, where it had been brought by Turkic frontiersmen, and along the Euphrates by 1574.[39] Some of the first European representations of New World plants, maize corn and hot peppers, appeared in German herbals (in particular that of Leonhart Fuchs in 1543) as plants from the eastern ends of Eurasia, with maize called "Turkish grain" and chilis labeled "Calcutta peppers" (figure 2.4). This may indicate that these New World plants entered Europe from easterly ports as well as from New Spain and Lisbon, for maize may have been cultivated on a larger scale first in the Ottoman Empire and only later in Europe. We know that it was already being grown in Spain on a small scale in 1530, only a few years after it had first been brought to Europe. In the Ottoman Empire, maize was called *Mısır buğdayı* (Egyptian wheat) as it was first introduced into Egypt sometime after the Turkish conquest in 1517 and later reached Anatolia and Istanbul via Syria.[40] As for "Calcutta peppers," the Portuguese took New World peppers to India, where they came to be cultivated and regarded as native and then flowed west as one component of "curry." Whatever the source of Fuchs's plant names, it is clear that food plants traveled in advance of recorded contact between peoples, carried by sailors and other anonymous intermediaries (figure 2.5). Their cultivation in new soils must have occasioned much experimentation by their growers both in the field and at the dinner table. As in the case of "Turkish corn," the very name of the material at times indicated this movement. Take, for example, the words "china" in English for porcelain and *cini* in Turkish for decorated ceramics, or the north Indian word *cini* ("Chinese") for sugar, which also seems to have derived from local reference to that ubiquitous trade object porcelain.[41] In the case of ceramics produced in Iznik in northwestern Anatolia, called *Iznik* çinileri (Iznik chinoiserie), a mix of Ottoman and Chinese decorative elements, the name itself reveals a long itinerary of this "material complex" from East Asia through Central Asia, via Persia, to Anatolia. Similarly, the instruments of war and the trappings, ceremonies, and techniques of staging power, such as the royal hunt carried out with noble animals and raptor birds, also spread extensively throughout Afro-Eurasia (figure 2.6).[42] Techniques of governing may also have flowed across this space: Roger II (r. 1132–1154) of Norman Sicily gathered information and informants from all parts of the world, and his plan for training members of the court in three stages of education may have been based upon his knowledge of Chinese practices.[43] Common mathematical techniques for dividing up partnerships and profits for the purposes of commerce traveled far and wide, as well as the problems of calendar making, and the tables needed to construct star charts, astrological horoscopes, and for figuring the direction of Mecca. Story problems for teaching rules of calculating the relative speeds of animals and planets chasing one another are found in China in the first century, in India in the seventh century, and in Latin—in their simplest form—in the ninth.[44]

Shared patterns of consumption emerged as well, Christopher Bayly has argued, that constructed the value of unique, rare, and unusual objects, part

FIGURE 2.4. Calechutischer Pfeffer (Calcutta pepper). Leonhart Fuchs, *New Kreüterbuch* (1543). Signatur 19 257 4°, Stadtbibliothek Ulm.

FIGURE 2.5. Provisioning of water for ships by slaves in Barcelona. Christoph Weiditz, "Trachtenbuch" (1529). Fols. 73–74, Hs. 22474, Germanisches Nationalmuseum Nürnberg. Germanisches Nationalmuseum, digital library. Public domain.

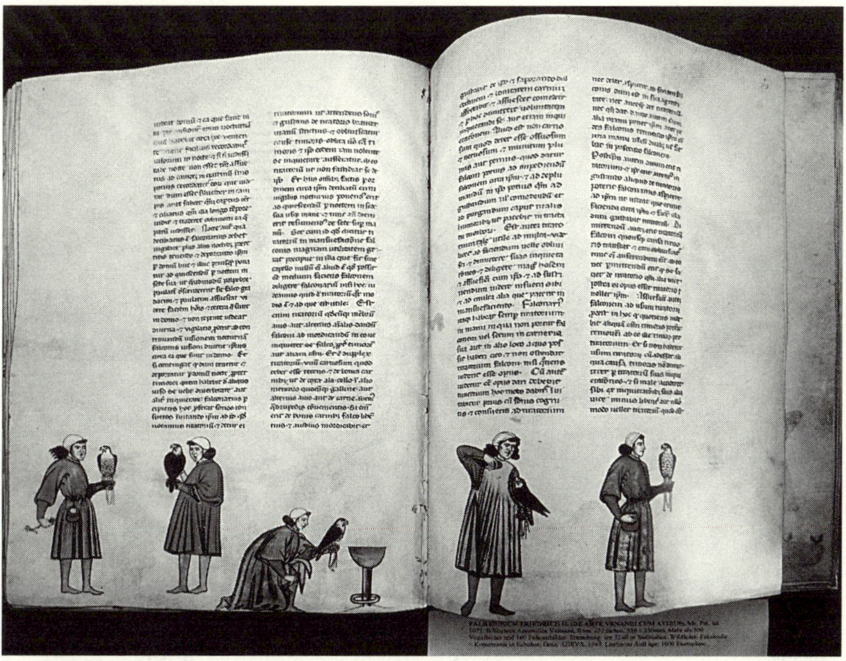

FIGURE 2.6. European traces of the royal hunt. Frederick II, Hohenstaufen, *De arte venandi cum avibus* (thirteenth century). Vatican, Ms. pal. lat. 1071.

of a material imaginary that could make real and tangible qualities of distant parts of a realm. The simultaneously social, moral, and exchange economy in which such valuable wondrous objects were embedded extended downward and outward through the courts of regional nobles (figures 2.7 and 2.8). At the same time the consolidation of the major religions—with their similar practices of prayer, sacrifice, and pilgrimage by which devotees pursued the signs and sites of gods scattered throughout the world—brought into existence a large infrastructure of transport, food, credit, and international trade to support their practices of devotion and their extensive systems of pilgrimage (spice for incense and Brahminical food restrictions, for example, and salted fish for Catholic fasting).[45] The individuals traversing these various pilgrimage and trade routes shared a common preoccupation with the relation of body and spirit—which sometimes converged in similar views of health and a common range of medicines, even when understood differently. Historians have debated the similarities and translatability of central medical concepts—such as "life-breath" and "winds" (*qi*, *pneuma*, and *prāna*) that can be found in ancient Chinese, Greco-Roman, and South Asian medical writings as well as the Greco-Roman-Islamic system of the four humors and the Chinese system of yin/yang. Some historians have argued that all these systems have origins in a Eurasia-wide notion of balance and the tempering of contrary forces in order to bring the body to a healthy state.[46] Historians have also debated what role these conceptual systems played in the variety of medical practices and the often similar substances and remedies that informed care of the body throughout Eurasia. At the most general level, an individual's constitution was viewed as influenced by the heavenly bodies, and many of the materials by which bodies could be tempered and balanced were transported over very long distances via the same trade routes as were traveled by weapons and rarities. Spices (wholly integrated into the health framework) and all manner of materia medica—including medicinal plants (such as dragon's blood), animal parts (such as rhinoceros horn), precious stones (for example, turquoise for eyesight and red coral to stop blood flow), and mineral-based concoctions (such as large quantities of sulfur, which flowed by the ton into Tang China from Indonesia and Japan between the seventh and tenth centuries CE, used both for industrial processes and fireworks and to temper the constitution)—all possessed medicinal virtues and commanded high exchange values throughout Eurasia.[47]

A literature on "wonders," which began to codify and canonize the marvelous qualities of these objects and simultaneously to stimulate desire for

FIGURE 2.7. Mother of pearl and inlay (Gujarat, India, sixteenth century), casket with northern European mountings. Staatliche Kunstsammlungen Dresden. Photo: Jürgen Karpinski.

them, grew up around the exchange in rarities and medicines described above. In this literature India was identified as the primary location of wonders and marvels, as well as the source of precious stones, a role that, along with Sri Lanka, it had played at least since 400 BCE.[48] While the literature on wonders extends back even to Greek texts, the view that wonders originated in India, Berthold Laufer argues, emerged out of an aristocratic Arabic-Persian milieu.[49] In Europe the very first of 129 wonders that Gervase of Tilbury lists in his *Otia imperialia* (1210) is the magnet, "an Indian stone," and William of Auvergne (ca. 1180?–1249) regarded "parts of India and other adjoining regions" as containing "a great quantity of things of this sort, and on account of this, natural magic particularly flourishes there."[50] Muzaffar Alam and Sanjay Subrahmanyam note that medieval travel literature throughout the world includes "wonders" or "aja'ib" (in other words, marvels, astonishing things or "mirabilia"), most often told in first-person narrative, like that of tenth-century Persian author writing in Arabic, Buzurg ibn Shahriyār, in *Kitab 'ajāyib al-Hind* (The wonders of India, ca. 960), which recounts marvels of all sorts purportedly told to him on board merchant ships, including many involving fearsome snakes, future-divining lizards, and crayfish that, for example, fell into the Sea of Senf, turned into stones, and were subsequently "carried into Iraq . . . and used in making an ointment to cure eye complaints."[51] Such travelers' reports and sailors' yarns (best known perhaps in those of Sinbad the Sailor and the *Arabian Nights*) spread widely through the "Maritime Silk Route," forming the stuff of "folk tales" across Eurasia.[52] On land, troubadours "blended traces of Sufi and Islamic mysticism with Manichaean Christian narratives," as in the story of Barlaam and Josaphat, which tells the tale of a prince's renunciation of all worldly pleasures and material belongings. This version of a Manichaean tale began as a recounting of the life of the Buddha, but it emerged again in the eleventh and twelfth centuries in Islamicate and Christian lands, apparently carrying other Manichaean beliefs along with it, and perhaps even contributing to the Manichaean revolt in the Northern Song in the twelfth century and the Cathar crisis in Languedoc.[53]

Even this brief and cursory survey of the pathways of exchange that coursed across Eurasia since prehistoric times shows that the flows of goods and of the peoples who mined, sourced, worked, produced, transported, and consumed them helped to constitute the "material-knowledge complexes" that both structured and fostered this exchange. In a reciprocal process between humans and their material environment, the mundane production of material things informed and transformed theoretical formulations of knowledge. The transfer of practical techniques across space fostered new ways of controlling both nature and other people. The movement of goods and their trade gave rise to people telling oral tales and forming and reforming conceptual, spatial, and intellectual imaginaries, some of which were compiled, written down (only sometimes by identifiable authors), and translated into various vernacular and learned languages in texts that also moved across great distances, where they stimulated the itineraries of yet further travelers, objects, and ideas.

Figure 2.8. Interior accoutrements of casket, created from Indian Ocean naturalia and set in northern European mountings. Staatliche Kunstsammlungen Dresden. Photo: Jürgen Karpinski.

Part 2

Entangled Itineraries

Modes of Approach

Chapter 3

The Silk Roads as a Model for Exploring Eurasian Transmissions of Medical Knowledge

Views from the Tibetan Medical Manuscripts of Dunhuang

Ronit Yoeli-Tlalim

At the beginning of the twentieth century, Wang Yuanlu, a Daoist monk in the western frontiers of China accidentally discovered a cave full of manuscripts near the Chinese town of Dunhuang in Gansu province. The cave, which had been sealed for nearly a thousand years, contained several tons of manuscripts. This cave, now known as Cave 17 or the "library cave," was sealed in the early eleventh century for reasons that are still being debated by scholars.[1] Following this discovery, a race began between the great nations of the time, to acquire as many manuscripts as possible. Today these manuscripts are dispersed among libraries in Paris, London, St. Petersburg, Tokyo, Beijing, and elsewhere and are currently being united on the Internet as part of the International Dunhuang Project, based at the British Library.[2]

The Dunhuang manuscripts are of enormous significance for Buddhist, Central Asian, and Chinese history. Their significance for the history of science and the history of medicine has only recently begun to be explored in European scholarship by Vivienne Lo, Chris Cullen, Catherine Despeux, Ming Chen, and others.[3] Observed in their overall context, the Dunhuang manuscripts are a bit like a time capsule, providing traces of what medicine was like "on the ground," away from the main cultural centers, at this particular geographical location. Being in manuscript form they preserve the benefits of unedited texts, revealing more diverse forms of healing and telling different stories than medical canons preserved in print.[4]

Analyzing the medical material from Dunhuang allows us to observe several continuities, such as the fuzzy borders between what we might term in English today "medicine," "ritual," "religion," "divination," and "magic."[5] The Dunhuang medical material also provides interesting data with which to analyze continuities between center and periphery. Although the Tibetan medical manuscripts from Dunhuang are singular manuscripts found on the outskirts of both the Tibetan and the Chinese Empires, we do find continuities with later printed texts.

Finally, it is the Eurasian continuity that is the focus of my ongoing research. Two main methodologies are at hand when analyzing Eurasian continuities. One is looking at European and Asian material in a comparative way, and another is trying to uncover specific interactions. Although studying interactions is a complex and difficult matter, particularly because of the nascent state of research into so many relevant areas of exploration, I believe this endeavor is worthwhile. This overview presents some of the directions my research is taking in exploring the Tibetan medicine from Dunhuang as a case study of interactions of medical knowledge along the Silk Roads.

The Meaning and Metaphor in "Silk Road/s"

The term "Silk Road" both in the singular and plural was coined by the German geographer and traveler Ferdinand von Richthofen at the end of the nineteenth century.[6] Richthofen used the term to refer to the routes along which Chinese silk moved from the Han Empire to Central Asia. Richthofen himself was interested not only in geography but also in the greater historical and cultural importance of these trade routes. Since Richthofen's day, the term "Silk Road" has evolved to refer to the many ways and exchanges connecting China and Europe, along which people, stuff, and ideas were moved and exchanged. The people were traders, missionaries, soldiers, brides, artists, and diplomats, as well as various combinations of these. The stuff they carried was much more varied than just silk and included (among many more) horses, paper, musk, musical instruments, and dumplings. Genes, germs, and immunities were also carried along the way.[7] The ideas included the many religions that traveled across Asia—Buddhism, Islam, Eastern Christianity, Manichaeism, and also some Judaism; medicine and other fields of knowledge that we might term scientific; stories, arts, and technology—including papermaking and printing.

When thinking, researching and writing about the Silk Roads, scholars were initially focused primarily on the two extremes, China and Europe. This focus on terminus points is understandable considering the sources at hand—that is, the sources that were on hand from earliest time to the twentieth century. All this changed when the great finds along the Silk Roads were excavated at the beginning of the twentieth century. Particularly thanks to the great archaeological discoveries along the Silk Roads of the twentieth century, scholarship gradually shifted to the great expanse in between, slowly uncovering traces of cultural exchanges.[8] These traces, which were buried for over a thousand years, are now slowly being studied. Archaeology of the Silk Road has also been relatively slow because of a combination of political and environmental reasons. But as archaeological research on the Silk Roads progressed, it has become clear that the exchanges mediated by the Silk Roads are much older and more extensive than has been thought in the past.[9]

While I take Khodadad Rezakhani's point on the overuse, misuse, and abuse of the term, I would argue that by dismissing the entire historiography of the Silk Road and the term itself we end up throwing the baby out with the bathwater.[10] Indeed, within the huge bulk of literature dealing with the Silk Road, one can find problems of various sorts, as Rezakhani ably demonstrates, but there have been of late some game changers in this respect, most notably Valerie Hansen's *Silk Road: A New History* (2012), which came out after Rezakhani's article and addresses most, if not all, of the problems Rezakhani raised with regards to the problems of dealing with the "Silk Road." The particular case discussed here—that of Tibetan medicine—serves precisely to amend the overemphasis on China and Europe that Rezakhani is criticizing.

Hansen in her illuminating account of the history of the Silk Road has demonstrated how trade on the Silk Roads was not on a grand scale, neither geographically nor economically. Unlike popular perceptions of vast amounts of silk and other precious items going from China to Rome, with devastating effect on the Roman Empire, the nature of trade as it emerges from the documents unearthed in the oases around the Taklamakan Desert is much smaller and more local. The interactions of trade, as of knowledge, happened at the in-between: between Gandhara and Kucha; between Kucha and Turfan; between Turfan and Dunhuang; between Dunhuang and Chang'an. And, of course, among the vast variety of peoples who dwelled and moved among these sites.

We now know that mediating cultures whose languages were once *lingua francas* of multicultural empires or spoken by long-distance travelers are as important for our understanding of history as the more well-known cultures of the termini points. The term *lingua francas* in this context refers to languages that are "contact languages," which "facilitate communication among people who do not share the same mother tongue."[11] Another significant point regarding *lingua franca*s is that they are transitory and unstable, intertwined with transitory and unstable power and prestige. They are useful as foci since they are, as Jocelyne Dakhlia has pointed out, languages for *communication* rather than languages of *identification*.[12]

And so, when we attempt to study the history of Eurasia, sources in languages such as Tibetan, Sogdian, Uighur, Tocharian (A and B), Bactrian, Khotanese, Hebrew and Syriac, as well as certainly those in Persian and Arabic, can help us fill the huge gaps in the puzzle that is Eurasian history. In some of these cases, entire languages and cultures were buried in the Taklamakan Desert for over a thousand years, with no followers or scholars to speak their word. Our knowledge of them now relies on a small and precarious group of scholars who are scattered around the world.

Toward a Eurasian Approach in the History of Science and Medicine

In 2006, at the International Conference of Tibetan Studies, I presented a paper discussing close similarities between Ibn Sīnā's *The Canon of Medicine* and an early text of Tibetan medicine, the *Zla ba'i rgyal po* (Lunar king). Many people (including myself) were astonished to discover that there *were* such close similarities.[13] Today, after much collaborative work, I am more astonished that we were astonished.[14] In Central Asia during the second half of the first millennium cross-cultural transmissions and mélanges were happening everywhere and were more of the norm than the exception. So, alongside our delving into the multiple ways of cross-cultural transmissions, we ought to also reflect on why we were astonished. I shall return to this question later.

The notion of "globalized medicine" is usually associated with modernity. The study of ancient medicine teaches us, however, that medicine needs to be treated in a much more "globalized," or, perhaps, interconnected way than was previously assumed. Treating ancient medicine globally, however, is still a rarity in medical history. Just as the treatment of five thousand years of world systems presented in global history publications such as *The World System: Five Hundred Years or Five Thousand Years?* has been, as its authors call it, an "enlargement of scale" of the now more common treatments of five hundred years of world systems, it is also high time for an equivalent "enlargement of scale" in the history of medicine.[15] At the core of these endeavors is the idea that methods and perspectives of "global history," which have been very fruitful in historical research dealing with the last five hundred years, can also be just as fruitful when dealing with the medieval and ancient world.

As in the context of the five thousand years of world systems, the term "global" or "world" needs to be aptly qualified: we are obviously not speaking of the "globe" or even the "world" as we know it today but, rather, of the "known world" of the time in question. One of the important contributions of this perspective (as discussed in the work of Gunder Frank, Gills, and others) has been showing the limitations of the focus on "civilizations" and bringing to the fore a more united "Eurasian" history and an inquiry into historical *connections* between peoples, places, and ideas. Rethinking dichotomies of "Eastern" and "Western" is key for the way we approach the history of ancient medicine. We have been gaining more and more evidence that allows us to question the constructed dichotomy between "Eastern medicine" and "Western medicine," or "Asian medicine" and "European medicine." We need to think of medicine in a more connected way, in line with current approaches in global history, the history of science, and the history of ideas.

Intensified communications between certain cultures tend to come and go. Some of the material "fossilized" in the manuscripts found on the Silk Road provide a fresh view into some of the interactions, exchanges, and influences that were in one way or another later written out of printed sources. The emerging focus on "exchanges" rather than "cultures" necessarily brings to the fore knowledge stemming beyond cultural elites, with a greater focus on locations where knowledge is more easily subject to change. For this purpose, it is particularly useful to study manuscripts such as those of Dunhuang.

How can we begin to untangle any history of transmissions of medical knowledge along the Silk Roads? I would like to suggest two types of starting points: first, we can follow some of the entangled itineraries of specific texts and key people as they traversed boundaries of language and culture, and second, we can analyze the specific collections of nodes of convergence in which medical manuscripts were found. Taken together, these can help us begin to delineate a picture of the transmission of medical knowledge, a picture that, although no doubt partial and lacking, is important. What I provide here are a few indications in this direction.

Entangled Itineraries of Texts and People

One illuminating example where we can trace an itinerary of a key text as it moves and gets translated into different languages and cultures is that of the *Siddhasāra*. This text is a comprehensive medical manual that includes the main theories and practices of Ayurveda that was written or, more likely, edited by Ravigupta of the seventh century.[16] It originated from India, possibly Kashmir, and had widespread influence in various parts of Asia, both in Central Asia and West Asia.[17] Primarily thanks to the work of Emmerick, we can trace the movement of this text across languages, as it traveled both northeast and west. In the ninth century it was translated from Sanskrit into Tibetan as well as into Arabic.[18] Around the tenth century it was translated from Tibetan into Khotanese, and sometime before the thirteenth century it was translated into Uighur.[19]

A Khotanese translation of the *Siddhasāra* was found in Dunhuang. Khotanese manuscripts were found by Stein in Dunhuang and in the Khotan area. The language was unknown at the time. Khotanese is an Iranian language contemporary with Middle-Persian and Sogdian, written in Indian Brāhmī script, with an extensive vocabulary borrowed from Sanskrit. Following Stein's discoveries, Khotanese was deciphered primarily by Hoernle.[20] Khotan was ruled by Tibet from the late eighth century till the mid-ninth century, but it appears that the Tibetan language remained in use in Khotan even after the end of direct Tibetan rule of Khotan.[21] During the tenth century the contacts between the Chinese, Khotanese, and Uighurs intensified as a result of marriage alliances.[22] As a result more members from these ethnic groups settled in Dunhuang, as is attested by the high number of Khotanese texts from this period found in the Dunhuang library cave.[23]

The introduction to the Khotanese translation of the *Siddhasāra*, not found in any other version, gives us an inkling of the process of translations from language to language to language at the time. The Khotanese introduction states that the text was translated from Tibetan, but as the translator found the Tibetan version problematic, he also consulted the Sanskrit original. The introduction also tells us the reason it was translated, suggesting that before this medical knowledge arrived there were many people dying because of the lack of medical knowledge.[24]

The material history of this manuscript provides further dimensions of this text's transmission. The recto of the Khotanese *Siddhasāra* from Dunhuang is a Chinese text (Pelliot chinois 2893). This manuscript was brought to Dunhuang from Khotan by Daoyuan, a Chinese Buddhist monk who traveled to India and, on his way back, stayed in Khotan for ten years before returning to China in the year 965.[25] Interestingly, we also have evidence of the owner of the manuscript. His name is written in Sogdian script on folio 156 of the manuscript. He has been identified as Zhang Jinshan, an ambassador to Dunhuang from the Kingdom of Khotan.[26] Analysis of Khotanese manuscripts from Dunhuang has shown that the king of Khotan maintained a kind of embassy in Dunhuang, where Khotanese princes lived, sometimes with their Chinese wives. It appears their library of Khotanese manuscripts was given to the monks who collected the manuscripts found in the library cave (Cave 17).[27] So here is a Sanskrit medical text translated into Tibetan, then into Khotanese, carried from Khotan to Dunhuang by a Chinese monk on his way back from India and, finally, in the possession of a Khotanese ambassador in Dunhuang who signed his name in Sogdian.

The *Siddhasāra* was also held in high esteem by Persian and Arabic scholars. Al-Rāzī (b. 864/5, d. ca. 925), from Rayy near modern-day Tehran, incorporated many passages of the *Siddhasāra* into his *Kitāb al-Ḥāwī fī ṭibb* (Comprehensive book on medicine).[28] The *Kitāb al-Ḥāwī fī ṭibb* was translated into Latin in the thirteenth century, and there are many remnants of the *Siddhasāra* in its many Latin versions.[29]

That al-Rāzī incorporated Indian medical knowledge into his writings should not really surprise us. Alī b. Sahl Raban al-Ṭabarī (fl. ca. 820–860), a Christian physician who later converted to Islam, is considered to be al-Rāzī's teacher whether directly, as tradition has it, or indirectly via his writings. He was a native of Marw (Merv) in Central Asia, one of the world's most cosmopolitan centers of learning at the time.[30] Its libraries were so famous that we are told the Middle-Eastern scholar Yāqūt al-Hamawi traveled halfway through Eurasia to access them and spend three years working in them.[31]

Alī b. Sahl Raban al-Ṭabarī wrote an encyclopaedic work on medicine, the *Firdaus al-ḥikma* (The paradise of wisdom). The *Firdaus* includes a long section dealing with Indian medicine. On the whole, this work is considered a compilation of Greek and Indian medicine showing an influx of Persian drugs.[32] His discourse on Indian medicine reflects knowledge of the main Ayurvedic classics, the *Caraka*, *Suśruta*, and the *Aṣṭāṅgahṛdaya saṃhitā* of Vāghbhaṭa. According to Siddiqi, these were translated from Sanskrit either directly into Arabic or via Persian during the early Abbasid period.[33]

In addition to the traces left by specific texts such as the *Siddhasāra*, we can also trace conditions and figures that facilitated such translations and transmissions. Such, for example, is the case of the Barmaks, as analyzed by Kevin van Bladel.[34] Barmak, the father of the Barmakid family, was an educated

Buddhist official from the city of Balkh in Tokharistan, in the valley around the upper Oxus. Thanks to Bactrian documents deciphered in the last two decades, we know that this was an area where Buddhism and its related Sanskrit sciences flourished at the time the Arabs arrived there in the late seventh and early eighth centuries. The Barmak family then became highly influential in the Abbasid court in Baghdad, and the Barmaks' grandson, Yaḥyā, became the tutor and then the powerful minister of the caliph Hārūn al-Rashīd (r. 786–809). Van Bladel has demonstrated how, as a result of his Buddhist roots and his family ties with Tokharistan and Kashmir, Yaḥyā facilitated a substantial translation enterprise from Sanskrit to Arabic in the caliph's court. A major outcome of this enterprise was the monumental translation of the Indian medical classics into Arabic: the *Suśruta*, the *Aṣṭāṅgahṛdaya saṃhitā* of Vāghbhaṭa and the *Siddhasāra* of Ravigupta, mentioned above. These same texts were also translated into Tibetan, a short while later, and thereafter became core texts in the Tibetan medical tradition too. Recent analysis has shown that Muslim physicians' and natural philosophers' interest in Asian works of natural knowledge at this time was not a passing episode but, rather, resulted in long-term influence.[35]

Multilingualism and Multicultural Aspects of Dunhuang

The study of multiculturalism on the Silk Roads has not received the attention it deserves yet, but the last few years have seen an immense growth. My discussion within the sphere of the history of medicine follows explorations of multiculturalism as they are studied in other Dunhuang and Silk Road areas of study, such as history of art, history of religions, literature, numismatics, economic history, history of ideas, and genetics.[36] In addition to tracing transmissions of key texts and key figures, we can also gain some insights on the processes of transmission of medical knowledge by analyzing specific locations of medical manuscripts in entire collections, coming from locations that served as nodes of convergence, such as Dunhunag and other Silk Roads sites.

The oasis of Dunhuang, where the northern and southern branches of the overland routes around the Taklamakan Desert met, was an important nexus of international trade routes in the ancient and medieval world. For most of its history, Dunhuang was a Chinese town. During the eighth century it became part of the Tibetan Empire and remained under Tibetan rule for about seventy years (781–848). The changing ethnicity of the rulers of Dunhuang had linguistic and cultural implications. In the second half of the first millennium, the cities of the Taklamakan oases were cosmopolitan seats of sophisticated cultures. Understanding the cosmopolitan nature of medieval Central Asia, where east, south, central, and western Asia interacted in multiple ways, is important for understanding not only the history of Asian medicine but the history of medicine and science at large primarily because of the input of West Asian medicine—exemplified in the works of Ibn-Sīnā (Avicenna) and al-Rāzī (Rhazes)—into Europe. The fascinating letter exchange between al-Bīrūnī (973–1048), born near the Aral Sea, and the slightly younger Ibn-Sīnā (980–1037), from Bukhara, reflects the existence of a large, competent, and interconnected community of scientists and thinkers.[37] Bukhara at this time was an important center of learning and culture.[38] Central Asia at this time was home not only to an assemblage of scientists and thinkers but also religious scholars, poets, artists, and musicians. This was truly an "Age of Enlightenment," which lasted several centuries, when Central Asia was the intellectual hub of the world.[39]

The search for Eurasian cultural interactions was one of the initial main driving forces for Aurel Stein's expeditions. In the application he made for financial support from the British government in India in 1897, Stein promised to supply tangible evidence of cultural exchange in ancient times.[40] The "library cave" that he discovered as a result—and its "polyglot" nature, as Stein put it—indeed supplied some of the evidence he was seeking. Stein himself realized the importance of the "polyglot" nature of the Dunhuang collection. Although Chinese is the most important language of the collection, there are also many documents in multiple languages (including some that were unknown at the time of their discovery): Tibetan, Sogdian, Khotanese, Sanskrit, Uighur, and Judaeo-Persian.[41] Other languages from nearby sites include Syriac, Turkish, Arabic, Tocharian A and B (Agnean and Kuchean) and others.[42] These languages and the cultures that they represent reflect both peoples who resided in Dunhuang and peoples who passed through Dunhuang such as traders, diplomats, and missionaries.

But does multiplicity also mean interaction? With the advance of research in the various languages of the manuscripts of Dunhuang, we are able to ascertain that the different cultures represented in the manuscripts found in the caves of Dunhuang and of other Silk Roads sites were not only present in the same locations but also interacted in multiple ways. Analyzing which texts were found where and in which languages can help us reconstruct processes of transmission: based on an analysis of the locations and numbers of Sanskrit medical manuscripts excavated in Kashmir, Haḍḍa, Kucha, Kyzil, Tuyuk, and Dunhuang, Ming Chen has reached the conclusion that the route followed by Indian medicine was predominantly from Kashmir via the Turfan area to Dunhuang.[43] The Dunhuang collection, more than being simply a "polyglot" library, mirrors some of this multiculturalism. It includes translations, transliterations, bilingual texts, glossaries, as well as "Berlitz style" phrasebooks for travelers—all pointing to active cross-cultural exchanges. Further evidence for cross-cultural interactions can be derived from evidence of diplomatic embassies and the documentation of traveling monks and of itinerant merchants.

Tibetan as a Mediating Language and Culture

Tibet was an important point on the trade route between India and China, not only for Buddhist missionaries but also for Arab and Jewish traders.[44] The main attraction of Tibet was its commercial goods—many of them lucrative to the point of being legendary, such as gold and musk. The case of Tibet as a cultural mediator stems from its key position and size during the time of the Tibetan Empire (seventh–ninth centuries), bordering with Tang China in the east and the Abbasid Caliphate in the west, as well as from the role of Tibetan as a *lingua franca* in and around Dunhuang. As a result of the Tibetan domination, the use of Tibetan prevailed among Tibetans and non-Tibetans and many of them became bilingual or multilingual. Tibetan apparently remained the most widely used second language among various local ethnic groups in and around Dunhuang including Han-Chinese, Khotanese, and Uighurs, until the early eleventh century.[45]

Multicultural Aspects of Tibetan Medicine in Dunhuang

Tibetan medicine developed as a synthesis of Indian, Chinese and Graeco-Arab medical systems.[46] The multicultural character of Tibetan medicine is emphasized in Tibetan medical histories starting from the earliest extant exemplars of this genre, the Tibetan medical history by Che rje zhang ston zhig po, dated to the early thirteenth century. Che rje sets medical knowledge within what he terms "The Seven Schools" (*lugs bdun*), referring to both divine and human realms.[47] Within the human realm, the list refers to medical systems from India, Kashmir, Urgyan (in present-day Pakistan), Nepal (*bal po*), Arabo-Persian (*stag gzig*), Dol po, Uighur (*hor*), Tangut/Xixia (*me nyag*), Khotanese (*li*), Byzantine (*phrom*), Chinese, and Tibetan.[48] Variations of this list become practically standard in subsequent Tibetan medical histories. Medical histories portray the earliest stage of Tibetan medicine as its most multicultural. Is any of this reflected in our earliest extant Tibetan medical sources: the Tibetan medical manuscripts from Dunhuang?

Mentions of Foreign Places in Tibetan Medical Manuscripts

There are two types of foreign locations mentioned in the Tibetan medical manuscripts from Dunhuang: locations associated with specific medical knowledge and places associated with specific materia medica or other imported goods. The foreign locations associated with imported goods are Persia (*ta zig*), Khotan (*li*), and possibly Kashmir (*kha che*). The manuscript Pelliot tibétain 127 includes a mention of Persia (*ta zig*) as a source of paper, saying: "If there is a bleeding from the nose, use paper from Persia."[49] Another Dunhuang manuscript, IOL Tib J 756, includes a mention of "sugar from Khotan" (*li*).[50] The same manuscript also includes a mention of a particular kind of silk, *kha che dar*—possibly referring to Kashmiri silk, or perhaps Persian brocade, which is to be used in the medical case of having a foreign body stuck in one's throat.[51] The Tibetan medical manuscripts from Dunhuang also contain many foreign names of materia medica, derived from Chinese as well as Persian.

The Tibetan medical manuscripts from Dunhuang mention a number of foreign locations as sources of knowledge. One is a mention of Turks (*drug gu*), referring to cauterization or possibly to bloodletting, saying: "The Turkic [*dru gu*] method [using] iron for cautery [? *sur phug*] is also suitable."[52] In the Tibetan Dunhuang manuscripts, the term *drug gu* or *dru gu* refers either to "Turks" in general or more specifically, particularly from the ninth century onward, to Uighurs.[53] More intriguing is the colophon of Pelliot tibétain 1044, which says: "This type of method comes from a land of the/an Indian king."[54] Interestingly, however, the moxibustion methods described in Pelliot tibétain 1044 are not known to have been used in classical Indian medicine.[55] The next line of the colophon explains that this technique derives from *ha ta na bye*, a Tibetan transliteration of the old Khotanese name *hvatana* for Khotan. Chinese and Tibetan accounts concerning the foundation of Khotan associated the kingdom with the son and ministers of Emperor Aśoka, hence "the Indian king." A number of Khotanese texts within the sphere of popular ritual medicine bear resemblance to Chinese texts, including one that has

Khotanese and Chinese parallel texts. The knowledge reflected in Pelliot tibétain 1044 is predominantly Chinese. So it appears that what is represented in this manuscript is a Tibetan adaptation of a Khotanese medical text, which preserved Chinese notions.

The third mention of a foreign land as a source of knowledge is the reference to Zhang zhung.[56] This appears in the colophon of Pelliot tibétain 127: "This text [*yig*] on medical practice [*dpyad*] is not even [to be found] at the archives [*phyag sbal*]. It is a compilation of all traditions of medical practice [*dpyad yig thams cad*], in addition to being compiled according to the indigenous [?*phugs pa*] medical practice [*dpyad phugs*] of Zhang Zhung."[57] Among all the mentions of foreign places in the Tibetan medical manuscripts from Dunhuang, there are no references to China. How can we explain this lack? One plausible explanation is the omission of the obvious. The medical culture of Dunhuang that transpires in these Tibetan texts can be best described as being of Sino-Tibetan nature. A comparison between the Tibetan and Chinese moxibustion texts from Dunhuang reveals striking similarities in both form and content. The Tibetan and Chinese texts are structured in similar ways: a list of symptoms, the location on which to apply moxa-cauteries, and the number of moxas to be burnt. The descriptions of ailments that are to be treated by moxa-cautery are also similar. Both the Chinese and Tibetan materials deal primarily with wind (T: *rlung*, C: *feng*) related ailments; genito-urinary and reproductive disorders as well as digestive and abdominal illnesses. Some of the descriptions bear almost literal similarity. Such, for example, is the description found in the Chinese manuscript S6168 dealing with "wind in the face as if there are insects on the face," which bears almost literal similarity to what we find in Pelliot tibétain 1044: "[If] due to wind [*rlung*] illness swellings are forming on the face, and there is itching like a walking insect."[58] It is interesting to note here that the Tibetan measurement word *tshon*, which is a loanword from the Chinese *cun*, is mentioned quite frequently in the Tibetan moxa-cautery texts from Dunhuang.

Another key point of similarity between the Chinese and Tibetan moxa-cautery texts is their links with time reckoning and divination. The Dunhuang manuscript Pelliot tibétain 127 includes not only a medical text on moxibustion (recto, lines 78–184), but also several divination and calendric texts. These different texts appear to have been written by the same hand. We also find such juxtapositions of medical and divinatory texts among the Chinese texts.

The link with Chinese notions is reflected in Tibetan divination texts from Dunhuang that refer to Confucius (rendered as Kongtse) as their author.[59] Among the Tibetan manuscripts from Dunhuang we also find Tibetan translations of Confucian maxims.[60] Indeed we know that Confucian doctrines were taught in the prefectural Dunhuang school and that special rites for his worship were conducted twice a year at the time of the equinoxes, in which the physicians of Dunhuang participated.[61]

One of the texts of Pelliot tibétain 127 is believed to be the earliest Tibetan delineation of the sexagenary cycle. This type of calendric cycle is used in Pelliot tibétain 127 (verso, lines 1–9) for divinatory/astrological purposes.[62] The text preceding the medical one on the recto is also a divination text (recto, lines 1–77). This divination text begins with the words: "Formerly, the gifted man of magical faculties [*'phrul gyi myis*] established this text of divination [astrology, sciences (*gtsug lag*)] as a model [*dpe*] for future generations. It deals with the positive and negative [aspects] of the level of prosperity [*dbang btang che chung*], years of life [*lo srog*] and power [*mthun*]."[63] These categories, which are derived from Chinese divination, are very similar to the ones we know from later sources in Tibetan divination, such as vitality (*srog*), body (*lus*), destiny (*dbang thang*), and luck (*rlung rta*).[64] The mention of the "gifted man of magical faculties" (*'phrul gyi myis*) probably refers to Kongtse.[65] This section of Pelliot tibétain 127 is very similar to IOL Tib J 748, another Tibetan divination text from Dunhuang, ending with what appears to be either a very early Tibetan reference to the Yijing or to divination in general.[66] The two texts are probably two copies of the same original.[67]

This evidence from the Dunhuang manuscripts of an early point in time when Chinese divinatory ideas were brought into Tibetan culture is further corroborated by two other types of evidence. The first are Tibetan narratives ascribing the transmission to the Chinese princesses Wencheng and Jincheng, who married Tibetan kings. The second comes from linguistic data analyzed by Berthold Laufer. Based on the form of the names of the trigrams as they appear in Tibetan in comparison with the Chinese, Laufer has shown that the Tibetan transcriptions have partially preserved the ancient initial consonants and the ancient finals of the Chinese, hence concluding that the transmission occurred during the Tang period.[68]

A key issue regarding the Tibetan texts, which is shared with the Chinese moxibustion charts, is their relation to the notion of "channels." Vivienne Lo has shown that the Chinese moxa-cautery charts from Dunhuang bear no explicit relationship to the concept of circulating *qi* through an integrated network of channels.[69] This characteristic of the Dunhuang Chinese material is shared with the Tibetan moxa-cautery texts. In this sense the Dunhuang Sino-Tibetan texts appear to be more similar to what we find in Tibetan received texts (e.g., the *Gyushi*) than what we find in Chinese received texts.

With the increasing penetration of the Tibetan script and language in the Han-Chinese society of Dunhuang during the period of Tibetan rule, a practice gradually developed among some local Chinese inhabitants of using the Tibetan script rather than Chinese characters to write Chinese. The influence of the Tibetan rule culminated in the creation of bilingual Tibeto-Chinese communities. Among the Han-Chinese of Dunhuang there seem to have been some who were fluent not only in their native Chinese but also in speaking, reading, and writing Tibetan. There is also evidence that these "Tibetanized" Chinese formed associations or communities, and the practice of writing Chinese using Tibetan script not only was carried out during the Tibetan rule but was being maintained in the tenth century under the Return-to-Allegiance Army.[70] This has resulted in the survival of various kinds of Chinese texts transcribed in Tibetan. Broadly speaking, these consist of Buddhist scriptures, Buddhist eulogies, songs and poems, primers, and a multiplication table.

Within the vast enterprise of sutra copying that took place in Dunhuang during the Tibetan rule, the majority of those responsible for copying the Tibetan texts were in fact Chinese. This would have required sufficient knowledge of Tibetan, or at least of the Tibetan script. The copyists were not simply copying Tibetan graphs mechanically, but they had considerable knowledge of the Tibetan language as well. These Sino-Tibetan aspects bring us to reflect on the nature of the Tibetan medical texts. Rolf Stein, who studied the Tibetan texts of Confucian maxims from Dunhuang, has pointed out that they are not straightforward "translations" but rather "variations." It is impossible, Stein noted, to know whether the texts we have at hand were composed by Tibetans or whether they translated or paraphrased a Chinese text. We are faced with a similar problem regarding the Tibetan moxibustion texts from Dunhuang. But perhaps rather than trying to label them as "Tibetan medicine" or "Chinese medicine in Tibetan" we should regard them as "Sino-Tibetan medicine" or "Dunhuang medicine."

I would like to return to the question of why we were astonished to discover connections between Tibetan and Arabic/Persian medicine with which I began this chapter. Victor Mair, in his brilliant introduction to *Contact and Exchange in the Ancient World*, provides an illuminating discussion of what he terms the "academic pathology" of "extreme indigenousness": a bias to disregard the plethora of data indicating contact and exchange among early people. He considers this bias a result of two factors, one political and the other disciplinary. The political, he points out, is particularly true of the second half of the twentieth century, the outgrowth of a surge in nationalistic consciousness, when "it became impolitic to assert that any significant element of culture needed to be borrowed."[71]

Although he locates this political factor within the twentieth-century nation-state, we can also see it at work in accounts of the history of knowledge. Although many—perhaps the majority, or even all—cultures in different periods regularly adopted and adapted knowledge, in most cases these foreign influences were not acknowledged. Indeed, as Arun Bala has argued, when well-developed cultures undergo intellectual change through knowledge transmission from outside, they often deny the significance of the external culture by ascribing the changes to their own traditions.[72] What Bala's analysis shows is that what matters more than the adoption of knowledge itself is whether a culture—at any particular time and place—also cares to acknowledge it. Similarly, when we look at the early history of medicine and analyze the correspondence (if any) of *being* multicultural and of *declaring* yourself as such, we find that in some cultural contexts foreign knowledge is absorbed into the main culture without any reference to its origins while in other cases we find elaborate descriptions of the sources of foreign medical knowledge.

The multicultural character of Tibetan medicine is emphasized in Tibetan medical histories starting from the earliest extant exemplars of this genre—the Tibetan medical history by Che rje zhang ston zhig po and its numerous variations thereafter. The colophon of Pelliot tibétain 127, speaking about "a compilation of all traditions of medical practice" in addition to being according to the medical practice of Zhang Zhung can be viewed as an early precursor of these accounts from Tibetan medical histories emphasizing the multicultural nature of early Tibetan medicine. Acquiring medical knowledge from different cultures appears to be celebrated in the Pelliot tibétain 127 colophon, and the results of this endeavor is regarded as being superior to what might be found "at the archives." This is quite different from what we can gather from the Chinese material. Illuminating in this respect is the preface to a Chinese medical manuscript from Dunhuang, Pelliot chinois 2675. As analyzed by Lo, Pelliot 2675 is a moxa-cautery canon that is an abridgment of moxa-cautery techniques of a number of teaching lineages.[73] It was produced in the capital with the purpose of providing a practical text for those who lived in the "outlying regions" and could not obtain sophisticated drugs. The text hence situates itself as disseminating a simplified, practice-oriented knowledge from the capital to the provinces. Here, the knowledge of the "outlying regions" is viewed as inferior to that of the center.

In the Tibetan Pelliot tibétain 127, on the other hand, the location of Dunhuang—and the availability of a variety of medical traditions there—appears to be regarded as advantageous for the author of this manuscript. In similar later accounts this attitude vis-à-vis foreign medical knowledge continues. The references to foreign knowledge in Pelliot tibétain 1044 and Pelliot tibétain 127 are indicative precursors of two distinct characteristics that are key to Tibetan medical history: the influence of foreign knowledge and the acknowledgment of it.

The Tibetan accounts of the sources of its medical knowledge often contain mythical elements, and so while we cannot read them as straightforward historical narratives, we can—and should—take some cues from such texts.[74] These accounts often serve as pointers to strata otherwise forgotten or else rewritten by later historical accounts. Literary narratives of universal histories of knowledge such as we find in the Tibetan medical accounts exemplify different ways of managing relationships between foreign and local knowledge and of negotiating cultural differences. The organization of knowledge from and about different peoples has been a powerful political tool, making claims to comprehensiveness and unity across multiplicities of locales.[75] Such accounts may be seen as a subcategory of "origin narratives" as discussed by Sonja Brentjes: they speak of the origin(s) of a field of knowledge, the motivation for "inventing" or "establishing" that kind of knowledge, and the field's subsequent development.[76] The Tibetan medical manuscripts from Dunhuang reveal the value of looking at sources in the "bridging" languages of the Silk Roads, such as Tibetan, Sogdian, Uighur, Tocharian (A and B), Bactrian, Khotanese, Syriac, and Persian. Through focusing on mediating cultures and languages, multicultural locations and collections as well as specific key texts and figures—such as the ones I have discussed above—we can begin to address the great puzzle of Eurasian history of science.

Chapter 4

Things (*Wu*) and Their Transformations (*Zaowu*) in the Late Ming Dynasty

Song Yingxing's and Huang Cheng's Approaches to Mobilizing Craft Knowledge

Dagmar Schäfer

Historians Timothy Brook and Craig Clunas, among others, have shown how manufactured goods—from food to craft products—acquired an important role in the late Ming dynasty.[1] As society became commodified and markets flourished, an increasing number of wares and raw materials became available from all corners of the empire. Browsing such markets within and outside Ming territory, the elites, rural gentry, and merchants discovered commonplace (*bao* 寶) or precious (*gui* 貴) and sometimes even exotic (*qi* 奇) "things" (*wu* 物) from places that were beyond the civilized world of "all under heaven" (*tianxia* 天下). Economic, social, and intellectual values and practices all changed. For instance, in response to these new opportunities local officials, elites, and the state all had to enhance the methods they used to verify an object's origin and follow its passage through time and space. Owners began marking their possessions, such as the calligrapher Zhou Tianqiu 周天球 (1514–1595), who had even his favorite chair inscribed.[2] Compilers of "Local Gazetteers" (*Difang zhi* 地方志) became increasingly meticulous about documenting regional tax and tribute wares. All of these practices could themselves add market and social value to a commodity: ceramic teapots were considered most precious if they originated from Yixing district, for example, whereas wine from the shores of the Shaoxing River was preferred.[3]

As such research has also shown, this growing mobility of goods, people, and materials conferred increasing significance to items as markers of cultural, social, and geographic belonging, as knowledge about an object's origin and history of use gained currency. Moreover, scholarly debates about authenticity and fraud can be read as a response to the challenge posed by such movables to moral, social, and material orders. In fact, intellectual orders were also on the move in this era. Two prime examples of this are the literary works of the artisan scholar Huang Cheng 黃成 (active ca. 1567–1572) and the minor official and scholar Song Yingxing 宋應星 (1587–1666?). Both men stand out in Ming scholarly literature because they assigned an epistemic function to processes of material production and use. This approach rocked the cosmological foundations of that era's approach to the "ten thousand things" (*wanwu* 萬物), in which arts and crafts and their products did not usually play any role.

The focus of this chapter is on the kinds of order that these two authors envisioned and how they reflected upon the changes to movable things and their movements. The authors are rarely discussed together, since Huang and Song worked with crafts in different ways. Huang, in *Xiushi lu* 髹飾錄 (Record on lacquer decoration, with the preface by Yang Ming 楊明 [active ca. 1621–1627] in 1625), tackled only one craft—namely, lacquer. Song, in *Tiangong kaiwu* 天工開物 (The works of heaven and the inception of things, 1638), in a rather more comprehensive way engaged practically in eighteen fields as different as agriculture, textile production, metalworks, mining, yeast production, and pearl diving. Both authors' approaches were equally and squarely situated within a cosmological view on material complexes. They were both searching for universal patterns, ultimate truths, and authenticity in the practical arts rather than focusing on consumption or on averting fraud or fakery.

In this chapter I begin by briefly locating each author's approach in their philosophical and historical context. This is followed by a comparison of their views on cosmology and crafts and an exploration of the role that each of them assigned to spatial and geographic characterization in the making of things. By examining how their shared cosmological framework and approach to transformative processes and creation affected their views on the corporality of various products and their movement in time and space, I hope to offer new insights into the relationship that Chinese intellectuals drew between geographic variance, local specificity, and claims to universality.

The Ten Thousand Things in the Late Ming

Song and Huang were pondering the intellectual role of art and craft products in a period where ruminations on materials and material complexes philologically and philosophically were generally undertaken in the context of the "ten thousand things" (*wanwu* 萬物), a term that exemplified the multifarious nature and forms of the world that people lived in, which comprised every organism (including ghosts), mineral, and human. The commentator Kong Yingda 孔穎達 (574–648), for instance, explained things as "the external environment" (*waijing* 外境), which he set out in a functional contrast with a person's inner self, their "heart-mind" (*xin* 心).[4] Late Ming philosophers would indeed consider the search for one's heart-mind as the royal way (*da dao* 大道), following the Song philosopher Zhu Xi 朱熹 (1130–1200) who identified things as being everything outside the self: "all the phenomena one can observe when looking up into the sky, as well as those methods observable on earth when looking down" (仰則觀象於天, 俯則觀法於地).[5] Zhu Xi also emphasized the importance of "investigating things and broadening knowledge" (*gewu zhizhi* 格物致知), a maxim that Jesuit missionaries such as Matteo Ricci (Li Madou 利瑪竇, 1552–1610) appropriated when accommodating European *scientia* into the Ming court in the mid-seventeenth century.[6] Although there were many overlaps among all these thinkers, those who mobilized things within a cosmological viewpoint were not always in agreement with those who moved things for the benefit of society and state—the civil servants who were following the minor way (*xiaodao* 小道). Both philosophers and civil servants would largely have agreed that much could be known *about* crafts. They would also have concurred that little was revealed *by* them.

Song and Huang departed from this contemporary mainstream stance. Their main works tackled the otherwise widely ignored subject of craft production as a series of epistemic processes. Huang's *Xiushi lu* and Song's *Tiangong kaiwu* share other features too. For instance, the two authors based their theoretical approach on *Yijing* 易經 permutation studies and discourses of *qi* (氣), *li* (禮), and the five phases (*wuxing* 五行).[7] Both authors considered the "making of things" (*zaowu* 造物) to be indicative of "transformative processes" (*zaohua* 造化) that again revealed moral and epistemic orders. By comparison, scholars from the next generation such as Wang Fuzhi 王夫之 (1619–1692) and Li Yu 李漁 (1611–1680) discussed manufacturing within the arts as a matter of human creativity.[8] Chinese historians have described this as a shift from the idea that "utensils were used to convey the *dao*" (*qi yi zai dao* 器以載道) to the notion that "the utensils [were] for substance, the *dao* for function" (*qi ti dao yong* 器體道用).[9] With this return to a utilitarian view of crafts in the wake of a political change from the Ming reign to the Qing reign, scholars seem to have become even less interested than before in Huang Cheng's and Song Yingxing's tracts on things. Both monographs, which drew theoretically on the *Xici* 系辭 chapter of the *Yijing*, had disappeared from China by the early eighteenth century, even though they contained information on arts and crafts that was not discussed elsewhere.[10] They survived mainly through their transmission to the Japanese Tokugawa period. I have argued elsewhere that when Song Yingxing connected *Yijing* change closely to material change this may have confused the scholars of his time.[11] The same then might be applicable to Huang Cheng. Although many readers may have shared similar ideas about the materialist grounding of a cosmological principle (and a philosophical theory), such a practical approach may not have appealed to most intellectuals of this time and may even have repelled them, resulting in the readers' merely extracting information about art and craft products from the publications and ignoring their cosmological approach.[12]

Despite their commonalities, these two works are rarely discussed together. As the only treatise on lacquer in the Ming, the *Xiushi lu* is mostly considered an expression of late Ming connoisseurship, whereas Song Yingxing's *Tiangong kaiwu* has long been venerated for its encyclopedic style.[13] In this chapter I aim to address this oversight.

Recording Crafts

> Good utensils are like the four seasons. Beautiful materials are like the five phases. As the four seasons move along and the five phases are completed, things are brought into being. The four good virtues [*deyi* 德義] unite, the five colors [*wucai* 五采] are prepared, and finery [*gongqiao* 工巧] is created.
>
> **Huang Cheng, *Xiushi lu*, preface**

> Each manufacture [*zuo* 作] that generates utensils and things imitates the creative transformations [*zaohua* 造化] of heaven and earth. Thus the sagely [*sheng* 聖] and divine [*shen* 神] all effectively apply methods [*fa* 法]. For this reason, innate, talented work [*lianggong* 良工] is good for utensils.
>
> **Song Yingxing, *Tiangong kaiwu***

What unites Song and Huang is that they found a solution for the challenge of changing material orders by identifying order in the processes of material production and use, whereas most scholars of that generation tried to stay as far away as possible from such diverse material, instead finding solace in classic texts and abstract contemplation. Rather than elaborating on philosophical and moral grounds, Song and Huang straightforwardly recounted the materials and the working processes of mundane tasks. Both grounded their works theoretically in *Yijing* permutation studies, notions of *qi*, *li*, and the five phases (*wuxing* 五行). That such theoretical underpinnings were more than rhetoric (and more than a means to elevate the value of craft expertise) is evidenced by the fact that the concept of change permeates the argumentative structures and contents of both books. The lacquer master Huang Cheng explicitly divides his book into two sections, one on *qian* 乾 and one on *kun* 坤, the hexagrammatic equivalents of heaven and earth as complementary forces within *Yijing* permutation studies. For Song Yingxing, the *Yijing* provided the structural scaffold for a chronology of eighteen different crafts, while the interchange between yin and yang explained why and how to smelt the five metals.[14]

These authors analyzed manufacturing processes in strikingly similar ways, but their approach differed in focus. Huang Cheng only considered lacquer and its procedural logic, whereas Song Yingxing identified the logic of yin-yang permutation within and across various material practices.[15] It could be said that, for Huang Cheng, yin-yang explained the procedural logic of lacquer production, and for Song Yingxing, any craft production could reveal the workings of yin-yang and *qi*.

These differences in approach and objective are evident in the way that each volume is constructed. Huang Cheng distinguished between naming conventions and working procedures—that is, naming the linkages and assistances (*mingming fuzan* 命名附贊) on the one hand and the format and design, materials, and usages (sorting action, *fenlei jushi* 分類舉事) on the other. These two parts (*ji* 集, further divided into 18 *zhang* 章, 186 *tiao* 條) were complementary, the two sides of a coin, because names and working procedures were subject to *qian*, the heaven, while format and design were related to *kun*, the earth. Although Song Yingxing explored a universal pattern of permutation across things, finding order within their manufacture, he arranged his work around a classic view of civilizational change, from the production of food to the casting of bronze bells. For Huang Cheng, a thing's "becoming"—that is, its transformation into a utensil—constituted the structural framework of its existence.

Huang Cheng compiled his work almost a century before Song Yingxing, although it is worth noting that his tract rose to prominence mostly during Song Yingxing's lifetime, when Yang Ming published a new, annotated (*zhu* 注) version.[16] Chinese scholars have speculated that Huang may be associated with a story in the *Jiaxing fuzhi* 嘉興府志 (1685), which records a lacquer craftsman who came from Xitang 西塘 in Jiaxing prefecture with this name. Xitang, by the Yuan period (1279–1368), had already earned a reputation as a center for lacquer production, and it was the birthplace of Zhang Degang 张德剛 (active ca. 1403–1424), who regularly signed his works in conspicuous places. Zhang is known to have worked with Yang Mao 楊茂 (active ca. 1350–1380), who mastered the art of lacquer carving. One piece signed by him is now held in the National Palace Museum, Beijing. Later authors were thus clearly attempting to situate Huang Cheng within a locally defined and anchored trajectory of lacquer crafts, rather than trying to identify Huang's native origin for authenticity reasons as this was done at the same time in European crafts. We can see how both authors, although equally grounding their claims to universal principles in the *Yijing*, vary substantially in their view of regional differences.

The *Xiushi lu*'s Organizing Principle and Lacquer Production

> Today I name the nearby and supportive [*mingming fuzan* 命名附贊], elaborating it as *qian* [first hexagram of the *Yijing*, i.e., heaven, yang] collection. *Qian* is at the beginning, giving birth to the ten thousand things, and lacquer tools and working regulations [*gongze* 工則] that are the original *qi* of [all] finery. The deed of *qian* is indeed great.
>
> **Huang Cheng,** ***Xiushi lu***

By revealing the chronology of production, Huang Cheng was following standard accounts of agricultural and craft works, yet he opened up an unprecedented literary space for the topic of making and carving lacquer wares.[17] Throughout all the steps described, he quoted the *Yijing*, thus explicating crafts as transformative processes. However, he never mentioned the origin of his raw materials or where the products were eventually traded or brought to use. By comparison, in his *Tiangong kaiwu*, Song Yingxing mainly quoted the *Yijing* in the introductory passages to each chapter, thus emphasizing the structural qualities of crafts and their role for society and state. Song's arrangement of topics directly challenged traditional frameworks of scholarly debate. For instance, his colleagues would have considered sericulture an agricultural (*nong* 農) topic, but Song viewed silk as part of the discourse on "clothing" (*naifu* 乃服). Song's chapters followed a logic progressing from raw material to the markets for finished products, succinctly identifying their places of origin, production sites, and market places, too.

A perfectly valid explanation for such a different awareness of regionality would be the social conditions in which Song Yingxing and Huang Cheng grew up and lived. Although both grew up away from the politically influential centers of Ming power, they originated from slightly different social strata: Song from a scholar-official family and Huang from an artisanal or merchant-official background. Both authors and their scholarly works can indeed be situated within a broader intellectual atmosphere of commercialization, in which markets were expanding and products were being traded over ever-increasing distances while elites pondered the management, organization, and ordering of goods and people within the society and state.[18] But Huang Cheng came from a vibrant market place, Huizhou prefecture in Anhui province, whereas Song Yingxing was from a rather rural place in Nanchang, south China. Very little is known about Huang Cheng except that he was an expert on lacquer craft in a region of state-owned lacquer production. He hailed from one of the newly prospering book centers; by the late Ming period Huizhou was brimming with book-carving businesses.[19] Hence, this place naturally also produced high quantities and qualities of ink, while changes in salt taxation turned Huizhou merchants into some of the richest inhabitants of China, often operating on a global scale. Huizhou merchants "traveled throughout the country in search of profits," bringing the world to Huang Cheng.[20]

In contrast, Song Yingxing lived in a rather small place. The markets of Nanchang were less exuberant than those of Huizhou, although they still seem to have contained a panoply of nonregional products. In comparison to Huang, who had the world's goods delivered to his doorstep, Song may himself have traveled more. He made his way to Beijing six times for imperial exams and was appointed to posts both near the coast and north of China. Such divergent experiences of mobility are one good explanation for the varying degrees of geographic awareness reflected in these authors' works. Song was well aware of the geography of his empire in terms of raw materials and production, as well as both imperial and tributary trade, even though he may also have drawn some of this information from books or from his colleagues. Certainly Song's framework of endeavors encompassed many more crafts than just lacquer, as in Huang's case, so he acknowledged a vast landscape of production, use, and trade in terms of both the country (*guo* 國) and all under

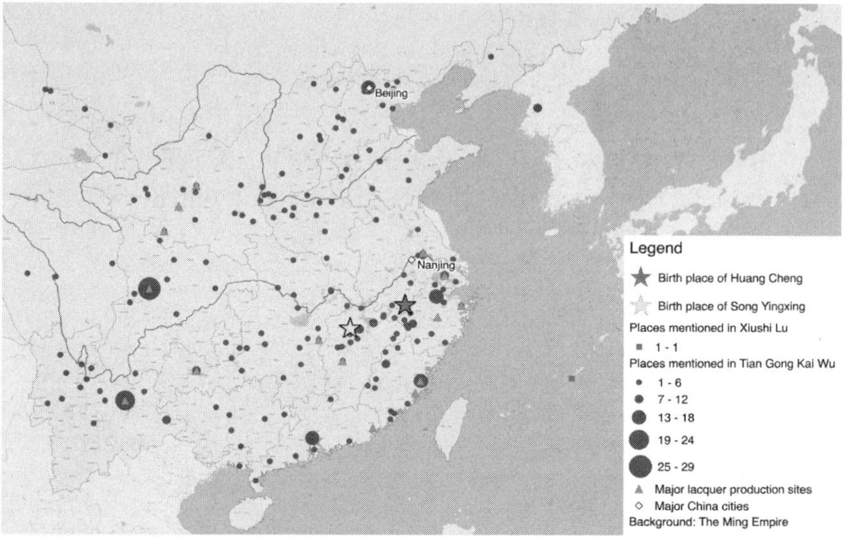

FIGURE 4.1. The geographic mobility of crafts. Huang Cheng, author of the book on lacquer *Xiushi lu*, grew up in Huizhou, a major marketplace during the late Ming. Song Yingxing, who wrote the compendium of crafts *Tiangong kaiwu*, was born in Fenyi, located in China's rural south. Huang Cheng did not mention any of the places renowned for lacquer production during the Ming and only once mentioned a specific locality, the marginal Ryukyu Kingdom (now Okinawa, Japan). Song Yingxing, by comparison, was acutely aware of the Ming's geography of production and use, although it seems he gathered most of his information from his peers and from texts. The map shows the major lacquer production sites of the period, and the over 100 localities mentioned in the *Tiangong kaiwu* as well as the single mention in *Xiushi lu*, at the extreme right of the map (on the islands).

heaven (*tianxia* 天下). He noted, for instance, that wild silkworms could be found in Qingzhou and described Gaozhou's minting technology. Both places are some distance away from his birthplace as well as from the posts to which he was appointed while compiling his *Tiangong kaiwu*.[21] In fact, Song mentions a whole range of places that he never visited but may have found out about from books or in conversation with colleagues in local, regional, or even higher central state posts, such as the sponsor of the *Tiangong kaiwu* Tu Shaokui 涂紹煃 (1582–1645?). Huang, by comparison, rarely mentioned other places at all, not in terms of divergent lacquer techniques of products nor in terms of possible origins of raw materials or markets, even though regional variance was huge and such variance was certainly available on Huizhou markets too (see figure 4.1).

We can gather from this that for Huang Cheng regional variety was neither important nor a challenge to universal claims; or that such variance was at odds with his claims. Clearly, his aim was to establish his trade, lacquer, as a series of transformative processes in which the combination of tools and working rules represented *qian* and *kun* corresponding to design (i.e. also *shanggu* 上古, archaism), defining "the creation of phenomena as *kun*, and its effectuating methods as *qian*" (成象謂坤, 效法之謂之乾). He argued that, "with the manufacture of utensils the phenomena [of heaven, i.e. the ten thousand things] are esteemed." He suggested that lacquer tools were wrought through the transformations of ten thousand things (*wanwu zaohua* 萬物造化) to serve as the original *qi* or "raw material" for the composition of lacquer wares. *Qian*, however, also included repeating procedures and following examples, as portrayed in the term "effectuating methods" (*xiaofa* 效法).[22]

For Huang Cheng, the processes and tools of production—whether using a brush to paint patterns or a knife to carve decorations—lay at the heart of classifying design along the yin-yang spectrum. He argued that materials (*zhi* 質) constituted the yin, and decorations or "patterns" (*wen* 文) the yang. The intricacies of designs could also be derived from the *Yijing*, which could refer to schemes such as the magical patterns of writings from the Luo River (*Luoshu* 洛書) and the river map (*Hetu* 河圖)—that is, a genre of powerful objects and magical texts represented by diagrammatic forms.[23]

Huang Cheng believed that cosmological principles exerted an influence over human creativity and skill in making an object. He suggested "the skillful [*qiao* 巧] method of transformation" as one of three ways (*sanfa* 三法) to produce and use lacquer wares (*qiqi* 漆器).[24] "Every utensil and thing transformed by a worker likened the transformations of heaven and earth, some so as to contain strangeness, others so as to contain divineness. All successfully use methods that enable innate goodness to be used as ritual vessels. Then, superior utensils are likened to the four seasons and the beauty of their materials likened to the five phases" (凡工人之為器物, 猶天地之造化. 所以有聖者有神者, 皆示以功以法, 故良工利其器. 然而利器如四時, 美材如五行).[25] Hence, Huang Cheng considered superior utensils and beautiful materials two basic requirements for the transformation into objects. Therefore, skillful work (*gongqiao* 工巧) was a condition of creating luxury goods.[26] His interpretation extended the notion in the *Kaogong ji* 考工記 (Artificer's records) that "the material is beautiful and the work skillful."[27] Regional variance had no role in this, as Huang was trying to establish universal claims, not list where lacquer could be created, produced, or bought and sold.

Tiangong kaiwu, the Order of Crafts

Whereas Huang Cheng used the *Yijing* as a way to frame lacquer production and thus challenge scholarly views about what comprised a legitimate intellectual topic, Song Yingxing opposed scholarly learning when he translated the *Yijing*'s scheme of human civilizational logic into a chronology of types of crafts. Song argued that the *Yijing* justified a progressive sequence of crafts, beginning with the production of food, ploughshares and other agricultural tools, followed by clothing (dyes were an addendum), transportation (boats and carts), and the production of tools (such as pestles and mortars) and ending with weaponry, pearls, and gems. Such a civilizational logic had its geographic center of gravity in China (*zhongguo* 中國), where the general principles were understood properly, even though in principle the logic itself was evident everywhere.

Song's listing of the practical arts concurs with the *Ru*-Confucian ideal of "sustenance before luxury." His corresponding classification of content, however, subtly challenged traditional ideals. For instance, Song placed sericulture and silk textile production in a section entitled "clothing" (*naifu*), a topic that was usually considered an agricultural theme in the genre of "agricultural books" (*nongshu* 農書), lumping silk in with other fibers, including felt and fleece. Although *naifu* is not an unusual term in Chinese literature, it is not used as a separate heading in any other work.

How Song's view on the role of regions diverged from those of his colleagues is particularly evident in comparison to other writings that viewed sericulture as part of agronomy. Francesca Bray has brought attention to the high status attained by the science of agriculture—agronomy—in Ming culture, when it was considered to be a moral obligation of the state and rulers and a venerable scholarly field of study.[28] This was because sericulture was understood to concern people's well-being, so dynastic rulers commissioned publications that described silk technologies and explained the fabric's social and political functions. In such books, as well as various other kinds of writings, treatises, collections, and monographs, sericulture featured mainly as a cultivation technology, sometimes appearing as a periodic task within seasonal processes, sometimes organized into chronologies of production. The state-issued collection *Nongsan jiyao* 農桑輯要 (Essentials of agriculture and sericulture, 1273), for example, juxtaposed sericulture with agriculture in its title and structure. Consequently, learned men, the gentlemen (*junzi* 君子), were required to engage with a theme that had substantial social impact and conveyed political values and ideals, such as the gender-based distinction of production techniques in which "men till, women weave." This notion was regularly traced back to Confucius, who was increasingly invoked by Song dynasty scholars as sericulture became a major occupation of noble households and an important source of income for scholar officials. By the Tang era (618–907), however, silk reeling and weaving, for instance, must have become viewed as male work, since they were pursued in (male-dominated) Buddhist monastic settings.[29] Wang Zhen 王禎 (1271–1368), whose agricultural treatise featured a broad range of tools and implements, also covered sericulture, discussing it in a procedural order from cultivation to processing. The few texts focusing on silk alone, such as the *Canshu* 蠶書 (Treatise on sericulture) by Qin Guan 秦觀 (1049–1100) or the much later collection *Cansang jiyao* 蠶桑輯要 (Compilation on sericulture) by Shen Bingcheng 沈秉成 (1823–1895) likewise depicted silk production from raw material to end product.[30]

All of these works described a procedural operational chain that aimed to spread knowledge about efficient techniques to other regions. The writers emphasized regional specificity (often using this to explain why they had written a book), frequently taking great pains to detail the local accessibility of materials, goods, people, knowledge, and skills. It is unsure who the intended recipients of this information were, but, as no one else could read, it is likely that the authors were writing mainly for their peers, such as the Chinese poet scholar Yang Shen 楊慎 (1488–1559), or landlords and officials such as Chen Hongmou 陳宏牟 (1696–1771).[31] For Song Yingxing, this procedural chain underpinned the logic of *qi* and thus indicated that a universal order existed that was valid in realms beyond sericulture and silk. Since he considered validating these principles as the primary purpose of his work, he was less concerned with providing information about locally specific materials or crafts in a systematic fashion.

His contemporaries were writing about silk in order to disseminate technological expertise—that is, detailing how to plant mulberry trees, raise silk worms, spin, reed, and weave—but Song meticulously noted down the silk processing techniques and specific products that explained the subtle balance of needs and demands in this technical system, claiming that its inherent causal nexus revealed underlying universal schemes. Thus, he described the structure of a loom and included an overview of different fibers, felt, and furs in one section of his *Works of Heaven*. He did not record things such as where to find the best loom carpenters or describe what distinguished silkworm breeding on the plateaus of inland Sichuan province from the methods used in the coastal regions of Jiangnan.

A further apparent difference is that Huang Cheng concentrated exclusively on production and cosmological issues whereas Song Yingxing took the raw materials, final products, and markets into account as well. This difference is also reflected in the two men's divergent use of classical literature. The practitioner craftsman Huang Cheng mainly referred to the *Yijing*, while Song Yingxing's more scholarly approach also referenced statecraft mythology such as the Yellow Emperor (*huangdi* 黃帝), Yao 堯 and Shun 舜, who "handed down upper and lower garments and thus the world was properly ordered."[32] Song also invoked classics such as the *Lunyu* 論語 (Analects), to legitimate the relevance of topics such as dyeing fabrics: "A gentleman does not wear facings of purple or mauve, nor does he use pink or roan in his underwear. In hot weather he wears an unlined gown of fine thread loosely woven, but puts on an outer garment before he leaves the house."[33] In identifying clothing as a signifier of status, Song thus ambivalently venerated traditional values and clothing as markers of social identity, in line with the philosopher Xunzi 荀子: "The nobles were given garments, decorated with sparkly mountains and dragons to govern all under heaven.[34] The lowly people wear coarse hempen jackets or woolen clothing, in winter to protect themselves against the cold, and in summer to cover their body from nakedness so as to distinguish [themselves] from animals" (貴者垂衣裳, 煌煌山龍, 以治天下, 賤者裋(shu)褐, 枲(xi)裳, 冬以御寒, 夏以蔽體, 以自別于禽獸).[35]

Song continued by admonishing his readers to be modest, asserting that the sole function of clothing was to provide protection against the cold and for defined purposes that "this material is brought forth by cosmic creational forces" (*zaowu* 造物). "Zaowu" is a general term that embodies natural creative forces without excluding humans. This creative force is manifest in the spontaneous and self-regeneration process of all cosmic activity.[36] Instead of categorizing silk as an advanced method of producing cloth (the Yellow Emperor was said to have bought silk to replace furs and feathers) and emphasizing the thread's quality as a fiber, he grouped silk with furs and pelts "of animal origin": "[Materials] subordinate to plant origin are hemp, *meng* hemp, and *jie* hemp. Of animal [literally: Song distinguishes between wild beasts and insects] origin are furs, wool, silk, and spun silk. Assuming that each kind [of source] provides its share, there is sufficient clothing [屬草木者為, 麻, 萵, 葛; 屬禽與昆蟲為裘, 褐, 絲, 綿. 各載其半, 而裳服充焉矣]!"[37]

Song's interest in local differences focused on techniques, while his approach to the origin of materials seems to have been topic specific and not part of a consistent taxonomy. He remarked, for instance, on the local origin of metals. Yet, in listing dye ingredients, he concentrated exclusively on recipes. In the chapter on stone calcination, he noted that some materials were substitutive: wherever stones could not be calcined for lime, "heaven creates [*tian sheng* 天生] oyster [shells] as a substitute."[38] In principle, therefore, he asserted that all settings offered the same conditions, whereas local differences resulted from different techniques of transformation. For example, he noted that cocoons appeared in different colors in Sichuan, Shanxi, Shaanxi, and Henan provinces and were white in Jiaxing and Huzhou, because "when a white male is crossed with a yellow female the offspring will make light brown cocoons."[39] Localities provided different material components and compositions because transformative processes of yin and yang were locally diverse. Techniques could originate in one local context—for instance, Song claimed that vegetable candles were first created in Guangxin prefecture—but then developed in diverse ways locally. However, he held that the underlying cosmic principles at work were the same everywhere.

Accounting for Production, Accounting for Things

Over the last two decades, historians of science and global historians have been equally enthralled by the spatial turn in academic discourse. When "mapping the large-scale networks in which practitioners of the sciences are involved" within global accounts on goods and raw materials exchange, historians of science have shown how, whenever people, goods, and ideas escaped the confines of certain territorial logics, then knowledge and information were circulated, altered, or exchanged.[40] An evident emphasis has been on far-flung sites and on comparisons between East and West, hence the histories of translation, incommensurability, adoption, and circulation that have emerged seem somewhat to imply that the degree of alienation in this era increased inevitably over geographic distance.[41] At the same time, such accounts hold true that geographic distance became less relevant when sociopolitical systems such as the Chinese empires were able to spread their material and intellectual culture and economic practices. Thus, a scholar in the northern city of Xi'an would feel close to Fenyi (Song Yingxing's birthplace, modern Jiangxi) but far removed from the oasis of Dunhuang in Central Asia, even though they are both almost equally far away in metric terms.

Historians of China investigating travel and its purposes have complicated this metric view by suggesting that there is a more nuanced history of "time-space convergences" (as Donald Janelle has termed the impact of travel times on notions of distance) than has hitherto been acknowledged.[42] Moving beyond geographies to examine mobility as a historical practice and experience opens up exciting views on interchange and knowledge exchange within imperial confines. From this perspective it becomes clear that, although Song and Huang were both inhabitants of the same sociopolitical system and although they lived equally far away from the Ming capital Beijing, they may have encountered mobility in different ways and, thus, also understood geographies in different ways.

Although there was no substantial reduction of travel time through technical change in the late Ming era, Ming politics turned travel and goods exchange into an everyday experience.[43] Craftsmen had to commute annually to complete their corvée duty at the state-owned workshops. Merchants, civil servants, and examinees populated official and trade routes, rivers, streets, and canals on business trips. Pilgrims and literati leisurely wandered through the landscapes. Even if an individual refused to move at all, the world would pass by in front of them, in the form of people, ideas, and the transport of goods.

Song and Huang reflected quite differently on such a world in motion. Song explicitly remarked on this changing time-place "distanciation" (as Thomas Giddens has termed the familiarity created by the stretching of sociopolitical systems such as the Chinese imperial state across time and space).[44] Places that had once seemed remote and outside imperial rule could now be easily accessed. Goods from all four corners of the civilized world were available on the markets, while Chinese porcelain, tea, and silk were being traded to countries all across the known world.[45] Huang, in comparison, did not remark on such concerns at all. Whereas Song thus saw a relationship between the empire as a geographic representation of civilization and his universal claims, such imperial issues did not play any such central role for Huang.

Huang Cheng's and Song Yingxing's operational accounts from production to use have been called "managerial," to underline their difference from the view of a producer or maker but also to show how their perspectives may have been affected by their relationships to statecraft and administrative tasks.[46] Clearly, Song Yingxing and Huang Cheng were not interested in things and affairs as indicators of regional specificity, as administrators of their time were. This distinction becomes apparent, for instance, when comparing their work to genres such as local gazetteers or reports from local administrators such as Sun Pei's 孫珮 (fl. 1690) *Suzhou zhizao ju zhi* 蘇州織造局志 (Record of the local weaving and dyeing bureau of Suzhou), which meticulously list wares and goods.[47] In fact, in contrast to such executive approaches (that were, in fact, often pursued by local officials), Song and Huang both emphasized that "things" themselves were always universal. Huang Cheng ignored regional identifications entirely (which would have been important for any administrator who had to calculate taxes). Song, in contrast, emphasized the regional origins of raw materials and the local diversity of products. He also identified regional differences in techniques, to emphasize that the same principles could be applied everywhere.

In their explanation of procedures both Huang and Song combined the speculative with the operational, believing that the organizational principles of "doing things" revealed the nature of knowledge—that is, its dynamic character. In other words, both authors considered the manufacture of things (*zaowu*) to be on a par with (or even the same as) transformational processes (*zaohua*). This meant viewing the production process as embracing all the actions through which materials were transformed into items, including their use as artefacts (*wenwu* 文物) or products (*chanpin* 產物). Both authors were aware that every element and detail of production needed to be in place in order to reach a satisfactory conclusion. Place, however, did not matter as a "characteristic," since the same principles were valid everywhere. Huang Cheng never addressed geographic origin. Song Yingxing did remark on local specificity in various contexts; his thoughts were rather unsystematic, however, when compared to agricultural accounts of this time. Song clearly did not pay attention to the qualities of materials either, or discuss properties in terms of local origin per se. In this sense, transformative processes (that is, human or natural processes of creation) revealed important sequences of events that made things work. While Huang Cheng's tract identified the properties and transformative processes of production that were revealed through yin-yang *qi* (an issue Song Yingxing explained in one of his other works, *Lun Qi* 論氣), in many respects Song Yingxing went one step further in the *Tiangong kaiwu*, mainly using the yin-yang dichotomy to underline the organizing and civilizing principle revealed by the variety of crafts.

Finally, the writings by Ming scholars such as Song Yingxing or by identifiable artisan scholars such as the lacquer master Huang Cheng must be distinguished from a connoisseur literature that was mainly concerned with space and locality as an issue of authenticity and fraud. Huang and Song concerned themselves much more with the structures leading to specific mechanics and the logics of production than with the origins or place of things in time and space. Although both authors itemized skills, practices, and procedures in a similar way to material registries or managerial guides, their references to the *Yijing* imply that they were more interested in revealing the universal principles inherent in the underlying structural contingencies, patterns, methods, and procedural logic of things and their production. Therefore, the two tracts also need to be differentiated from the works of Qing civil servants such as Chen Hongmou, who compiled books on practical crafts with the aim of promoting a prosperous and morally intact state economy. Song and Huang, by comparison, shared the same aim—to situate objects, their creation and their intrinsic characteristics, as indicators of a wider (in time and geographical terms) and deeper (in more than just tangible things and people) universal order that encompassed all things, movable or not.

Chapter 5

Curative Commodities between Europe and Southeast Asia, 1500–1700

Tara Alberts

It was near midnight in the Portuguese colonial city of Malacca (Melaka, Malaysia) and Juliana Coutinha was enduring a difficult labor, which had brought her close to death.[1] Three days had passed since the child had died in her womb. No remedy had worked to bring about her delivery; the pains were excruciating, her debility total. Unable to help and seeing that the end was close, the midwives abandoned her. Although Juliana had already called on the "remedies of heaven" to no avail, she asked her son-in-law Manoel Carvalho to go to the nearby Franciscan convent to request they send over the cincture of a lay-brother, Luís da Cruz, who had died a few years ago, in 1622.[2] The friars were accustomed to lending the belt to the sick. They obliged and also gave Manoel a small piece of Luís's habit. Belted with the cincture and with the morsel of habit beside her, "at the same moment she pushed out the infant, already fetid, swollen, and skinless, whose aspect, contrasted with the mother's feeling of certainty that she would recover, excited such a tumult amongst the bystanders that, despite the fact it was the middle of the night, they began to yell at the top of their voices, 'Miracle! Miracle!' calling on everyone who heard them to give thanks to the clemency of God for such mercy."[3] As with many stories of health and healing, miraculous or otherwise, this account of Juliana's suffering is the only snapshot we have of her experience of ill health, accident, or misfortune. Such brief schematic accounts are often the only opportunity to catch glimpses of the protagonists' lives and of their strategies to obtain cures and regain their health. This fragmentary evidence demands deconstruction and contextualization. What remedies and healing techniques were available? How did individuals determine the relative efficacy, reliability, and authenticity of available cures? How did certain healers, substances, and practices gain the reputation as having health-giving, curative qualities? Why were certain materials or practices used during moments of crisis?

When exploring these experiences against the backdrop of the shifting populations of colonial port cities and the wide range of healing practices and substances available in these diverse areas, where individuals from varied backgrounds converged and interacted, an additional layer of complexity is added. How far did a commodity's perceived value survive a voyage between distant shores or over cultural frontiers? Attempting to understand these issues through the fragments of human experience shared in these narratives necessarily invites further reflection on the methodological complexities of the sources. Read sensitively and contextualized carefully, Juliana's story and others like it can be deconstructed so as to analyze the strategies employed to assert claims of efficacy; to illustrate how beliefs and practices connected to healing were created and transmitted; and to explore anxieties surrounding the legitimacy, orthodoxy, and broader connotations of a range of curative commodities, materials, and techniques. What emerges is a clearer picture of the range of complex, sometimes apparently contradictory factors that underpinned understandings of value, efficacy, and mechanisms of healing.

Stories of Healing

Juliana's miracle was one of hundreds attributed to Luís da Cruz reportedly experienced in Malacca, Macao, Goa, and other locations throughout Portuguese Asia. Manoel Carvalho recounted the story before a tribunal, which was held in Malacca between 1622 and 1627 to gather evidence to support the canonization of Luís as a saint in Rome.[4] Three manuscript volumes were assembled, containing over one hundred witnesses' testimonials. There was clearly considerable interest in Malacca in supporting the cause of a putative local saint. The evidence collected by the tribunal was eventually examined by the Congregation of Rites in Rome in 1647. Although the cause for sanctity failed to proceed beyond this stage in the canonization process, selected parts of these tribunal records were later reproduced in print for circulation in Europe. By the late seventeenth century, Luís's miracles had become a significant feature of the institutional memory and identity of the Franciscan order in Portugal.[5]

These tribunal records and their subsequent printed elaborations and reiterations can be rich sources of incidental information. They open a window on lived experiences, on the material culture of sickness and healing, and on institutional anxieties about the legitimacy of cures in colonial Malacca. Yet the limitations of this type of source material require some methodological reflection. Stories such as Juliana's can help us to consider some of the challenges that face historians in attempting to gain access to understandings of the efficacy of remedies and hierarchies of value in complex sites such as this colonial city.

First, the content of the testimonies is inevitably shaped by the nature of the tribunal eliciting them and the conventions of the documentation recording them. In each account, only the most salient details of experiences that would further the canonization agenda survive the journey from oral to written testimony. Between the investigative work of the tribunal in Malacca and its documentation, the witnesses' testimonies were molded into the formal narrative structure of a tribunal report, beginning with a formulaic impression of the witnesses' interrogation as to their moral character, and whether "the religious of the said [Franciscan] convent [of Madre de Deus in Malacca] had paid them to make any testimony in this tribunal about Brother Luís da Cruz, lay brother of the said order."[6] The statement then invariably proceeds through a brief, schematized account of the alleged miracle and concludes with an affirmation of the truth of the testimony.

Scant identifying information is given about each witness, beyond his or her name, social status, and occasionally place of birth. The witnesses seem to represent a broad cross-section of the city's population: "Bengali" and "Cafre" slaves, Portuguese *fidalgos*, secular priests and members of religious orders, local converts, natives and residents of the city, and passing Indian and Chinese merchants. Yet the Tuscan Italian of the scribe universalizes the experiences of these groups: not only do their stories follow similar patterns of tribunal reports and miracle narratives occurring the world over but experiences and even their names are transformed in the tongue of the holy city. Juliana becomes "Giuliana," cured by the "corda" of "Frà Luigi della Croce." Juliana's experience thus reaches us through many layers of translation: through her kinsman's testimony and through the tribunal scribes' recording, reordering, and translation and the narrative exigencies of the texts. How far can we read an individual's experiences and understandings of the materials she used or the healers she consulted through the distorting lens of such a process and such documentation?

The testimonials are metronymic in the predictable rhythm of the witnesses' experiences and homogenized and familiarized by the language of their authors. However, reading them closely allows us to gain some insight into the variety of spiritual cures on offer in the city, indeed the variety of cures attributed to Luís. Not least, these texts are often the only way we can catch a glimpse into privileged domestic spaces in which decisions about health and healing were made.

For those who could afford them, a range of curative substances was available in Malacca from a variety of practitioners. Like Juliana's midwives, some of these figures appear as supporting characters during the miracle narratives. In addition to the royal hospital, Franciscan and Jesuit pharmacies, apothecaries, and the European physicians and surgeons employed by colonial authorities, cures could be sought from individuals versed in various systems of healing. These included Malay, Chinese, and Indian doctors; ritual specialists such as Malay *bomoh* and Catholic priests who provided drugs, incantations, spirit propitiations, and other spiritual remedies; various specialists in bone-setting, moxibustion, bleeding, massage; various midwives including informally trained, experienced older women and traditional birth attendants such as the Malay *bidan kampung*.[7] How did individuals choose among the types of remedies available and decide, if necessary, which specialists to approach for assistance?

Childbirth in the Colonial City

The miracle tale collections of Portuguese Asia provide one possible entry point to explore a number of issues related to these choices about health and healing and the beliefs underpinning them. First, the preponderance of miracles related to women, and especially childbirth, reflects the broader concerns of the colonial and ecclesiastical authorities. Behind the reiteration of exemplary tales of licit, correct use of certain curative commodities and spiritual remedies, we can see a number of colonial anxieties and fears about what else was being birthed in these intimate female spaces of pain, danger, and physical toil. Ecclesiasts fretted about the sexual morality of Portuguese *moradores*, some of whom maintained harems of slaves and often had numerous illegitimate children. Later, there would be increasing official disquiet that children born of European fathers and non-European mothers would form a new "mestizo" class, blamed for exacerbating the "degeneration" of Portuguese domestic mores and habits in colonial cities.[8] Even before this, though, and even when the child was born within a church-sanctioned marriage, childbirth in the colonial city could be a site and a moment of tension and threat.

A large number of the miracles related to Luís da Cruz feature women, both as patients and as decision makers around the sickbed. This was especially true of the numerous childbirth miracles attributed to the friar, which were often selected for inclusion in the printed accounts. In Soledad's *Historía Serafica*, for example, we read of Anna da Costa, who used a section of the Luís's habit dipped in water as a miraculous cure to save the life of her slave Aldonça, who was also dying in childbirth. We also hear of Luiza das Neves, wife of Antonio Coelho, suffering "from that terrible distress experienced by those who cannot bring forth the fruit of their womb," who obtained no relief from either "physicians of the land, [or] from intercessors of Heaven." God only permitted her delivery once she sent for Luís's tunic, "which had already been the instrument of marvels." As soon as it touched her, "without any violence she delivered two living children."[9]

In some accounts Juliana is not named, and the story of her salvation becomes emblematic of Luís's many childbirth miracles. In Paolo da Trinidade's *Conquista espiritual do Oriente*, for example, other, more notable or prominent women are named, but Juliana becomes merely "a woman who was suffering greatly in labor with great pains and afflictions" (*Uma mulher que estava muito mal de parto com grandes dores, e afliçöes*), cured by the application of a fragment of the friar's habit.[10] Only the distressing details of the three days passing since the child died and its foul appearance and smell (*toda esfolada e que cheirava já mal por haver três dias, como dito é, que estava morta*) remain to link the miracle to Juliana's story.[11]

Juliana, now rendered anonymous, becomes a symbolic composite figure, fittingly perhaps as we know so little about her beyond the miracle recounted at the tribunal. She may have been a native of Malacca or of some other part of Southeast Asia. She may have originated in Macao or Goa, she may have been of Eurasian descent. Portuguese colonial authorities were initially keen to encourage relationships between male colonists and local women. Marriages and temporary partnerships with Europeans were also encouraged by many communities in Southeast Asia as a means to further trading links.[12] The role of these relationships, and of the resulting offspring, in the creation and maintenance of colonial and commercial power has been explored in detail by a number of scholars.[13] Yet, as a number of historians have lamented, the experiences of the women themselves can be extremely difficult to uncover.[14] Women often took Lusophone names on baptism, rendering their origins and pasts obscure. Their voices are rarely heard unmediated in the archive.

How would Juliana's background, origin, and even the languages she spoke have affected her and her attendants' selection of and experience of remedies for her suffering? Juliana's case invites us to consider how commodities, practices, and technologies first obtained reputations within these spaces of healing as efficacious curative substances. Why did they come to mind in times of crisis? How were hierarchies of curative value understood and shared? We should bear in mind that domestic spaces in the colonial city were often culturally heterogeneous: Portuguese *casados* with Asian wives and slaves and servants from a variety of backgrounds. As Angélica Morales Sarabia has argued in reference to households in New Spain, domestic spaces could "function as microcosms of the broader culture." Stratified socially and racially, they were places in which exchange and conflict between culturally diverse household members meant that "new practices were generated and subsequently conserved."[15]

As some historians and anthropologists studying childbirth in later periods have pointed out, Southeast Asia was unusual in that in many cultures around the region there was a relatively widespread occurrence of male midwives, and men often played other important roles in birth customs. However, in the cases encountered in miracle tales and other colonial records, the birth attendants (if mentioned) were described as female. In colonial cities at least, birth occurred—and was managed, for the most part—in female-dominated domestic spaces.[16] The female medical knowledge systems and practices that dominated these spaces are alluded to by the Dutch physician Jacob de Bondt (1592–1631) in his account of the medicines and medical traditions of the Malay Archipelago.[17] He translates local knowledge into the categories and terminology of European medicine ("In facilitating the birth it [turmeric] is a sovereign remedy: in complaints of the uterus it is specific") and empirically verifies it through his own investigations ("to confirm this opinion by my own experience, I have in reality found nothing so beneficial in all the disorders above mentioned as this most excellent remedy").[18] Yet it is clear that women were his first source of the botanical knowledge needed to treat a range of conditions. His account of the Malacca stone (a bezoar from the stomachs of pigs and porcupines), for example, reports how he was "told by the Malayan women" about its abortifacient properties. His account of turmeric similarly emphasizes women's agency in its formulation that "in female disorders no medicine is so celebrated by the Malayan women as borbory [*boberri or borri-borri*, e.g., turmeric]."[19] Laboring women were attended by a range of helpers (relatives and midwives), each potentially inculcated into varied systems of female knowledge about the process and remedies for when things went wrong.[20]

In scholarship concerning the circulation of ideas, materials, and technologies, experiences of pregnancy and childbirth are often overlooked or, as Lenore Manderson asserts, "taken for granted or treated as static and 'natural.'"[21] This disregarding or universalizing of the experience of childbirth in part echoes the evidentiary base, where the concern of colonial or religious authorities created records detailing specific issues related to governance and control. The itineraries of materia medica, technologies, knowledge systems, and practices relating to childbirth are especially obscure in the archive, only coming into focus at points of crisis or contention.

This also holds true even for modern research into beliefs and practices surrounding parturition. A recent meta-analysis of Southeast Asia ethnobotanical studies reveals the hitherto unexplored diversity of plants used by women today around the region, finding around two thousand different species used to treat "female disorders," with some long-standing remedies widely used by disparate communities. As the authors point out, this "traditional" knowledge, which is often shared orally between women, is mostly overlooked by researchers. It can also be difficult to analyze because of "limited means of diagnosis and the resulting cryptic descriptions of use" in which it is couched.[22] Yet, in order to understand the development of beliefs about efficacy and the hierarchies of value of various curative commodities, we must try to catch sight of these systems of knowledge and practice that governed the domestic spaces in which the majority of decisions about health and healing were made.

Shared Values and Concepts of Efficacy?

We could speculate about how far those present at Juliana's delivery would hold commensurate concepts about pregnancy and childbirth, perhaps drawing from shared, if differently understood, underpinnings in humoral theory. Many Ayurvedic, Chinese, Arabic, European, and some local Southeast Asian medical traditions contained similar concepts of fundamental bodily humors and of the balance of opposing (hot/cold, dry/wet, etc.) elements as the conceptual basis for understanding the pathology and management of diseases and other vulnerable conditions. Such commensurable ideas, however, do not necessarily guarantee a shared understanding of pregnancy. As

Manderson has discussed, for example, there is considerable variation around Asia about whether pregnancy is understood as a "hot" or "cold" state and, therefore, about what diet and therapies are recommended.[23] Indeed, a simplistic hot-cold dualism does not adequately capture the complexity of beliefs about pregnancy, its effect on the humoral system, or its proper management in many traditional Southeast Asian medical systems.[24] Similarly a wide range of local practices and understandings about childbirth were found around Europe, as recent research on early modern midwifery has revealed.[25] These complexities are important to consider in order to understand what was at stake when Juliana called for Luís's cincture. At the crisis point of the obstructed birth, in a hybridized colonial domestic sphere, we can consider which traditions were drawn upon and which could win out.

Several key questions emerge. First, how did Juliana think of her attendants? Were they known to her as *parteiras, bidan* (traditional Malay midwife) or *daya* (midwife, Malay borrowing from Arabic)? They are called *parteyras* in the published Portuguese account, but it seems more likely that she would have called them daya. Manoel Godinho de Erédia wrote of the skill of the daya, in his 1613 account of Malacca. He described them as "female physicians who are excellent herbalists, having studied in the schools of Java Major" and who prepared their plasters, syrups, and pills "with regard to the humour of the infirmity, and with regard to the age of the moon, following the doctrine of Galen, Book 3 chapter 6."[26] The term "daya" is still used today for midwives among the Portuguese-descendant Kristang community of Melaka.[27] What power does a name confer? As Andaya points out, one Malay law code affirmed the high status and authority that should be accorded to daya, describing them as "rajas" within birthing chambers.[28] Would a *parteira* command the same respect and would her remedies be as efficacious?

Second, what practical traditions, theoretical understandings, and technical knowledge did her attendants share? Did they massage her according to custom, or prepare fabric with which to wrap her abdomen after birth (*melenggang perut*)? Did they practice the ritual of "rocking the womb" as described in some treatises of Malay custom?[29] Did they prescribe turmeric, or other remedies to be eaten, inhaled, smoothed onto the skin, or bathed in? Did they practice the "sympathetic magic" of avoiding knots, closed doors, or lying counter to the flow of local rivers?[30] Childbirth was a moment not only of physical danger: the spiritual peril of perinatal mortality was a common theme in folk beliefs across Europe and Southeast Asia. Those who died in childbirth could return as vengeful spirits, tragically doomed, and now dangerous to the wider community.[31] Birth customs and practices aimed at mitigating the physical and spiritual risks were essential knowledge and could prove extremely enduring.

Finally, which practices and commodities caused anxieties for colonial and ecclesiastical authorities? Which remedies did they hope to replace with suitable understanding and practices such as Juliana's exemplary faith in Luís's cincture? Colonial authorities frequently attempted to stamp out childbirth beliefs and practices in domestic spaces that they perceived to be "backward," superstitious, or threatening. In Goa, for example, the Provincial Council of 1567 banned women from using traditional birth attendants.[32] The Inquisition of Goa punished a number of women who were accused of using "superstitious" rites during deliveries, and numerous edicts to ban other "heretical" childbirth ceremonies were subsequently issued, all apparently to no avail.[33] By the eighteenth century many European colonial authorities would take more serious steps toward the formal regulation of midwives as a matter of social reform.[34] In the nineteenth and twentieth centuries, British and Dutch colonial commentary again was dominated by condemnations of "ignorant" or "superstitious" childbirth practices throughout their possessions in Asia, which were seen as undermining colonial attempts to "modernize" public health and reduce infant and maternal mortality.[35] Yet, despite the continuing efforts of reformers, traditional female birth attendants often remained repositories of valued knowledge and spiritual power.

Materia Medica in Malacca and Beyond

In addition to attempting to regulate practitioners, ecclesiastical and colonial authorities were concerned about perceived misattribution of curative power or the "misinterpretation" of the mechanisms of efficacy of curative substances. The fight against "superstitious" or otherwise heretical beliefs in the efficacy of certain substances entailed that certain materia medica were banned outright because of their connection to non-Christian belief practices. In Goa, for example, betel leaves, areca nut, turmeric, and certain flowers were prohibited.[36] Similarly in Spanish America peyote was forbidden. The bans were enforced in each case by the Inquisition. However, even where bans were not enacted and the substance in question was widely and openly used, certain remedies could have complicated and contested connotations that affected their value and perceived efficacy. To further illustrate some of these points of tension let us consider in more detail the range of commodities available and used by healers and practitioners in Malacca.

Descriptions of Malacca by sixteenth- and seventeenth-century travelers provide fascinating insight into the variety of materia medica converging in the port from markets near and far. With the capture of Malacca in 1511, the Portuguese had seized a complex and well-connected port city used by merchants from all over the Indian Ocean region. It was an important clearinghouse for the valuable substances found in the forests of the hinterlands, including aromatic woods, gums and resins, valuable spices and herbs, and bezoar stones and other animal products. It was one of the key trading ports for the most lucrative spices of the Moluccas—including nutmeg, mace, cloves, sandal, benzoin, and aloes-wood. Malacca was also a node of exchange for a number of long-distance trading routes, with commodities brought from South, Central, and Eastern Asia, Africa, and the Mediterranean.

Some commentators, seeking to promote the city as the key asset of Portuguese India, were especially keen to demonstrate the variety of medical commodities on offer. Duarte Barbosa reported that "musk, rhubarb, pepper, incense, saffron, quicksilver, opium, . . . rose water from Mecca, . . . gall-nuts from the Levant, cloves, white sandal wood, benzoin, camphor, aloe-woods . . . nutmeg, mace" came in on the trade winds to the city.[37] From Erédia's account it is clear that daya (female medical practitioners) were also employing a wide range of local and imported commodities to make their medicines. He reports that they "ordinarily" use "clove, nutmegs, pepper-corns, long pepper, *betre* [betel], saffron-root, saffron flowers, ginger, *lancoas* [galangal], another and hotter kind of ginger, *conchor* [probably *kenchur*—aromatic ginger], *bancalê* [Cassumunar ginger], *dringo* [sweet flag], *pulacary* [alyxia cinnamon], canaphistola [golden rain tree], *tamarindi* [tamarind], cayoular [Indian screw tree], *cayotay* [stinkwood?], and innumerable other roots about which a special treatise could be written."[38] Some of these ingredients could be gathered locally, but others originated from further afield.

Exploring some examples of materia medica, we can start to identify features that would differentiate certain commodities as being *particularly* valuable within various medical systems. Take, for example, three commodities mentioned by Barbosa: gallnuts from the Levant, rosewater from Mecca, and aloes-wood. The first two commodities remind us of the flow of materials that could have a medicinal function from the trade routes of the Red Sea and Persian Gulf, and the continuation—indeed the later resurgence—of this trade despite the Portuguese occupation of key trading cities.[39] The third commodity, aloes-wood, is representative of the rich, rare, forest produce of Southeast Asia, which was widely prized and sought after throughout Asia, the Middle East, and Europe. The substance is formed when the interior of various types of the aquilia tree is infected by a mold that causes the tree to produce a deep red resin deep within the wood. It is still extremely valuable. Aloes-wood was known by a wide variety of names, including, for Europeans, variations on the appellations agarwood, eaglewood, lignum aquila, calambac, and garuwood. For much of the sixteenth and seventeenth centuries, European authorities debated the nature of this substance, how it was obtained, whether there was more than one variety, and where the best quality could be sourced.[40] All three commodities, like many products with "medicinal" applications, had a range of other uses. All three also had a range of cultural associations and symbolic meanings that could shift and take on new weight as they moved in and out of different contexts, perhaps also as they moved into different domestic spaces.[41] We will see that, with these commodities, lingering connotations and concerning associations could on occasion cling to, and be exacerbated by, the commodity's itinerary. These associations could subtly affect its value and sometimes create disquiet among colonial and ecclesiastical authorities.

Origin, Quality, and the Complex Connotations of Commodities

The first feature that could factor into estimations of value and efficacy was the origin of the commodity. Nutgalls, for example, are formed on a range of plants and are widely used in medical preparations, to make ink, and to tan leather, but those from the Levant and, in particular, those from Mosul (in modern-day Iraq) were most widely prized as being of the best quality.[42] Levant gallnuts (*Gallae Levanticæ*) are formed by the action of the gall wasp (*Cynips gallae tinctoriae*) on the Aleppo oak (*Quercus infectoria*) and were widely used in a number of medical traditions as an astringent.[43] It is clear from a number of accounts that the stated origin of these gallnuts acted as a guarantee of quality, although in European accounts at least there was no consensus as to why this was the case. "The reason why the Levant gall-nuts are of superior quality to the European does not appear to be well understood," one nineteenth-century encyclopaedia entry averred, summarizing the rather hazy scholarship of past generations on the matter. "Perhaps the insect which produces them is a species of cynips different from ours. Perhaps too, the warmth of the climate of the eastern countries promotes the concoction of the specific juice, which is secreted after the puncture made by the insect."[44]

Whatever the reason for the accepted superiority of the Levantine gallnut, there is no sense in any of the Portuguese accounts that this product was viewed with any suspicion per se due to its origin and trajectory, which was mainly along trade routes plied by Persian, Ottoman, and Gujarati Muslim merchants.[45] After all, it had long been acknowledged in European herbals that the primacy of the Levantine product both as a source of superior ink and the most efficacious tannic acids for use in medicines. Yet it was lamented that control over the trade of such sought-after and effective products was not in Christian hands. In 1622, the same year that Luís da Cruz died, Portugal had suffered its first devastating imperial defeat as the city of Hormuz in the Persian Gulf was captured by an English-Persian coalition.

Afonso de Albuquerque had seized a jewel when he took Hormuz in 1507. According to one early seventeenth-century depiction of the city, it had been known as "*emporium mundi*" by the ancients, because there a merchant could acquire all medicines of the world. And, the author asserted, the luster of Hormuz had not diminished over the centuries: "The neighboring Persians and the natives say that the world is a ring, and that Hormuz is the precious stone set in it, due to the great confluence of merchants who come from all over the world and to here come all health-giving things and medicines. Here from Persia come the best rhubarb, nutmeg, mace, many aromatic herbs, perfumed water, and the best bezoar stones that there are in the world, worth three times as much as those from Malacca."[46] Yet even before the loss of the city, the Portuguese were unable to control trade in Hormuz: according to some accounts the majority of merchants using the Portuguese port were Muslims who profited at the expense of the Portuguese Crown. The Augustinian friar Agustinho Azevedo (in Hormuz between 1589 and 1600), for example, warned the king that "Turkish" trade through the port was increasing. He directly linked the fate of Christianity to the flow of commodities through these networks: "along with the merchandise they [the 'Turks'] take many purchased or deluded Christians, and take them back to Turkey where they make them Turks . . . from which comes much harm to your Majesty's treasury, and danger to India, because these renegades know the ports, and know those who can give advice about how fleets can come to India, at the cost of your majesty the Turk makes himself more powerful with all the riches of India."[47]

Moreover, renewed trading and diplomatic activity between the Ottomans and Malacca's Malay neighbors—especially the sultanate of Aceh—saw a revival in the Red Sea spice trade via Surat.[48] Sultan Ala'ad-din al-Kahar of Aceh sent an embassy to the Ottoman sultan Suleiman I in 1562 requesting that regular ships be sent with trade goods and military assistance including artillery, cannon founders, and gunners to support his fight against the Portuguese and against rival Malay sultanates. In return, Aceh offered the sultan "the commerce and trade in all the drugs and spices of the Moluccas, Banda, Java and the other islands of that archipelago, assuring him that he would gain innumerable riches from it."[49] A further formal alliance was agreed with the Ottoman sultan Selim II in 1567, which furthermore meant that henceforth attacks made by the Acehnese with Ottoman support on the Portuguese in Malacca acquired the character of a jihad.[50] By the 1580s, according to Portuguese accounts, four or five ships "from Mecca" were arriving in Aceh each year, bringing gold, silver, slaves, rosewater, glassware, silk, and munitions.[51]

So there was a complex—and for the Portuguese, uncomfortable—context for many of the commodities originating in the Middle East that ended up in the markets of Malacca. The commodities were serving to enrich and strengthen the hands of Portugal's enemies and to bolster diplomatic, commercial, and military alliances against them. If we turn to the second commodity mentioned by Barbosa (the "rosewater from Mecca"), we can see a further way in which the origins of these curative materials might both enhance their perceived value and effectiveness for some customers and raise profound concerns about their utilization for colonial authorities.

Rosewater and attar of roses were used in a wide range of medical traditions in Europe, Asia, and the Middle East.[52] They could serve as the base of medicines using other ingredients: one seventeenth-century Malay recipe, for example, recommended mixing the dried gall of the slow loris, from the forests of the archipelago, with rosewater and rubbing the resultant concoction on the abdomen of pregnant women to ease delivery.[53] Rosewater could be distilled from various rose varieties and was fairly easy to produce, but some markets nevertheless favored the product of certain regions of renown: higher prices were obtained in many parts of southern Europe and the Middle East for rose oils and waters distilled from *Rosa damascena* in Persia, in the Ottoman province of eastern Roumelia (now part of Bulgaria), and in later centuries in Taif, near Medina, for example.[54] Rosewater was widely used in cookery, in perfumes and medicines, and for religious or ritual purposes throughout Southeast Asia. Yet the stated origins of the substance in Mecca add another dimension to the rosewater listed by Barbosa. In many ways it is unsurprising, given the Islamic identity of the sultanate and the city's extensive connections to the Middle East through trade, scholarly exchange, and the *hajj*, that rosewater from there would be sought after and prized. In the nineteenth-century medical treatise *Tayyibul Ihsan*, written for a Malay audience, rosewater and oil of roses (*ward*) were among seventeen items of materia medica distinguished by their connection to Mecca: the author explains that the roses were not grown there but imported and then distilled by experts in the pharmacies of the holy city.[55]

Rosewater was associated with some Islamic practices in the Malay world: being used for cleaning ritual spaces, in the sanctification of mosques, during religious ceremonies, and by pilgrims on the *hajj*. It is also possible that other forms of curative water were also brought from Mecca for sale in Malacca: containers of water from the miraculous Zamzam well in Mecca, for example. Erédia described the use of rosewater as one of the characteristics of Malay self-presentation and deportment. He described the Malays as "all Serracenos or Moriscos" (that is, Muslim) and when characterizing their manner of dressing explained how they made "extensive use of precious perfumed unguents, rose-water, sweet-scented leaves and cloves."[56] How far did the Islamic associations and qualities endure during the Portuguese colonial period? Did "Muslim" curative commodities endure in "Christian" Malacca because of these connotations? The non-Christian spiritual, religious, even political connotations of some commodities could be concerning for the colonial and ecclesiastical authorities, especially as their use was not limited to the non-Christian population of the city. As Sarah Abrevaya Stein has demonstrated in her discussion of the topic in other times and other climes, the evocative scent of rosewater would have complex and shifting cultural connotations, depending on the location of its use.[57] Could the ostensibly unremarkable use of rosewater by recent converts mask a surreptitious continuation of non-Christian beliefs and practices? Even materials widely used in Europe could have different meanings and value in the colonial city.

Managing "Spiritual" Dangers

Especially given the context of fears over the Ottoman-Malay alliances, it is unsurprising that there were heightened anxieties over the possibility that non-Christian healers and curative substances could serve as Trojan horses for spiritual warfare. Concerns expressed by missionaries and priests about the behavior of some of their converts and parishioners hint at an ongoing ontological project on the part of some religious leaders to establish the boundaries between the categories of licit and heterodox cures. Yet could a line be drawn when even common ingredients for medicinal preparations could gain their curative qualities and value from implicitly understood "spiritual" or "supernatural" characteristics?

There were several ways in which these broadly defined "spiritual" or "supernatural" dimensions of commodities could cause anxieties for colonial and ecclesiastical authorities. First, the location and collection of valuable forest products such as aloes-wood, for example, often required esoteric skills and knowledge that were outside the control and understanding of colonial and ecclesiastical authorities. The forests in which these resources were found were haunted not only by dangerous animals but by ghosts and spirits.[58] As work by Peter Boomgaard and Terenjit Singh Sevea has shown, well into the nineteenth and twentieth centuries ventures in these hazardous liminal spaces required the assistance of spiritually powerful miracle workers: the *pawang*, *dukun*, or *bohmoh* who had the spiritual and practical expertise to cope with these dangers.[59] "Aboriginal" (*Orang asli*) groups living in forests were also often important vectors by which valuable and efficacious forest products were brought into the cities. Andaya cites, for example, the encounter between a group of Dutch men and some "Banua" (forest dwellers) just outside Malacca in 1642. The latter were bringing "roots for pregnant women experiencing difficult labor" to sell in the markets of the town.[60]

These groups made claims to spiritual know-how and efficacy that could compete successfully with "orthodox" healers and commodity purveyors. They could therefore become vectors for dangerous practices and spiritual peril. Erédia describes how "certain apprentices of the wild Banuas act as doctors, for they understand the properties of all the miraculous herbs and plants." Yet their practice was not merely rooted in knowledge of the natural world: "in curing maladies they make use of incantations, as though their medical treatment depended on magic and diabolical arts, for at the first glance they recognise the interior ailments of the human body."[61] Moreover, among their number he describes sorceresses, "who are enchantresses, witches, and cheiromantists, apprentices of the wild Banuas from the cave of Gunoledam," who kill children and babies in the womb, charm the creatures of the forest to do their bidding, or change their form and turn themselves into animals.[62] He relates how Jorge de Santa Lucia, bishop of Malacca, "wished to stop the great injury done" to the city and so, in 1560, excommunicated a number of these "enchantresses" and "wild Banuas from the interior, who changed themselves from men into tigers."[63] This brought an end to tigers coming out of the forest attacking people and helped to persuade the city's inhabitants of the superior efficacy of the church: "the infidels and Mouros were astounded at this miracle and many Chelias, idolatrous natives of Malaca, were converted."[64] Yet the fact that the church had to intervene repeatedly so as to "defeat" these individuals and to "disprove" their heterodox knowledge about the collection and use of precious curative forest commodities suggests their continued role and value to consumers in the colonial city.

Second, patients' understandings of the virtues or effects of certain substances could be problematic. Aromatic woods such as aloes-wood played an important part in many early modern medical systems; their odor and flavor could indicate their innate properties and virtues. In Europe, while Galenic theorists considered flavor and odor as secondary properties, writers influenced by alchemical theories often emphasized the heuristic importance of these sensory qualities.[65] In many systems these properties could also be understood to have both "natural" and "supernatural" effects, and understandings of the exact mechanisms through which they operated could be varied. In many Chinese medical systems, the strong and pleasing scents of aromatic woods indicated their medicinal value and their efficacy against demonic agents. Aloes-wood was prized as an ingredient in medicines to drive out internal pain and demons and as a material from which to make incense, Buddhist rosaries, religious images, and auspicious, valuable clothing toggles and amulets.[66] Similarly, Bontius described how aloes-wood was offered as a "sacrifice" to the gods by the "Moors and the gentiles" around the Malay world. In Vietnam, where the most valuable varieties were found, it was offered to the deity Thiên Y A Na, who was also propitiated before wood hunters entered the forest to find precious woods.[67]

These divine relationships would have sounded familiar to well-educated Europeans. Pliny explained the virtues and value of trees bearing aromatic woods and resins with reference to their "early history," when they "formed the first temples of the gods, and even at the present day, the country people, preserving in all their simplicity their ancient rites, consecrate the finest among their trees to some divinity."[68] Garcia da Orta referred to a similar "fable" often repeated concerning aloes-wood: that it originated in the "terrestrial paradise," which explained its heavenly scent.[69] How were ecclesiastical authorities to ensure that beliefs about medicinal efficacy were based upon an understanding of natural virtues of substances, and not attributed to supernatural or "spiritual" qualities, or that they had a natural effect on the body rather than some ritual purpose? Was it even possible to disaggregate these effects and meanings? Such understandings and the urgency with which authorities tackled them could be dependent on context: the use of turmeric or areca as a remedy might be judged licit in Malacca but highly problematic in Goa, for example.

Similarly, there was also considerable concern among missionaries in the city about whether "Catholic" cures were correctly understood by the local population, especially given the similarity of many of these practices and materials to those offered by non-Christian ritual experts. A variety of curative water, for example—holy, scented, blessed, or sourced from sacred springs—was used by different communities in Malacca.[70] Catholic holy water was distributed by missionaries and churches, but other faiths had similarly effective remedies. Holy water bowls, inscribed with verses of the Quran could be obtained from China, India, and other parts of the Malay Archipelago. Alternatively, the afflicted could create a cure by writing holy words from various religious traditions on a piece of paper or in red sandal powder on the bottom of a bowl and then create their own cure by adding water. How could missionaries ensure that the laity understood the distinction between the licit Catholic version of these practices and the illicit, "superstitious" use of cures attributed to the "wrong" gods or spirits? All the witnesses in the Luís da Cruz enquiry reportedly attributed their healing to his intervention, resulting from the application of his relics or related material. Yet this unified impression of orthodox understanding of the action of the saints perhaps masks a more complicated picture.

Hierarchies of (Spiritual) Value

One complicating factor in resolving this issue was the sheer variety of cures offered by the church—from sacramentals such as holy water, prayers, *nominae* or scraps of paper with other holy words, blessed candles, Agnus Dei, religious medals, and saints' relics to the supposed healing benefits of engagement with the sacraments, and to the drugs and simples dispensed as an act of charity by missionaries to the indigent sick. Even within the testimonies of Luís da Cruz's cures, the variety is striking: some people are cured by his touch, others by praying to him, others by kissing his image, others by touching various items that had belonged to him. Several, such as Juliana Coutinha, were cured by applying his cincture to their bodies.

Moreover, sacramental cures too could be increased in value and efficacy through various means, creating a hierarchy of curative religious materials, which echoed the diversified market for other curative commodities. A number of cases in the Luís da Cruz miracle collection, for example, saw the afflicted cured not simply by holy water but by a type of "super-charged" holy water, fortified and turned into a tertiary relic by having a piece of the friar's habit dipped into it.[71] Intriguingly, in many of the testimonies there seems to be an effort to stress the efficacy of the Luís da Cruz's cures even over and above other Catholic sacramental remedies. Just as gallnuts from the Levant were better than other varieties and rosewater from Mecca was more highly prized than local or Indian imitators, sacramental cures originating with the friar won out over other invocations of the saints and religious remedies. Thus we hear from the priest Emanuel dos Anjos that one of his parishioners, Manuel Lopez, had been tormented for four months by demons to the extent that he no longer knew himself and was driven to violently attack his servant. The possessed man did not respond to any other cures: "he did not get better with the recitation of words of the Gospel, or with other images [of saints] that were placed on him," reported Emanuel. So the priest took a small piece of Luís's habit, tied it up in a handkerchief and tied it to the patient, who immediately quietened down.[72] Similarly Juliana was cured by Luís after her other appeals to heaven had failed.

Conclusion

The case of Juliana Coutinha and her fellow *miraculés* reminds us of the need to consider the wide range of available materials and practices of healing, including the deeper connotations and implicit hierarchy of spiritual value which might affect the perceived value of a range of commodities. It raises a number of questions about the interaction and overlap among imported trade commodities, whose value depended on factors that served as proxies for quality and efficacy—including origin or place of manufacture—and both locally created and imported materials valued for spiritual efficacy. Commodities and practices could derive effectiveness from origins in unseen, distant, famous, or religiously significant lands or from the unknowable (and perhaps spiritually significant) depths of the forest. Equally, however, "local" materia medica and the relics of "local" saints could be also held in high esteem as being uniquely suited to the humors of those who dwelt on the same land. Numerous witnesses to Luís da Cruz's miracles were at pains to demonstrate that God had granted their prayers only when they turned to Malacca's own putative saint: their miracle would be a homegrown one.

Yet the labeling of curative commodities with an indigenous (or indeed, a "foreign") identity could often be a complex process, reflecting much more than simply their origins or place of manufacture. In her study of "Korean" ginseng, the historian Soyoung Suh demonstrates how, in early modern Chinese and Korean botanical texts, for example, the authors carefully differentiate between the virtues and uses of various types of locally grown and foreign ginseng.[73] Modern botanists draw no distinction between these two types of the plant: perceived or constructed differences in the text are born of a range of socioeconomic, cultural, and political imperatives rather than of fundamental physical or morphological differences between two distinct commodities.[74] Similarly, merchants, healers, and other commentators arrived at different conclusions about how the origin of a particular substance could affect its efficacy and value. In some cases the reasoning seems somewhat circular. Subtle distinctions in appearance, weight, scent, density, and oiliness could be used both to grade valuable woods such as aloes-wood and sandal and to categorize a sample as originating in a particular locale.[75]

Contextualizing Juliana's delivery, we have seen how systems of values and beliefs about efficacy could interact in complex ways. We can often see traces of authorities' attempts to build identities or consolidate control through these webs of significance. Believing that a substance was more efficacious because certain regions were renowned for a particular produce or manufacture was acceptable. As we have seen, however, there was a need to ensure that these understandings of the hierarchies of efficacy did not slide into unorthodoxy or illicit practice.

The narrative of Juliana's cure raises some key questions, which need to be considered when we attempt to understand the flow of materials believed to have curative qualities across cultural boundaries. We have begun to see some of the factors—over and above any physiologically active compounds that may indeed have produced effects on the body—that could lead to certain commodities becoming especially valued as having curative powers. Its origin, the expense, and the reputation of its purveyor could all, unsurprisingly, affect the value of a commodity in ways that are fairly evident in the archive.

More difficult to trace are the implicitly understood connotations of certain substances and the ways in which even seemingly worthless items such as scraps of fabric, a rope belt, or plain water could become invested with great value through spiritual transformation. This dimension means that any investigation cannot focus simply on the categories of commodity that were held to be curative but must also take into account the factors that govern *how well* various materials within this category were perceived to cure. Which *type* of gallnut—or rosewater, or other substance—used in *what manner* was considered to be the most effective? *Which* saint had the most effective relics, and how should they be used to maximize their powers? Why did these beliefs arise?

In this chapter I have demonstrated a few ways in which we might start to explore some of these issues through a case study that underlines the range of complex, interacting concerns that affected perceptions of efficacy and legitimacy, especially during fraught and sensitive situations such as childbirth. Future research will lead to a more detailed understanding of how substances gain value in the understanding of the consumer. Of course there is a further vexed question of how this (admittedly rather vague) notion of "value" maps onto questions of monetary value, and how we might understand fluctuations in the two.

It is clear that policing the use of curative materials and the understandings of their effectiveness could be important for colonial and ecclesiastical authorities. A homegrown saint was needed and promoted in Malacca so as to displace the continued power of non-Christian sources of healing—from the *bomohs* of the forest who were worryingly relied upon even by Catholics for natural and spiritual remedies to prized ingredients imbued with power because of their origin in Muslim holy lands, whose import was perceived to line the pockets of Portuguese Malacca's enemies and to undermine her mission of Christianizing the region. Materials and techniques of healing could be used to further much wider objectives. Yet we have also seen how mutable materials, technologies, and practices could be, and how difficult it was to control the ways they would be understood and the impact they would have as they moved across cultural, religious, and spatial boundaries.

Chapter 6

Translating the Art of Tea

Naturalizing Chinese Savoir Faire in British Assam

Francesca Bray

In the late 1830s the British began experimenting with tea growing in Assam to substitute for expensive imports from China. They transplanted Chinese tea plants, Chinese tea workers, and various other forms of Chinese skill and expertise into Indian territory, adapting them to this alien natural and social environment through various processes of naturalization that notably included applying "the skill and science of the Europeans."[1] New plants were bred, new curing processes were developed, campaigns were pursued to reeducate British tastes, science was invoked, and engineers were engaged to design specialized machinery. By the late 1880s the native Assam tea stock had almost entirely replaced the original Chinese bushes, now disparaged as insipid and inferior, and the elaborate Chinese skills of tea processing, initially learned with great pains, were likewise dismissed as inefficient. In contrast to the decentralized, smallholder-based Chinese tea industry from whose components it was initially constructed, the Indian tea industry was conceived and developed as a large-scale, capital-intensive enterprise in which individual estates stretched over hundreds or sometimes thousands of acres, processing thousands of pounds of leaves each day, wherever possible with machinery. Over the decades the model of large-scale, centralized production developed in the Assam gardens was successfully transplanted across the Himalayan provinces and the mountain ranges of South India and Ceylon, and then on to other British colonies in Natal, Kenya, and beyond.

The Assam Tea Company was founded in 1839, just six years after the dissolution of the East India Company's monopoly on tea imports to England. It was the first British commercial tea company, "not only the premier tea company of India, but . . . the genesis, the *ma-bap* (mother and father), of the

Table 6.1. Assam Company tea production 1840–1905. Based on "The Company's Statistics," in H. A. Antrobus, *A History of the Assam Company, 1839–1953* (Edinburgh: T. and A. Constable, 1957), 406–12.

	1840	**1843**	**1853**	**1862**	**1872**	**1895**	**1905**
Acreage bearing	264	2008		4727		9319	10184
Acreage total	1356	2768	2921	4847	5200	10009	10184
Output (lbs)	10212	87705	366687	1043010	1463391	3272662	4044012
Yield (lbs/a/yr)	39	44	-	220	-	351	397
Price (shillings)	3	2	1.95	2.2	1.9	20	90
Dividend (%)		3	6	25	17.5	20	8.5

whole Tea Industry outside China."[2] The official company history, published in 1957, is a rich (but frustratingly disorganized) source of information on the technical evolution of Assam tea cultivation and manufacture. The author, H. A. Antrobus, divides the fortunes of the company, and of the Assam tea industry more generally, into three phases. The years from 1839 to 1854 he presents as the experimental (and near disastrous) phase; from 1854 to 1870 was a period of normalization and expansion; and finally, by 1871, Assam planters had succeeded in ridding themselves of the "incubus of the China plant," the pioneering days were over, a market for Assam teas was opening up in Britain, and "the running of tea companies had become commonplace."[3]

Antrobus admits ups and downs in the early years of the Assam industry, but the bare statistics on the historical expansion of tea production in Assam (see table 6.1) appear to support a bolder narrative of steady improvement. This is the version that features in the forewords of Victorian tea planters' handbooks and in today's global tea company websites: a tale of progress, of thrusting British enterprise coupled with "the inventive genius of the Anglo-Saxon race," steadily improving, transforming, and displacing an ossified and inefficient Chinese craft, where cultivation and manufacturing were both small-scale, the harvest was restricted to a few short weeks, quality was variable, and the excellence of the tea hinged on the skill and reliability of ignorant and often dishonest artisans.[4] In this narrative British energy and acumen, science, and capital transformed a stagnant and wasteful craft into a modern industry based on large well-equipped plantations planted with bushes picked year-round, and processing with sophisticated machinery reliably produced huge volumes of tea, to uniform standards, at prices so low that even the poorest British families could avail themselves of the sober and cheering beverage.

Both the triumphalists and the more cautious Antrobus concur that, by the 1870s, Indian tea manufacturing had been made anew and owed nothing more to China. But how accurate is this account? How close was the Indian tea industry—by 1880, 1900, or even 1930—to realizing its proponents' long-cherished ambition of mechanizing every step of tea production? How much was tea production informed, as its boosters implied, by the latest science? Had the Assam model indeed discarded the last vestiges of any old-fashioned legacies from the original "Chinamen" who had worked side by side with the British pioneers in the 1830s and 1840s? In the wider context of Victorian-era industrial values, it is not surprising that the Indian tea industry consistently pursued improvement through mechanization and economies of scale. It is not surprising that the initial years were hard nor that, at several points during the first decades, the industry's boosters were near despair. Yet a closer look at the records shows that even post-1870, when Antrobus judged that "the running of tea companies had become commonplace," progress was still far from steady and confidence in the industry remained shaky.[5]

One reason was that Chinese competition had stepped up to meet growing Western demand. Almost all the Chinese tea exported to Britain came from the southern coastal province of Fujian (see below). Until 1842, with China's defeat in the First Opium War, all tea exports had to pass through government-approved brokers in Canton. The Treaty Ports opened to direct Western trade in 1842 following China's defeat included two Fujian cities at the heart of the tea industry, Fuchow (Fuzhou) and Amoy (Xiamen). Prices fell and quality improved once the tea could be shipped directly from the province where it was grown. Fujian tea growers stepped up production, and exports rocketed from 35 million to 92 million lbs between 1856 and 1885, boosted by the speedy new clipper ships. Only after 1895 was the preeminence of strong Indian teas so firmly established in Britain that Fujian's exports began to dwindle.[6]

From the very beginning of the Indian tea industry, its British boosters pinned their hopes on the application of science as one sure way to undercut Chinese competition. But this proved more challenging than expected. The path toward what shareholders and engineers regarded as efficient production was strewn with obstacles. Neither tea plants nor tea drinkers always behaved as expected; various key steps in tea production proved nearly impossible to mechanize; and many manual skills persisted well into the twentieth century as the (typically unacknowledged) bedrock of high-quality production.

The challenge I address here is how to map such uneven processes of technical translation and adaptation. The chapters in this volume illustrate a variety of ways, mostly textual, in which to trace the history of transfers or transplantations of skills, knowledge, and artifacts from one society to another. There is no single "best way" to track or reconstitute such itineraries, and each case offers a different constellation of resources for doing so. As a historian of technology, however, I am particularly interested in developing methods that highlight and clarify the concrete, technical dimensions of such transfers, the choices that are entailed and the complexity of the transformations that ensue when a mature system of material production is implanted into a new environment with different resources and expectations.

The tool I use to map such details in the long-term technical evolution of the Indian industry is the *chaîne opératoire* (procedural sequence). Although the use of diagrams and charts has rather fallen out of fashion in history, including the history of technology, they remain valued heuristic and communication tools in related disciplines such as archaeology, science and technology studies (STS), and the French school of anthropology of techniques. The chaîne opératoire is a conceptual and visual tool initially developed by anthropologists of techniques that is much favored by archaeologists (see figure 6.1). It lays out the core sequence of technical processes that go into the production of an artifact or the exercise of a skill and provides a clear framework for feeding in information about the inputs and outputs—material, intellectual, social, or ritual—specified at each stage. It is an organizing device developed specifically for mapping technical assemblages in motion.[7] Here I apply a chaîne opératoire analysis to a historical case of technology transfer—namely, the transplantation of Chinese tea-manufacturing skills to British India.

Global Commodities and Early Modern Technology Transfer

Global historians have identified import substitution as the driving force behind some of the most significant East–West technical transfers across Eurasia during the early modern period. In particular, they point to its significance in triggering the industrialization of European manufacturing through the late eighteenth and nineteenth centuries. Asian manufactures including cotton

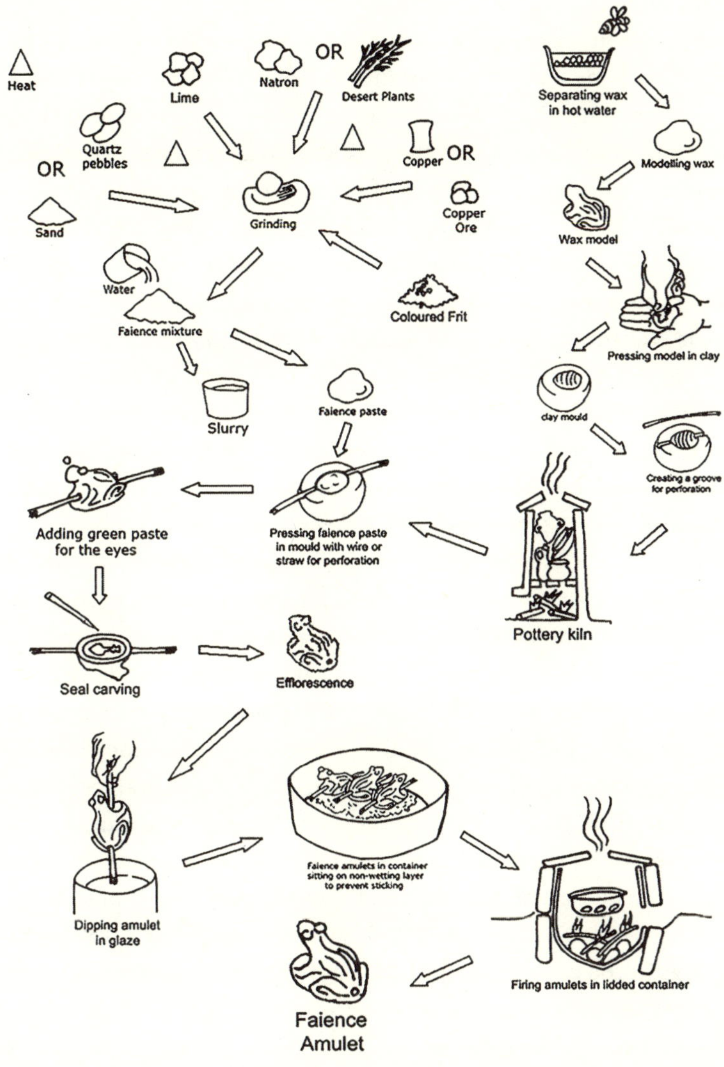

FIGURE 6.1. Chaîne opératoire for making a frog amulet. "The reconstruction of the 'chaîne opératoire' for the frog amulet revealed a complex sequence; it shows that the production of such amulets in the New Kingdom was part of an organised and controlled system. The choices of colours, the standard designs and the amuletic significance of these objects indicated a uniformed [*sic*] pattern, which was influenced by religious belief and traditional fashions. Also the scale of this production appears to have been large and highly organised—it involved sourcing some of the raw materials from areas beyond Amarna and Egypt with possible links to a wide trade network." Archaeology-Petrie Museum, Institute of Archaeology, University College London. http://www.ucl.ac.uk/archaeology/faience/chain.htm.

textiles, printed calicos, silks, porcelain, and tea were refashioned or re-created on European soil through complex combinations of borrowing, adaptation, reverse engineering, or pure invention.[8] These translations, typically involving the long-term transmutation of a constellation of artisanal crafts ("arts" or "savoir faire," in the language of the period) into a system of industrial, mechanized production, raise a number of fascinating questions for the historian of technology. Which elements of an alien technical system traveled successfully, and which did not? Which elements persisted, which were transformed, and which discarded, as the technical system was naturalized and reconfigured in its new setting? At which point could the imported technology be said to have achieved stability and sustainability (the industry was no longer at risk of failing)? In nascent or experimental industries, different methods and philosophies compete, so how and when was *closure* (consensus about desired outcomes and the best technological procedures for reaching them) reached? How closely did actual practice reflect ideals and stated goals?[9]

The Indian tea industry is also a case of import substitution involving the translation of a "native" savoir faire into a European industry. In contrast to the European re-creations of cotton-printing or porcelain manufacture, the whole process of British-style tea manufacturing was located outside Britain itself, albeit on colonial territory. The labor force (possessing significant if largely unacknowledged expertise) was Indian; the management, the acknowledged experts (agronomists, engineers, tasters, etc.), and the capital mobilized were almost entirely British, as were the consumers.[10] What is perhaps most unusual about the case of tea is that rather than reverse engineering or reinventing the original artifact, as in the case of European varieties of porcelain, or incorporating a subset of imported skills and materials into a well-established industry, as in the case of cotton-printing, the pioneering British tea growers directly imported the complete set of Chinese tea expertise (seeds, tools, and skilled workers) and set it up on Indian soil as a working system, a faithful copy of almost all the elements of a Chinese tea garden. The pioneers studied, documented, and initially transmitted the Chinese system, which lends additional interest to the technical choices that drove its mutations over the next century.

The implantation and transformation of Chinese methods is richly if unsystematically documented through the entire history of tea production in India. It thus offers a promising case with which to test the utility for historians of the chaîne opératoire. To illuminate the industry's evolution, I document and compare a series of chaînes opératoires, drawing on sources that begin in the 1830s and end one century later.

In evaluating and comparing chaînes opératoires, the devil is in the details. I compiled the chaînes opératoires in the form of tables and diagrams, most of which are unfortunately too large and elaborate to include in a book chapter. I have therefore included only a few indicative illustrations here, and the complete set of tables and diagrams, along with a detailed time line, are available online (www.francescabray.co.uk/tea) and referenced in the endnotes.[11]

Studies and Sources in Tea Literature

The history of tea, in China, in India, and in Europe, is booming. Almost every month sees the publication of an enticing new work, an article, monograph, or popular book, discussing tea as a Chinese drink that went global, as a gear-wheel in the constitution of the British Empire, as a quintessential vehicle of Victorian English or imperial Chinese culture and manners, or as the labor-intensive plantation crop in whose service the government of British India transmuted slavery into new forms of semi-servitude, notably the system of indenture. Themes that feature prominently in these studies include tea and empire, tea and colonial science, tea and taste, tea and class in Britain (viewed through consumption history), and tea, race, and class in India (viewed through labor history). In many of these works the techniques and technologies of producing, transporting, and marketing tea are an element in the argument. But despite the wealth of recent publishing on tea, no systematic technological history of the Indian tea industry has yet been written. There is no published study that collates and compares over time the rich but typically scattered technological information contained in the abundant sources.[12]

The flourishing field of tea history draws sustenance from an extraordinary wealth of source materials. The earliest Chinese work devoted to tea, Lu Yu's *Chajing* (Classic of tea), was completed in the Tang dynasty, in 760. Lu's text, and the taste for tea, took the medieval Chinese world by storm and started a flood of Chinese writings on tea that variously document the techniques of tea planting and processing, tea taxes and trade, myths and medical properties, the arts of preparation and appreciation, and local and regional specialties. Tea operas and songs, poems, paintings, wares, and utensils express Chinese tea cultures from the Tang through to the modern era. Some of the earliest works—notably Lu Yu's *Chajing* of 760, Han E's *Sishi zuanyao* (Compendium of essentials for the four seasons), probably compiled in the late ninth century, and Wang Zhen's *Nongshu* (Agricultural treatise) of 1313—provide rather complete chaîne opératoire–style accounts of how to cultivate and process tea.[13]

In the voluminous later Chinese literature on tea, however, it is rare to find a full and systematic treatment of the technical procedures of tea cultivation or processing. The focus of later works was mostly on varieties and grades of tea (of which there were many hundreds) and on the arts of tea brewing and appreciation. This was a literature of connoisseurship that found a wide readership. An eighteenth- or nineteenth-century chaîne opératoire based on Chinese sources would be heavily skewed toward the consumption end of the chain. We learn from Samuel Ball (see below) that tea merchants, working as they did between village and city, were directly acquainted with both the practices of production and the criteria of consumption and often combined the two complementary forms of expertise with an antiquarian interest in texts on tea. However, they apparently authored no texts on the subject or, at least, none that circulated beyond their immediate circle.[14]

In around 1850 two British writers, Samuel Ball and Robert Fortune, published three books documenting in graphic detail the skills and equipment of tea production commonly used in the tea regions of Fujian, Zhejiang, and Guangdong.[15] These areas produced almost all the tea exported to Britain, and both Ball and Fortune intended their accounts as contributions to the development and improvement of the nascent Indian tea industry.[16] Fortune's descriptions, scattered through two books written in the format of travelers' tales, were based on direct observation. Disguised as a Chinese merchant to avoid apprehension by the authorities, Fortune visited several provinces where tea was grown and manufactured; he even reached the fabled up-country tea regions of Fujian and Zhejiang where the very best export teas were grown.[17]

Ball, stationed in Canton between 1804 and 1826 as the inspector of teas for the East India Company, had discussed all aspects of tea production minutely with merchants, growers, and manipulators from the tea districts. Ball had studied Chinese and read voluminously in historical and contemporary technical and "statistical" works that his Chinese colleagues lent him. He set up a workshop where he had visiting tea makers from Fujian reproduce the most sophisticated tea-making procedures for him (albeit using inferior Guangdong leaves), and he conducted experiments, varying timing and temperature to try to produce standardized procedures for making the best teas.

Until well into the 1880s and even later, writers on tea in India continued to use Ball's and Fortune's works as references for assessing the quality of Indian teas and gauging the progress of the Indian industry. Since they document technical aspects more comprehensively and systematically than Chinese sources of the period, these works have also served as rich sources for historians of tea in China. I draw upon both Ball and Fortune to reconstruct the original chaînes opératoires of Chinese green and black export tea production from which early Indian tea manufacturing derived.[18]

The variety of sources for the history of tea drinking in Europe is equally impressive, if chronologically more compressed, since the strange dry leaves of the tea plant first reached Europe in the early seventeenth century. The documentation on early modern European tea consumption ranges from medical discussions to taxation acts. The sources are particularly rich for Britain, which by the mid-eighteenth century was the world's largest consumer of tea outside China. Not surprisingly, given that, in Britain, tea was handled and later also produced by large merchant companies, such aspects as shipping, marketing, and branding are minutely documented, which is not the case for the Chinese tea industry prior to the 1920s.[19] Another distinctive feature of the literature on tea in Europe, which again does not appear in China until the 1920s, are the scientific debates and reports on the chemistry and botany of tea, stimulated by research in London, Paris, or Berlin but probably more attended to in India than in the metropolis.[20]

Sources on the great English dream of producing tea in India first appear in the 1780s. The president of the Royal Society, the eminent botanist Sir Joseph Banks, argued eloquently for the national importance of cultivating valuable tropical crops such as tea, coffee, cinchona, and breadfruit in the British colonies. In pursuit of this goal an officer in the Bengal military, Robert Kyd, established a "Garden of Acclimatization," or Botanical Garden, for the East India Company (EIC) in Calcutta in 1787. Kyd and his successors as director of the garden—notably the botanist Nathaniel Wallich—worked closely with scientists at Kew on testing and developing suitable crop plants for India. In 1788 Banks sent a memorandum to the EIC chairman urging him to identify lands suitable for growing tea in the Bengal Presidency, in areas where the climate and altitude appeared suitable. Although attempts to grow the Chinese tea bushes sent to Calcutta by Lord Macartney's China embassy in 1793 failed, the project of growing tea in India steadily gained credence as being scientifically viable and economically desirable. So when travelers in the 1820s and 1830s claimed to have spotted native species of tea growing in the jungles of Assam, a state newly annexed into Bengal, the identification of the materials they sent to Calcutta fed into an already vigorous scientific and mercantile conversation.[21]

The Government of India established its first experimental tea gardens and manufactures in Assam in the late 1830s, "leaving the industry to be developed by private enterprise should its practicability be demonstrated." The first detailed report was published just two or three years later in 1839.[22] After a period of relative silence, reflecting some uncertainty about whether the tea project would indeed succeed, by the 1860s a critical mass of planters and technical experts was emerging.[23] Meanwhile in the British imagination the romance of the planter's life high in the misty ranges was taking hold—the public was eager to read of jungly hills and torrential rains, encounters with comical coolies, with elephants or tigers, and the evening ritual of manly conversation over a chota peg.[24]

From the 1870s the publication of tea literature both technical and anecdotal gathered momentum, eventually encompassing planters' reports, handbooks, diaries and memoirs, tea trade journals, and tea company correspondence and accounts. Although works calling themselves "a handbook," "a vademecum," or even "a textbook" start to appear in the 1870s, those that provided a systematic set of instructions about how to grow tea and/or run an estate were relatively rare until the last years of the nineteenth century. Most of the literature on tea produced in India unsurprisingly provides the planter's or company's view rather than that of the workforce, and any intrusion by government inspectors was typically portrayed as time-wasting and ignorant interference. For labor historians, the sequence of *Labour Enquiry Reports* of 1896, 1906, 1921, and 1931 "have become veritable bibles regarding the life and times of the Assam plantations." The centennial or similarly celebratory histories commissioned by big companies like the Assam Company or the Colombo Tea Traders' Association are also valuable even though they are often frustratingly incomplete and inconsistent repositories of information for the whole history of tea production in their particular region, often stretching into the years beyond Independence.[25]

Here I present and compare a chronological sequence of snapshots of the chaînes opératoires of tea production as extracted from some of the more significant texts of the moment. The initial focus is on the experimental implantation of a Chinese industry into an Indian context. My first technological snapshots come from the years around 1840, when Chinese plants, experts, and skills were first brought to the hills of Assam. The main source is a report prepared by C. A. Bruce, the manager of the Government of India's first experimental tea gardens in Assam, and there are some interesting contrasts and similarities between the chaînes opératoires that Bruce traces for green and black tea and those laid out in the accounts of tea production in China at the same period, written by Fortune and Ball.[26]

Then I trace the erratic progress toward a mechanized industry as it unfolded over several decades. The description is structured around two technological snapshots. In 1839 Bruce had already advocated designing machines to speed production, but it was not for several decades that conditions were ripe for mechanization of the industry. The second technological snapshot dates from the period around 1880, when (after four roller-coaster decades of boom or bust) experts assessed that the Indian tea industry's future was finally secured. Serious efforts to mechanize began, and a dense web of technical communications took shape around journals such as the *Tea Gazette.* Key texts for this moment are the fourth edition of Money's monograph *Tea—Cultivation and Manufacture*, published in 1883, and the *Tea Gazette*'s compilation *Tea Planter's Vade Mecum*, published in 1885.[27]

My third technological snapshot dates from a decade or so later, in the late 1890s, when Indian tea exports had finally surpassed China's. British tea drinkers had at last rejected the lighter China teas in favor of more robust Assam-style Indian black teas; engines were now presumed to be standard fixtures in every tea estate; and machine builders were pulling out all the stops to increase the rate and volume of processing. Crole's *Tea: A text book of tea planting and manufacture* of 1897 is the first work to call itself a textbook. However, contextual evidence suggests that most planters acquired their expertise through on-the-job training and that, in practice, they learned more from glossaries for communicating with the "coolies" than they did from textbooks. Crole, a retired Assam tea planter who joined forces with a London machine-building firm, was an unconditional enthusiast of machines for every stage of tea production, and his lists and assessments of different makes and designs offer a useful reference for subsequent progress in this domain.[28]

My fourth and final technological snapshot dates from the mid-1930s, a century after tea planting began in India. By this time tea planting had become routinized across India and had settled into an assemblage of manageable, dependable practices. My case study here is the cluster of Finlay tea estates in the Kanan Devan Hills of Travancore (the High Range) at the southern tip of India. Tea planting began here in the 1880s, following the Assam model. By 1895 a couple of dozen small tea estates were operating successfully, and they attracted the attention of the huge colonial firm Finlay Muir and Company. Finlay's bought up the lease on the whole area of the Concession ceded to Europeans for cultivation by the Maharajah and persuaded the planters on the ground to continue as managers for the company.[29] In 1922 an ambitious young lad from Glasgow called Alec Robertson (my father-in-law) started his apprenticeship in tea planting as an assistant manager in one of the High Range estates. He was sent to the head office in Calcutta for six years between 1924 and 1930; then his request to transfer back to the High Range was accepted, and there he stayed until Independence came in 1947.

In 1976 A. S. Robertson produced a memoir on tea planting that had been commissioned by the South Asian Studies Centre in Cambridge. The memoir includes a set of step-by-step instructions for developing, planting, and running a fifteen-hundred-acre tea estate, and this constitutes the chaîne opératoire of my final technological snapshot.[30] The aim here is to assess earlier ambitions and claims of unstoppable technical progress against the stabilized, standardized practices of tea planting in estates belonging to the huge tea companies that had steadily taken over the landscape in India and Ceylon, starting in the 1880s.[31] I also use this centennial case to assess what, if anything, survived of the original Chinese mode of tea production after a century of growth on Indian soil.

Using the Chaîne Opératoire

I use the chaîne opératoire (procedural sequence) as a device for organizing and comparing systems of tea production with their variants and transformations. My aim is to clarify what traveled and what was omitted, abandoned, or superseded during the transplantation of the craft of Chinese tea making to British India. The question is how the industry grew and branched from its Chinese origins as a constellation of artisanal skills and how to trace its complicated and far from linear or steady progress toward a modern mechanized industry.

The chaîne opératoire is a tool for mapping how the various elements are linked in a technical production process. At its simplest, a chaîne opératoire may trace only the sequences and conjunctions of material inputs and techniques that go into producing an artifact. Yet even this basic level permits archaeologists to deduce information about the systems of ritual and aesthetic values, and social organization within which the artifact is embedded (figure 6.1). As developed by anthropologists, the chaîne opératoire is more typically elaborated to incorporate a spectrum of inputs and outputs—political, social, economic, symbolic, or material—that go into producing a given human artifact.[32] Anthropologists may incorporate into their analysis of a chaîne opératoire, for example, the ritual offerings and the sociable sharing of betel nut that are an integral part of producing a barnful of ripe yams or the resolution of culture clashes between cosmologies and work disciplines that is necessary to operationalize a French-designed gazogene installed in Costa Rica.[33]

It is curious that the chaîne opératoire has seldom been used by historians of technology or by historians analyzing processes of technical transfer. Yet, I suggest, it has great potential for tracking the complexities of itineraries of knowledge and skill and the values attributed to techniques or technologies.[34] When it comes to how techniques, skills, and the multiple elements of a chaîne opératoire are documented or spoken of within the society where they are practiced, we must not look for the consistency of an engineer: some will be the object of attention, anxiety, or pride, others will be taken for granted and passed over in silence. We may be surprised or frustrated by gaps or obsessions, by the unexpected significance, value, or power attributed to some elements and the virtual disappearance of what with hindsight seem essential components.

By providing a template showing the technical sequences that must necessarily occur if an artifact such as processed tea is to be successfully produced, the basic chaîne opératoire can be used to pinpoint absences and presences in documentation and expertise. In our comparing of this basic and essential sequence of procedures against the recipes of the time, the chaînes opératoires envisaged and recorded in our sources highlight the explicit or underlying concerns and values expressed in these recipes or documents. Considering which media and genres (written handbooks, human expertise, material artifacts) containing information about a given technical practice such as tea manufacturing were produced in a given historical context, what informative value was attributed to them at the time, and what they can contribute today to the historical recovery of the technical practice can all offer further insights into what we might call the local culture of tea—and into the challenges or opportunities that this technical-cultural ensemble presented for the transfer of tea production or consumption from one society to another.

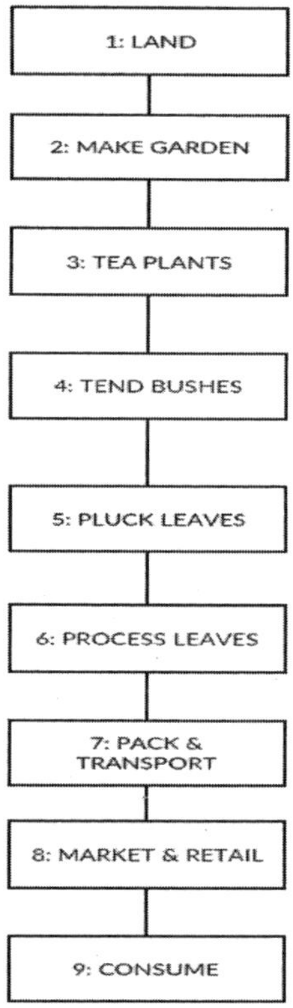

FIGURE 6.2. Basic chaîne opératoire of tea production. Author's image.

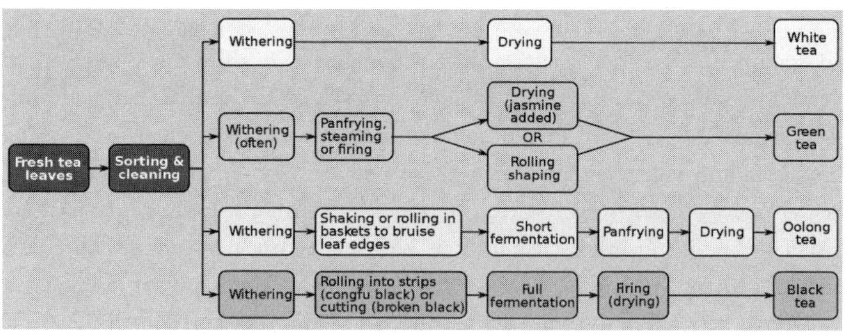

FIGURE 6.3. Processing sequences for green, white, Oolong, and black teas. https://commons.wikimedia.org/wiki/File:Tea_processing_chart_II.svg (accessed July 12, 2018). Creative Commons Attribution-Share Alike 3.0 Unported (author: Apis O-tang, 2008)

You cannot process, brew, enjoy, or tax tea without the physical product. Tea is a specialized crop, and the leaves require complex care and processing before they can be used as a beverage.[35] So at the core of the history of tea, necessarily, lie the material production and transformations of the tea leaf. Land must be cleared and prepared, seeds planted, seedlings transplanted and protected, the tea garden tended, the leaves plucked; then the leaves are variously fired or withered, fermented, rolled and refired before being sorted, graded, and packed, transported, sold, and finally brewed. Here I use a very basic chaîne opératoire. It comprises nine steps, from acquiring suitable land, through processing the leaf, to the final steps of brewing and drinking tea (figure 6.2). In my four snapshots I focus principally on steps 1–6, mapping the technical constellations and sequences of the tea garden and its factory. When they are reported as directly influencing upstream changes in technical choice, I also link in the subsequent stages of transport, marketing, and consumption. Step 6, "Processing," unpacks into different sub-sequences depending upon the type and grade of tea being prepared (figure 6.3). I compare the successive versions of the basic chaîne opératoire, and of the sub-sequences in step 6, so as to map continuity, variation, and evolution—both in the individual steps in the procedures of tea production and in the technical clusters or assemblages that characterized successive instantiations of Indian tea production.

As comparing figures 6.1 and 6.3 makes clear, there is no standard symbolic code or set of conventions for drawing a chaîne opératoire. While it can, in principle, be elaborated to chart the culture of tea in all its material, aesthetic, social, and political aspects, integrating production, consumption, and representation, as a first effort I necessarily restrict myself to the basics. For my analysis I used two very simple graphic versions, the table and the diagram, and (as already explained) in this chapter I translate the main features of the diagrams and tables displayed on the website into textual accounts.[36] Constructing and comparing the chaînes opératoires characteristic of my four different phases enables me to identify and contextualize the sets and sequences of technical change that progressively transformed a Chinese craft into a British industry.

Transplanting Chinese Savoir Faire

The initial prognosis for tea growing in India was unfavorable. Tea plants brought from China to the Calcutta Botanical Garden in 1793 by the Macartney embassy had withered and died. When British travelers in Assam sent seeds and plants of what they believed were indigenous tea trees to the Botanical Garden for identification in the 1820s, the botanists denied that they belonged to the tea species and made no effort to try them out. In 1826, however, the expert Dutch tea taster and merchant J. J. L. L. Jacobson began planting tea on Java using Chinese seeds, workers, and methods; although the tea was judged of less than excellent quality, here was proof that tea could be successfully transplanted from China to tropical zones.[37]

In 1834 Sir William Bentinck, the governor-general of India, persuaded the British Parliament to set up the twelve-member Tea Commission under his direction to investigate the viability of tea plantations in British India. The commission immediately began the mapping and assessment of potential Indian terrains where "the China plant" could flourish. The montane regions of the state of Assam, ceded by Burma to the British East India Company in 1826, looked particularly promising. No sooner was the Tea Commission established than the directors of the EIC, which had lost its monopoly on the China tea trade the previous year, decided to set up "an experimental establishment in Assam for cultivating and manufacturing tea."[38] But how could the British secure the knowledge, materials, and skills necessary to transform Assam's hills from jungle into tea garden? Of course, the start-up materials and knowledge would have to come from China.

At that time China was *the* world's tea country. The tea regions of China produced prodigious quantities of tea—green, white and yellow, black and red, scented and plain, fine and coarse, in loose leaves and in cakes. Most of this tea was consumed within China itself, where it had been the everyday drink of rich and poor alike for several centuries.[39] As the taste for tea caught on through eighteenth-century Persia and England, Morocco, Holland, the United States, and Russia, tea became one of China's most important export goods, rapidly outstripping even silks and porcelain as the biggest source of income for the British and Dutch East India Companies.[40]

Britain was by far the biggest market for Chinese teas. The millions of pounds of tea imported yearly into Britain by the 1820s—green teas and black, Singlos, Congous, and Boheas—all came from China. And yet these Chinese teas were a source of mystery: although tales abounded of tea trees so tall the leaves could only be harvested by monkeys, of different species yielding green and black teas, or of Chinese tea makers meticulously twisting each leaf separately between their fingers to produce the characteristic twist, nobody had much accurate knowledge of the tea plant itself, of how it was cultivated in China, or by what mysterious sequence of manipulations a fresh green leaf was transformed into the dry, crisp, fine twists that magically unfurled in hot water to produce a fragrant beverage.[41]

For Europeans the cured tea leaves were akin to what engineers and historians of technology call a black box, a technical device or product whose complex internal structures are a mystery to the user and cannot be deduced from the input or the output. For most of us, for example, our computers or cell-phones are a black box: we use them without the faintest idea of what is going on inside. But we do not need to understand the contents in order to use these devices competently, nor do we aim to build our own versions.

In the case of Chinese tea, the British invested copious capital—supplied to Chinese tea brokers in Canton, by EIC officers—to purchase futures in various types and qualities of tea. The money went up-country to the tea regions to finance the operations of smallholder tea growers and processors, but the EIC investors saw nothing of the tea until the crates of leaves were delivered to Canton. It was only then that they had any direct control over quality: experienced British tea tasters took samples, brewed them into fragrant green or brown liquid, and sipped them attentively to check the grade and quality and assess the value of the tea for auction in London.[42] Until the 1830s the British had grumbled at the expenditure involved in purchasing tea from China, and they had sought ways to restore the balance of trade.[43] But they had not interested themselves in how they might learn to grow tea themselves. Then, with the Chinese government threatening to cut off trade entirely, the British acquired an urgent interest in prizing open the black box, examining its contents, and learning for themselves the mechanisms or procedures by which cured tea leaves were produced. The question was how could they gain access to the contents?

The experts of the time were unanimous that the start-up materials and knowledge must come from China. Unluckily, however, both China and Japan (the only other tea-growing country at the time) were forbidden territory to Europeans. Except for the strictly controlled port at Deshima, Japan had been closed to Westerners since 1635. In China the Qing government likewise restricted access to a few ports where Western traders were allowed to deal only through officially accredited Chinese merchants. From 1757 foreign trade was restricted to the single port of Canton, and foreigners were strictly forbidden from penetrating Chinese territory beyond the city limits. Since "the Chinese guarded jealously the secrets which gave them the monopoly of supplying the world with tea," it was not easy, nor was it even legal, to gain access to the technical knowledge and materials needed to start up a new tea industry in India.[44]

Nevertheless, without Chinese know-how nothing could be done. Since Mohamet couldn't go to the mountain, the mountain must be brought to Mohamet. Bentinck, doubly empowered as Governor-General of India and Chair of the Tea Commission, argued that there were two essential steps. First, the lands under EIC control should be surveyed so as to establish where tea would flourish, and second, "concerted measures [should be taken] for obtaining the genuine [i.e. Chinese] plant, and the actual cultivator." The Chinese cultivator was a key figure: he alone possessed the understanding of the Chinese tea plant that would allow a successful implantation into Indian soil; he alone had mastery of the complex sequence of techniques for processing tea and could sow the seeds of this knowledge in the minds of British estate managers and the hands of their local labor force.[45]

No sooner had the Tea Commission convened than they sent out G. J. Gordon, an opium dealer with experience of China, to Canton via Macau, with orders to obtain tea seeds (illicitly) for India. Gordon set sail in June 1834. In December the botanists of the Calcutta Garden unexpectedly reversed their earlier judgment and confirmed that the tealike trees growing wild in Assam did indeed belong to the tea species. The Commission recalled Gordon but not before he had dispatched 200 lbs of tea seed, which landed in Calcutta in January 1835—50 lbs were planted at the Calcutta Botanical Garden, 50 lbs were sent to Assam, 50 lbs sent to Kumaon in the Western Himalayas, and 50 lbs sent to the Nilgiris in the Madras Presidency. Gordon brought three "Chinamen" back with him, two of whom knew how to make black tea; these men were the pioneers of a trickle of Chinese tea growers, tea makers, and tea chest makers who worked with the first British experimental tea planters in India from the late 1830s through into the 1860s, establishing a working industry.[46]

Whereas initial efforts to recruit large numbers of Chinese workers from Asia's port cities failed miserably, the Chinese tea makers recruited personally by Gordon, Fortune, and others appear to have settled in well. Ball mentions an estate of 118 acres under cultivation in Kamaon (Kumaon) in the late 1840s, with one overseer, ten Chinese tea makers, two Chinese tea box–makers, four sawyers, and an unspecified number of local laborers. McGowan's reminiscences, published in 1861, note that the Chinese tea makers brought to

Kangra ten years earlier had by now married local Muslim women and set up their own little village.[47]

Working closely with the British directors or managers of the tea gardens, the Chinese tea makers demonstrated and explained how tea was grown and processed. As experts they were well paid and well treated, and the managers respected their judgment. In addition to their own skilled work, they supervised individual tea gardens within the enterprise and trained "native" workers. The training ratio envisaged was one Chinese to train ten or twelve "natives," whether in the delicate art of rolling tea or in the more robust but equally essential skill of building lead-lined boxes.[48] The intention was that the expensive Chinese experts would be phased out once they had successfully trained sufficient "natives," who could be paid at much lower rates and treated with less deference.[49]

The Chinese tea workers threw open the lid of the black box of tea production for the British. But they did far more than simply transmit established Chinese methods to India. By constructing and comparing the chaînes opératoires laid out by Bruce, Fortune, and Ball I was able to disentangle the terminological ambiguities and complex processual strands interwoven in Bruce's report and thus to clarify the technological choices that went into selecting certain Chinese cultivation techniques, some for application to China plants, others to Assam stock, as well as the adaptation of processing methods to local conditions. If we examine Bruce's account in this light, we see that in addition to their success in establishing the China plant in Assam, Bruce's Chinese tea makers provided indispensable experience and support for the domestication of the native Assam bushes and for the streamlining of tea processing. Their flexible expertise helped lay the foundations for the eventual worldwide supremacy of Assam-style black teas and their production and processing model.

I first drew up the chaînes opératoires of tea cultivation and manufacture as extracted from Bruce's *Report on the Manufacture of Tea*, submitted to the Tea Commission in 1839.[50] As Antrobus remarks, they are "an account of a dialogue between Bruce and the Chinese black tea makers which took place in 1837 and which must have been Bruce's only manual of how to proceed." I then put together the chaînes opératoires of green and black tea cultivation and manufacture in China, as described by Ball and Fortune.[51]

If we compare the processing procedures recorded by Bruce with those detailed in the accounts given by Fortune and Ball, we glimpse an early hint of simplification, or at the very least, an indifference on Bruce's part to the subtleties of processing procedures that were so sensitively rendered by Ball. Fortune attributed the supposedly inferior quality of the teas initially produced in Assam to the fact that the first Chinese tea makers came from around Canton, where the teas were inferior to those of Fujian or Zhejiang.[52] Even so, it is unlikely that the rather rudimentary processing techniques Bruce sets out represented the best that his Chinese tea makers could do. It is more likely that Bruce and the tea makers between them devised a streamlining of procedures so as to accommodate the lack of skilled labor and the challenges of timing.

As Fortune emphasizes, in China, whether teas were being produced for the home market or for export, the quality and price of the tea were directly proportional to the time, care, and skill that went into processing. No complicated equipment was involved. Ball's sensuous descriptions show us how magic was wrought in a shed behind the farmer's house with nothing more than a pair of cast-iron pans, chopsticks, bamboo sieves, and above all, dexterity and judgment. For green teas the fresh leaves were first panned (tossed in a wok until they became soft and pliable). Then they were thrown onto a table of bamboo slats where three or four workers rolled big handfuls like "a baker working and twisting his dough" into balls, expressing the moisture and curling the leaves; this process was repeated two or three times, after which the balls were shaken out and the leaves set on screens of bamboo to air-dry. Then they were stirred and tossed in woks in a slow process of drying, before the final sifting and grading.[53]

Although the processes for making black teas differed in some details (see below), the general principle of Chinese tea processing was to expel the moisture from the leaves while retaining the flavor. This required different rates of drying and manipulation for different stages, flushes, or prevailing weather, and so on. During the brief plucking season all the labor of the household was mobilized day and night: processing must begin as soon as the leaves came in, and if the sequence was interrupted the tea was spoiled. Green teas required more careful cultivation and more processing than black teas and were more liable to spoil. Late-season, coarser leaves were typically used to make the lower-quality export teas. High-quality teas were panned and rolled several times before they were ready. In producing the very best green teas, the process of "To Ching" involved tossing the wilted leaves several hundred times by hand, patting, rubbing, and balling them, until their fragrance indicated that they were ready for roasting.[54]

This low-capital, skill-intensive way of producing tea worked well in China's tea regions, where tea was just one of several sources of household livelihood, combined with other crops, home weaving, and so on.[55] But it was not suited to large-scale, specialized plantation production, the model for the British Assam experiment, particularly when labor of any kind, let alone skilled, was scarce. Comparison of the British Assam procedures with the sophisticated, time- and labor-consuming chaînes opératoires detailed by Ball and Fortune makes it clear that both the green and black teas produced by Bruce's team were made with a processing sequence that was reduced to the bare minimum, using fewer tea workers, and with fewer repeats of key processes.

Even with procedures thus streamlined, Bruce feared that the more profitable green teas, preferred by the British public of the time, could not be produced successfully on the Assam estates. The solution he proposed was to follow common Chinese practice—namely, to process the green teas on-site to the stage where they were stabilized, then ship them to Calcutta (or better still, to Britain), where the complicated and intensive finishing processes could, Bruce thought, be mechanized. Bruce proposed shipping Chinese green-tea experts to London for a year to demonstrate the finishing procedures, after which "it might be left to the ingenuity of Englishmen to roll, sift and clean the Tea by machinery, and, in fact, reduce the price of the Green Tea by nearly one half, and thus enable the poor to drink good unadulterated Green Tea." Nothing came of this proposal. Instead, in 1843 the Assam Company decided to cease production of green teas, fearing it could never compete with China either for price or for quality.[56]

One criticism of the black teas Bruce first sent to London was that they were "rather coarse and strong," unlike the delicate Chinese teas they were supposed to displace. In 1840 Bruce was instructed to abandon "experimental teas" and concentrate on producing classic Souchongs, Hysons, and gunpowders.[57] But these teas, as mentioned, required concentrated and prolonged inputs of skilled labor; also, the leaves needed to be small and absolutely fresh when they arrived at the factory.

What with labor bottlenecks and delays in bringing in the freshly plucked leaves from distant tea tracts, Bruce and his tea workers found that often the leaves started to ferment before processing could begin. Furthermore they were using leaves from both Assam and China stock (*jhat*), but Assam leaves were naturally much larger and more turgid than leaves from Chinese jhat, so they did not respond as well to the classic processing. At the time, the spoilage and the unsuitable leaves were regarded as serious problems: they disrupted work schedules, produced inferior classic teas, and upset the Chinese tea makers.[58] But there were other forms of black tea made in China, stronger-liquored teas where the leaf was intentionally withered before treatment.

In China, although the finest black teas for home consumption such as Yen or Padre Souchongs were made of the most delicate leaves from the finest shrubs and "would be injured by any exposure of the leaves to the sun," it was recognized that coarser leaves (those from the main summer harvest, the typical ingredients for export teas) would not wither sufficiently in air and shade alone. They were therefore often exposed to the sun before panning to accelerate wilting. Sun-wilting was also useful for reducing excess moisture if leaves had been picked after rain. Huang traces the appearance of these partially fermented teas back to Wuyi (Bohea) in northern Fujian in the late eighteenth century.[59] Ball and Fortune, describing methods in Fujian and Zhejiang in the 1840s, distinguish between two kinds of black tea: the finer *luk-cha* was air-dried before panning; *hong-cha* (red tea) was exposed to the sun before panning; it was then panned, rolled, broken, and rerolled, then laid out in the air to ferment for two or three days after rolling and before the final firing. This produced warm, aromatic teas that later came to be known as oolongs.[60] Bruce's report does not include a description of the *hong-cha* method, but as it soon became the norm in Assam, it seems likely that his Chinese black-tea makers resorted to it as a way to rescue the "spoiled" leaves and maybe even recommended it as an alternative method, better suited than the classic fine-quality procedures to Assam leaves, climate, and labor resources.

Now let us turn briefly to the question of the Assam stock or jhat. Bruce was growing two completely different kinds of tea. Following the instructions initially of the government and later of the Assam Company, he grew the China plant either from seed imported from China or from seedlings raised in China or in the Calcutta Botanical Garden. The China seed generally did well, and it was sown, transplanted, and tended largely according to the orthodox Chinese method. Even once the wild Assam bushes had been certified botanically as true teas, they were clearly not domesticates and were generally viewed not as a potential source of leaves but, rather, as an indicator that the domesticated China plants could flourish in British India. But Bruce and his Chinese team set to work to domesticate the Assam stock, applying and adapting burning, cutting-back, and transplanting techniques that, although they did not feature in most Chinese written accounts, were actually common practice in China.[61]

However common they might have been in China, these vegetative methods of reproduction did not form part of the seed-based cultivation repertory approved by the Calcutta Botanical Garden and adopted by the other nascent tea regions in the Himalayas and South India. Outside Assam there was essentially no alternative to growing from seed: there were no wild teas to work with. But Bruce, in Assam, was charged with a twofold experiment: to raise viable tea plants and to establish viable tea-making procedures. For the latter he needed leaves, and no self-respecting Chinese tea grower would begin picking before a bush was at least three years old. So, while the China plants were maturing, Bruce and his team went out into the jungle to locate natural groves of the tall, vigorous, thick-leaved Assam tea trees from which they could harvest leaves that would at least approximate China leaves. They cleared the land using local Assamese slash-and-burn (*jihum*) techniques, and in as little as a year after the burn-off the tea bushes were ready for cropping. This method, however, could not be sustainable long-term since it produced an irregular scattering of small tea tracts that was difficult to harvest efficiently.[62]

In jungle areas with only a sparse scattering of tea trees, rather than harvesting on the spot the regenerated plants would be dug up and transplanted into the main garden. Because they were so much more vigorous than the China bushes, the Assam transplants were spaced in rows six rather than four feet apart.[63] Bruce also experimented successfully with growing the Assam plants from seed, but he found transplanting preferable because it reduced growth. Even so, the Assam bushes typically grew within two or three years to a height of six feet. The Chinese tea men recommended cutting them to the ground every three years (a procedure they said was necessary only every nine years back in China). While the low China bushes, kept to a neat height of three or four feet simply by plucking, were easily harvested by women and children, the six-foot Assam bushes were a challenge even to men. Radical pruning, which later became a core technique in India tea cultivation, was apparently never considered. This may have been because tea was seldom pruned in China, and the China tea makers had no expertise to offer. Bruce hoped that over the generations the bushes would grow smaller.[64] He was not, however, worried by the large size of the Assam leaves: "the size of the leaf matters nothing, in my opinion, provided it is young and tender; even [the China plants'] diminutive leaf, if one day too old, is good for nothing."[65] There is no indication in his report that the China and Assam leaves were processed differently, nor that different leaves were selected to make green or black tea. In this, Bruce and his team were following Chinese custom.

The quality of the teas that Bruce sent to London was deemed poor for several reasons. First, much of the tea went moldy in shipment. Second, the tea makers had not yet devised processing methods adapted to the distinctive qualities of leaves grown in Assam, whether from China bushes or from Assam trees. The heavy monsoon climate produced more turgid leaves, even in the China jhat, that resisted conventional Chinese processing. Finally, the Assam jhat leaves naturally produced darker, stronger teas, for which there was as yet no established market in Britain.

The initial reaction of the Assam Company was to insist on improving the processing of China jhat plants; Bruce was instructed to abandon working with the Assam stock. But London was a long way from Nazira and the message seems to have been ignored, fortunately for the long-term success of the Assam Company. It was not long before the advantages of the Assam stock became clear, at least to producers. Being larger and more vigorous than the China bush, the Assam plant yielded a greater weight of leaves at each plucking and flushed again more quickly and over a longer season. This promised great improvement in both output and productivity. Once production methods were improved and stabilized, not only would the acreage yield more tea overall but the longer season would allow planters to make better use of their capital investments: large buildings, access roads, permanent staff, indentured labor, and eventually, machinery. And then, as long as British tea drinkers could be persuaded to consume more Assam tea, the future would be bright.

"The Inventive Genius of the Anglo-Saxon Race"

In the 1830s and 1840s the Government of India and the Assam Company strove to implant Chinese tea bushes and Chinese methods as a prerequisite to establishing a successful tea industry in India. Yet from the start British ambitions were not restricted to mere reproduction of Chinese techniques. "Naturalization," the process advocated by Lord Bentinck, encompasses both transplantation and adaption: from the outset the British were convinced that they could significantly improve Chinese methods and produce superior tea. "We can scarcely doubt that when the skill and science of the Europeans, aided by thermometers etc should once be applied to the cultivation of tea in favourable situations, the Chinese tea will soon be excelled in quality and flavour," wrote one early enthusiast in a document included in Bentinck's report. Fifty years later an American observer of tea production in the Himalayan tea districts celebrated the rapidity with which Indian planters had transformed "crude" Chinese techniques, unchanged for centuries, into a new industry with "numerous machines cleverly contrived to save labor and better the teas, . . . a striking illustration of the activity, the energy, the inventive genius of the Anglo-Saxon race."[66]

This faith in the power of machines and science was typical of the age, and perhaps in London or Manchester such technical confidence seemed justified. On the ground in India, however, progress was typically erratic. If we construct a time line of innovations in techniques and machinery from 1836 to 1931 we see that inventions seldom succeeded first time round, for a variety of reasons.[67] One was the lack of local expertise and infrastructure. In 1839, for instance, Bruce ordered a steam-powered sawing machine to make planks for tea chests. Unfortunately, nobody knew how to mount or assemble it; after five years it was returned to Calcutta and sold. Only in 1853 was a sawmill successfully installed. Another factor was the lack of trust between different levels of management. In 1842 Assistant Manager Strong invented a rolling machine. If the machine had worked well, it would have been a vital contribution to improving the fermentation of Assam-style teas. Unfortunately, at that point the London board wanted only China-style teas. In order to avoid confrontations with London, Strong was told not to try out his machine. In 1859 Superintendent Mornay was requested by the company to develop a grading machine, while Company Engineer Gibbon developed a promising rolling machine. Both Mornay and Gibbon fell out with the company administration; they were fired and their inventions declared useless.[68]

Another factor that impeded successful mechanization in the early years of the Assam Company, between 1840 and 1870, was the instability of the tea industry and the lack of reliable markets, which discouraged expensive investments.[69] From the late 1860s the embryonic visions of mechanization began to take concrete, if only partial, shape. One fundamental innovation was the rolling machine successfully developed by Company Engineer Kinmond in 1868. The machine did the work of forty-four men and allowed a batch of black tea to be fully processed within daylight hours, achieving excellent quality. As thorough and speedy rolling of the leaves greatly improved fermentation, Kinmond's roller was a crucial contribution to the consolidation of Assam-style black-tea production. The company installed several Kinmond rollers over the next decade, and as mechanical rolling became standard other engineers from outside the Assam Company, notably the brothers Jackson, began working on improved rollers. The 1880s saw a surge in the mechanization of agriculture across the board; the tea industry was just one sector in a burgeoning market. Machine companies like Marshall Sons & Co. and Britannia—some based in Britain, some with branches in Calcutta—now started designing and building all kinds of new tea machinery.[70]

To gauge how the achievements and potential of mechanization were assessed at the time, I turn to the chaîne opératoire of tea production laid out by a respected expert, Lieutenant-Colonel Edward Money. Money had worked as a tea planter in Assam for many years and had visited tea gardens across India, from Chittagong to Darjeeling and the Nilgiris. His monograph *Tea—Cultivation and Manufacture* was originally written as an entry for a prize offered by the Agricultural and Horticultural Society of India in 1871 for the best work on tea. Money's book won the gold medal and ran into many editions; our chaîne opératoire uses the fourth, revised edition of 1883 where Money discusses the evolution of the industry over the last two decades.[71]

Information is sparse on how techniques of tea growing and tea making developed in India between Bruce's initial report and Money's work. Through the 1860s and 1870s, as existing tea regions expanded and new regions like the Nilgiris and Ceylon were opened to tea,[72] British money poured in, along with men hoping to make their fortunes. Yet we possess little technical information for the period, partly because there were no shared standards or general manuals. Money's book was especially valuable, he believed, because an alarming number of tea managers had no experience or knowledge; "each made [tea] in his own way, and often turned out the most worthless stuff." As another authority put it, "People who had failed in everything else were thought quite competent to make [tea] plantations." The *Indian Agriculturalist* hailed Money's book as "the standard work on the subject"; the *India Mail* called it a "thorough, clear, instructive directorium."[73]

Money's work was a milestone. It was the first work that systematically spelled out the steps of tea production and its variations, along with the principles of managing an estate. Furthermore, reflecting on changes in the industry since the first and fourth editions, the author gives us a considered evaluation of recent innovations and their potential. But Money's work was not written in a vacuum. Since the 1870s the future of the Indian tea industry had begun to look secure. By 1880 a community of practice had formed that included tea planters, engineers, labor recruiters, government inspectors, and specialist journals such as the *Tea Gazette* in India as well as auction houses, retailers, and machine factories in Britain. Their increasingly dense exchanges helped build consensus on norms and standards as the pace of innovation changed.

In his preface to the fourth edition, Money writes: "So much has been done in Tea since I last wrote [five years earlier in 1878], I found it impossible to embody all in the former book."[74] He therefore added six chapters, including one on "Making Indian Tea known in the United Kingdom" and one on "Tea Machinery." In mapping the main contents of Money's work as they related to the nine basic phases identified in figure 6.2, I aimed to show the many fields of competence and understanding that Money believed were necessary for a manager or an investor.

For each phase—from site selection and labor procurement, through costing and accounting, to the brokers' criteria for pricing grades of tea—Money lays out systematic instructions, while allowing room for variants depending on type of soil, plant, climate, and so on. He gives a standard sequence for cultivation, which adds manuring, composting, and pruning to Bruce's list of operations. Thirty years earlier George Williamson, superintendent of the Assam Company (1852–1858), arguing that Chinese cultivation techniques did not suit the Assam jhat, had introduced a regime of reduced early plucking and regular pruning, both intended to keep the Assam plant vigorous. When Williamson left, his regime was abandoned, but pruning was revived in the 1870s and was key to the system of "high cultivation" advocated by Money.[75] In climatically suitable regions a combination of manuring, skilled plucking, and severe pruning, Money claims, allows up to twenty-five flushes a year, almost a year-round cycle. (For comparison, Bruce's bushes gave three flushes and in China four flushes was the maximum.) Although not all regions, Money thought, were climatically suited to this intensive twenty-five-flush regime, he correctly believed that it had great potential in South India and Ceylon, as Robertson's account of tea-estate management in the High Range in the 1930s bears out.[76]

Money's experimental five-step processing sequence for making black tea from the Assam (or hybrid Assam-Chinese) leaf indicates that full fermentation was now fully developed. It includes withering, rolling, fermenting, and sunning before the final firing and sorting. He notes that these methods are designed to retain as much sap in the leaf as possible for a strong flavor; they are thus the antithesis of Chinese or green tea methods, where "strange to say [the sap] is purposely wasted." The five-step two-day sequence is designed to save both time and labor. Money highlights its advantages by contrasting it with a much an older twelve-step three-day process, including four rollings and two pannings, which closely approximates the Chinese *luk-cha* method described by Fortune and which Money admits had now been largely abandoned.[77]

By the time Money was writing it was clear that conditions in India were best suited to growing and making strong-fermented black teas using Assam-jhat leaves. Money's chaîne opératoire offered a stable and effective model for producing such teas. But English tea drinkers still preferred light China teas, and Money devotes the final chapters of his book to building up markets and reeducating tastes.[78] Lowering costs would also help Indian teas oust competitors from China, and here mechanization was the key.[79]

In the domain of machinery, Money felt so much had changed in a mere five years between 1878 and 1883 that he must add a whole new chapter to his fourth edition. Tea machinery "is on the road to perfection," and "if Indian Tea ever vie in quantity with China in the Tea consuming countries of the world, it will be due entirely to the economy effected by our machinery." As the long lists of machines in Money's own book and the *Planter's Vade Mecum* of 1885 show, inventors and machine companies responded enthusiastically to the rise of the new market. Machines proliferated; some worked, others didn't. "Though machinery reduces the cost of production, and in more than one case improves the quality of Tea, and planters know it, the difficulty before them to-day is to know which is the best machine for each operation. Unanimity on this point is not to be expected yet."[80]

Here too Money offers advice to planters that is intended to standardize methods and thus guarantee quality. Which machines does he think are needed, and which brands does he recommend? Reliability and affordability are prime criteria.[81] Money still recommends Kinmond's original rolling machine of 1868, which can be powered by animals, steam, or humans; more recent models he says are not as good.[82] But even the best rolling machines, Money holds, will never completely replace the necessity for hand rolling. Indeed many gardens that invested in rolling machines went back to hand rolling, or alternated between hand and machine rolling depending on the labor situation at the time, because machines couldn't give "the nice final twist" that added so much value. Firing, packing, and sifting would all benefit from suitable machines, Money thought. Variations on the Chinese wooden winnowing-fan were still the norm for fanning and sifting.[83] Because homogeneity of leaf size was crucial in determining the price of teas, Money suggested that the Agricultural Horticultural Society should offer a prize for a sifting machine to sort out the fine tips: this would be the most beneficial of any invention for the tea industry, Money believed.[84]

The *Vade Mecum* of 1885 (to which Money was a prominent contributor) sheds additional light on the technological culture within which mechanization was promoted, and the resources and forms of expertise required and available to support it. Machines need a source of power, and this was the age of steam. C. B. Fergus, "Mechanical Engineer and Tea Factory Mechanician," contributed a long chapter on "Steam engines in tea factories" to the *Vade Mecum*, followed by another on "Machinery." "Steam engines" begins by laying out the physical and mechanical principles of steam engines, then advises on installation, maintenance, different kinds of fuel, the suitability of different types to different scales of plantation, and so on. Planters need this basic mechanical knowledge, Fergus writes, because tea gardens are remote and calling in an engineer takes time. Estate managers therefore need sufficient knowledge to keep the boilers well maintained, to spot trouble, prevent explosions, and know when to send for an engineer.[85] This knowledge is all the more essential because the factory "enginemen" should never be allowed to do more than start and stop the engine, oil it, and keep it clean. They must not be allowed to tinker, even if they claim skills. "Remember that there are *no* Mohamedan fitters in Bengal, or very very few. If anything goes wrong with the engine or machinery, don't try to remedy it yourself unless it is something simple; don't let your engineman nor your amateur engineer-friend touch it, but send for the nearest engineer. It is better to be sure than sorry."[86] The chapter on machinery deals with the main categories of machine in use, and the chapter's introduction likewise explains basic principles: "wheels, shafts and spindles, cranks and eccentrics, drums and band-wheels, clutches and couple-gearing."[87]

Few plantation managers came from a technical background, and probably most were content to ignore the physics of the steam engine and the functioning of eccentrics, leaving the supervision of their boiler and machines to the local "engineman," who presumably would have learned on the job, either in a tea factory or in another industry down in the plains. But the technical ignorance of managers did not impede the steady mechanization of the industry.

By the mid-1880s the Indian tea industry was at last considered a reliable investment. Dividends had stabilized, and although they had been slow to be convinced, the British public had perceptibly shifted toward a taste for Assam-style teas. In 1888 tea exports from British India finally outstripped those from China. "The tea industry experienced phenomenal growth between 1880 and 1900: acreage under tea increased by 120 per cent, while production of tea quadrupled. . . . [Indian tea's] share of the British market increased from 27 per cent in 1880 to 54 per cent in 1901."[88] Spotting a winner, in the 1890s established British colonial companies like Finlay's entered the field, buying up tea plantations, injecting capital, centralizing operations within and across regions, and consolidating the Indian industry's long-standing trend toward large-scale, mechanized production.

The *Text-Book of Tea*, published in 1897 by a retired tea maker from the Jokai (Assam) Tea Company, D. Crole, interests us as an expression of the grandiose ambitions for science and engineering current at the time, though it met with mixed reviews at the time.[89] Crole lays out plans for factory buildings on a gigantic scale, recommends tramways to move the tea between buildings, and presents tea production as a field of applied science, setting out a syllabus in botany and chemistry that he "commends to the study of every planter."[90] While Crole was obviously as much a visionary as an objective observer, it is noteworthy that he took for granted that even isolated tea gardens would now have engines to drive machines, powered by timber, coal, petrol, or hydraulic turbines. Overhead wires should transmit electricity to all parts of the factory and garden.[91]

If machines had been an enthusiasm for Money, they were an obsession for Crole. As an earlier advocate had put it: "There is no reason why, because the ordinary machines have failed, planters should be sunk in the belief that that costly article the coolie must endure as long as Tea does." Observing that several processes are still carried out by human labor, Crole is led to "wonder why many branches of tea cultivation, especially those of an arduous nature that require a large expenditure of labour to accomplish, cannot be effected by the aid of steam and machinery, as is the case with other agricultural pursuits. Not many years ago such an idea might have been dubbed an idle phantasy, but surely it is well within the bounds of probability now; and if so, why is its practical accomplishment so long delayed?"[92] Crole runs through different makes and models of witherers and dryers, rollers, sifters, and "patent ball breakers." One notable point here is that broken leaves and dust are no longer discarded as worthless: for strong black teas "the more the leaf is broken up, the better is the liquor upon infusion." There was now a demand for machines to cut, tear, and shred the leaves, and established tea machine companies like Jackson's, Marshall's, and Greig's rose nobly to the task. But these machines performed only parts of the needed sequence: an integrated CTC (crushing, tearing, curling) machine was not developed until 1931—and unfortunately proved disappointing in performance. It was only in the 1960s that Indian engineers first succeeded in designing a workable CTC machine.[93]

Even Crole admitted that it was not every task that could necessarily be matched by a machine. Although Money and his colleagues on the *Tea Gazette* had been convinced that steam-plowing was the future for the clearing and cultivation of tea gardens, it proved impracticable and Crole did not raise the possibility.[94] And the experiment in mechanical plucking that Crole had observed was not encouraging: "Although one would be only too glad to welcome any successful invention in this line, yet one must, I fear, reluctantly confess that it does not seem possible for any feasible device to be forthcoming." Crole was more confident about the prospects for mechanizing the fermentation process. The endpapers of Crole's book advertise Arnolds Perfect Tea Witherer and Fermenter, a machine that his London company, Messrs Arnold & Crole, had just developed. The machine, the advertisement claims, is the ultimate solution to the common chemical problem of secondary fermentation. Fermentation is "THE operation in tea manufacture, for on this process alone the character as well as the quality of Black Tea depends. Machinery has gradually been adopted for every one of the other subsidiary operations, and the time has at length come when the crude methods till now in vogue for the fermentation of tea must give place to improved methods involving the use of science and machinery."[95]

Fermentation, fundamental to the quality of Assam-style teas, was a tricky and unpredictable process. The *Vade Mecum* devotes a section to how fermentation affects tea, the phases of fermentation and the effects of over- or underfermentation. The section is largely descriptive, consisting of empirical observation. The correspondence at the end of the section confirms that there was no consensus or standard procedure for fermentation. Crole tries to bring science to the rescue. In addition to a long section on the botany of tea, he includes a chapter on its chemistry, including fermentation. Although Crole gives his account a learned veneer of chemical terminology, citing recent research papers, it is hard to see how tea planters could turn this to use. One anonymous reviewer savaged this section, declaring it misinformed and useless. Crole himself concluded that "a thorough investigation of [tea fermentation] should be set on foot as soon as possible, so that planters may in future substitute a scientific basis for the rule of thumb that now always obtains."[96]

Despite dreams of total mechanization, probably the most significant technical innovation in the Indian tea industry after the mid-1880s was a refinement of manual skill. In the early 1890s, when competition with China for the fine-tea market in Britain was at its height, Indian planters introduced the practice of plucking two leaves and a bud or even one leaf and a bud, rather than simply picking off the top three leaves. "Fine plucking" allowed leaves from the earliest flushes to be made into superior teas.[97]

A Hundred Years On

Crole foresaw a rosy future in which machines would replace coolies and science would supersede rule of thumb. While Money and his colleagues at the *Tea Gazette* were perhaps more circumspect, they saw the future of the Indian tea industry in similar terms. How fully were these prospects realized? Let us finally turn to the 1930s, a century after the first tea was grown in India, when the British Indian tea industry, by now unchallenged on the world markets, had settled down into a comfortable routine.

By the 1920s big tea groups or corporations had largely displaced local companies as the main tea producers of India. Finlay's was the largest among them, a typical example of an imperial company with a broad portfolio.[98] Established in 1750 in Glasgow, James Finlay & Co. started as a textile company and by the 1860s had mills and offices across the world. Finlay's set up a Bombay branch in 1862, and another in Calcutta in 1870. In 1882 Finlay's began investing in tea; by 1901 the Finlay-managed tea estates in Assam, Sylhet, Cachar, Dooars, Darjeeling, and Travancore comprised more than 270,000 acres in total, with 77,000 acres bearing; they employed seventy thousand native Indian workers and a large staff of superintendents, managers, and assistants from Britain.[99] Finlay's is a textbook case of the advantages of scale: its percentage of tea acreage in India went up by 1 percent of the total between 1922 and 1938, from 33 to 34 percent; but its share of trading profit rose 5 percent, from 31 to 36 percent.[100]

In the 1890s Finlay's leased a large concession in the High Range from the Maharajah of Travancore; through the following decades the company steadily expanded the acreage under tea in the High Range. A. S. Robertson worked on several High Range estates, first as Assistant Manager, then as Manager, between 1922 and 1947. Over the quarter-century Robertson and Finlay's had to deal with serious fluctuations in the market (the Great Depression, the Second World War), the growth of nationalist and anti-British sentiment, the threat of a Japanese invasion, and the growing certainty of Independence. Labor relations and the husbanding of resources are constant concerns in Robertson's memoir; technological innovation, however, does not feature.

Robertson's accounts of learning on the job, supervising an established estate, and developing a new estate, together with his description of tea processing, confirm that after the frenzy of mechanical inventions of the 1880s and 1890s the curve of innovation flattened.[101] When Robertson began working for Finlay's in the early 1920s the Assam model of black-tea production, by now successfully implanted across South India and Ceylon, had reached closure. The technical practices of cultivation and processing were now standardized as routine across India; the industry no longer sought to transform production methods or to redefine the finished product; the taste for fermented black teas, the stronger the better, was firmly established among the British tea-drinking public; the mission of the Tea Research Departments in North and South India, and of Finlay's own scientific headquarters, was not to innovate but to maintain standards.[102]

The black-tea model was premised on intensive year-round harvesting and processing of Assam-jhat teas, grown on large estates averaging one thousand acres or more, each with its own factory.[103] A thousand-acre estate needed a workforce of around fifteen hundred; this included seasonal workers, with their families, brought in from several hundred miles away, and more permanent foremen, mechanics, and local craftsmen, all supervised by a British manager and his assistants. Estates were now typically owned by a large company such as Finlay's or Buchanan's, which provided financial, legal, managerial, and technical-engineering support, set quality standards, and undertook shipping, marketing, and sales of teas from different regions and of different qualities combined into its house blends. The risk for individual estates was thus considerably diminished, and the expectations of conformity heightened.

Were the technoscientific ambitions of the 1880s and 1890s fulfilled on the Finlay's estates of the 1930s? Let us consider Robertson's chaîne opératoire of picking and processing tea on the High Range estates in light of the practices and ambitions detailed by our earlier writers on tea. The time span is roughly a century, with Ball, Bruce, and Fortune describing practices from the 1820s and 1830s, Money detailing the state of the art and prognosticating for the future in the 1880s, and Crole expressing his utopian hopes for a fully mechanized and scientific industry in the 1890s.[104]

If we check off the processes that Money believed should be routinely mechanized against the actual assemblage of machines and manual skills that Robertson documents, the comparison suggests that normalized tea production in the 1930s fell well short of Money's or Crole's expectations. Machines were fully integrated, but a surprising number of procedures still depended partly or wholly on manual labor or on leaving natural processes to take their course. Robertson's factory used an automatic sifting machine, for example, but it was necessary to employ women to do the final winnowing and grading by hand. The withering room had hot-air machines, but they were only turned on in really wet weather. Despite Crole's fond hope of a machine that would eliminate the tricky fingertips decisions of how long and at what temperature to ferment, in the 1930s no machinery was involved in this crucial process. Instead, the fermenting room was kept cool by spraying water on hessian screens; the tea maker inspected the tea for color and smell at regular intervals; he knew fermentation was complete and the tea ready for firing when the leaf had "assumed a distinctive copper colour."[105]

It would probably not have been beyond the technical capacities of 1930s engineers to develop effective harvesting machines, nor to perfect the prototype CTC machine, as Indian engineers did in the 1960s. But we must presume that, in the highly competitive 1930s world of black-tea production, cost-effectiveness (calculated at both estate and company level) was a primary factor in technical decisions about how much labor to deploy at which stage of cultivation, what types or brands of machinery to invest in, when to let nature take care of drying rather than turning on expensive heaters, and when to substitute manual labor for machines. While labor supplies and relations remained a constant preoccupation for planters such as Robertson, he never mentions the refrain, so frequent fifty years earlier, that the industry will best prosper if it replaces the costly and troublesome coolie with a row of machines.

Robertson shows considerable respect for workers' skills, acknowledging the contribution they make to output and quality. Key moments include transplanting seedlings from the nursery, when the taproots must be gently loosened and dipped in a mixture of sharp sand and cow-manure slurry, and the skilled styles of pruning the bushes, to shape them, to encourage flushing

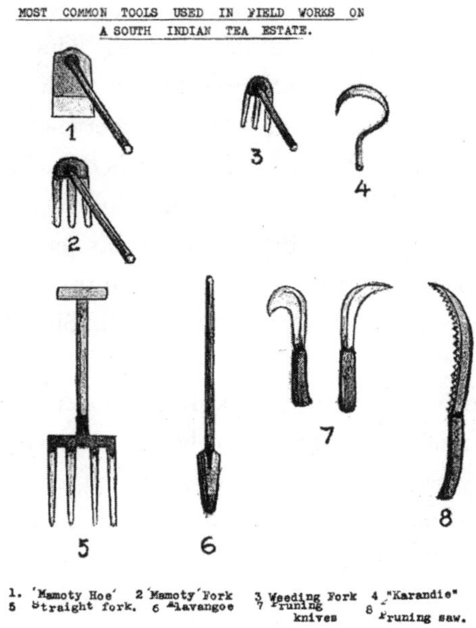

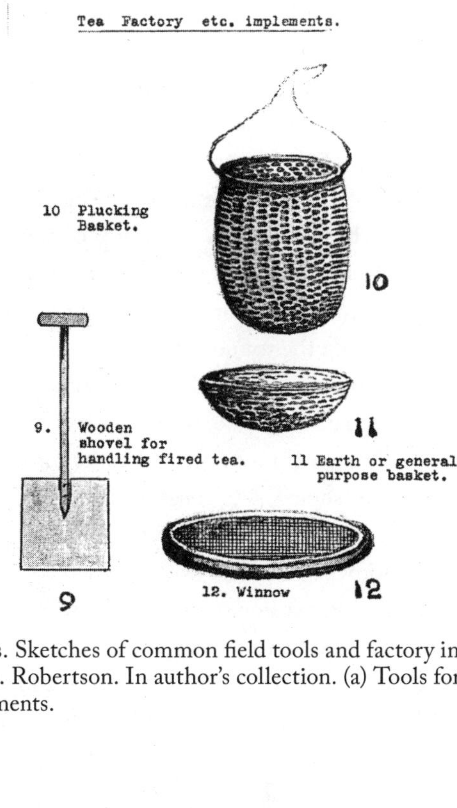

FIGURE 6.4A–B. Sketches of common field tools and factory implements, High Range, by A.S. Robertson. In author's collection. (a) Tools for fieldwork. (b) Factory implements.

or to restore old bushes to vigor.[106] Harvesting gangs were divided into three categories: first and second tipping and plucking. Tipping (earlier referred to as "fine plucking") was the careful selection of tender shoots from young or pruned bushes; it required much skill and experience and was the job of young and mature women. Plucking, which refers to picking throughout the pruning cycle, and especially toward the end, keeps the flushes of new shoots going through the year thus stabilizing yields; this was left to older women and children. Women's skills were also called for at the final stage of winnowing and cleaning the finished tea, an apparently humble task that determined the tea's final value.[107]

Simple tools such as hoes, pruning knives, wooden shovels, and bamboo sieves were as essential to the running of the tea estate as were the machines. The basic repertory had been introduced by the first Chinese tea men, and it had hardly changed a century later. Earlier authors sometimes give costings for these humble items.[108] Robertson went to the trouble of drawing, naming, and describing them (figure 6.4a–b): the effective and versatile iron hoe (*mamoty* or *mummti* in Tamil) "break[s] up ground speedily and with the least possible effort when used skilfully. . . . In my opinion the finest agricultural tool ever produced!"[109]

This is a good moment to reflect on the expertise of the Finlay's workforce, and on training. These had changed surprisingly little over the course of the century between 1835 and 1935. Let us start at the top of the colonial race hierarchy, with the British tea men. Company engineers in the 1930s were professionals, as they had been since the founding of the Assam Company. Estate managers and assistant managers were no longer just any old riff-raff who turned up: Finlay's selected them on aptitude, but they still learned most of their skills on the job, doing the rounds of the estate and factory. Assistants learned basic planning and costing from the managers. The care of machines was left to the Indian mechanics on the estate, with a company engineer on call for serious matters. Rule of thumb, general experience, local knowledge, and a keen eye were key for supervising most jobs and procedures; textbooks were of little use when so much depended on the weather or on details such as spotting the precise moment when the leaves turned the right copper color.

In fact, rather than issuing textbooks, companies were more likely to provide language handbooks to serve their new staff as a basic training manual. Language skills were essential both for giving orders and for learning what the orders signified. Finlay's representatives visiting tea gardens were told to assess the assistant managers on their knowledge of the vernacular. "It was a condition of the 4th year salary increase, and good language-skills brought a bonus."[110] A. M. Ferguson's Tamil guide, entitled *Iṅgē Vā*, was in its eighth edition in 1924, when Robertson bought his copy, which was much worn and copiously annotated when I inherited it. Many of the sequences of phrases it includes are complete chaînes opératoires in themselves:

> Spread the leaf on the shelves to wither, toss the wet leaves about a little, do not spread them too thickly, roll till the juice comes out, leave the balls covered for two to three hours to ferment, now break the balls and let them be covered up a little longer in a heap, put the withered leaf into the roller, turn on the water, turn off the water, too much water is coming, the belt has broken, bring a needle and some leather to mend it, put fire into the drier, spread the leaf on the trays and fire it off, don't burn the leaf, if smoke comes take off the tray at once.[111]

The Tamil workers—men, women and children—likewise learned on the job. They were supervised by British managers and Tamil foremen, who weighed and counted, foraged through baskets of leaves to ensure that the small leaves hadn't all been piled on top, scolded and sometimes praised. Naturally there were subversive practices. The *karandie* or scraper was a "home made implement which the labour liked but which was discouraged by the management of the estates because of its tendency to cause soil erosion on sloping land." *Iṅgē Vā* singled it out: "I have told you a hundred times not to use a scraper."[112]

A number of permanent workers who were recognized as having professional skills were employed in the factory, including mechanics and carpenters who learned through apprenticeship. Various other manual skills were also valued highly, and managers tried to keep these workers on long-term:

> With all parts of Tea manufacture it is well to employ the same men continually in each department, but above all, perhaps, should this be done in Tea sifting. A good sifter is a valuable man. He knows each kind of Tea by name; he knows what sieves to use, and the order in which to use them for each Tea; what the effect a larger or smaller mesh will have on each kind, &c., &c. In fact, he knows much more of the *practical* part of sifting than his master can, though the latter is, probably, a better judge of how far the Teas are perfect when made.[113]

Robertson likewise appreciated the skills of experienced workers and tried to arrange with his labor recruiters to persuade as many good workers as possible to extend their contracts.

We conclude by asking whether anything of the original Chinese tea system transplanted to Assam in the 1830s survived in the 1930s tea industry. Obviously the Assam model had a historical dynamic of its own. The vigor of the bush supported a progressive extension of the harvesting period, culminating in year-round plucking. The large, sapid Assam leaves were unsuited to Chinese processing methods, and their flavor developed more fully if they were broken instead of being kept whole. The characteristics of the Assam plant favored a dynamic of industrialization: mechanical treatment improved the quality of tea and increased the volume that could be processed, while an extended harvesting season and an extension of the acreage made such investment worthwhile. For the China bush, with its short season and its tougher leaves requiring more labor-intensive processing, the potential for mechanization was limited, and it is not surprising that the mechanization of the Indian tea industry coincided with the abandoning of the China bush.[114]

Yet some elements of the tea techniques originally developed by Bruce and his Chinese tea workers survived the transition. Growing tea from seed, using the same methods recorded by Bruce, remained the standard procedure in India until the 1970s.[115] Although the seeds were from the Assam jhat, not the China plant, the nurseries were made and tended according to methods common in China, interplanting shade trees and protecting individual seedlings with light matting. The repertory of hand tools used in Indian tea gardens and factories (*momaties* and sieves, for instance) was almost identical to that used in China—but such tools are common throughout much of Eurasia so we should not read too much into that. Although manual processes continued to play a key role even in the mechanized factory, most of them had evolved into something very different from the original Chinese processing techniques, which were designed to extract as much moisture from the leaf as possible and to keep the leaf whole. Only in Darjeeling were the Chinese techniques preserved, to make small batches of very delicate, highly seasonal tea.[116]

The long-term legacy of Chinese tea making in Assam is, however, more complex and less obvious, and it is the technique of comparing chaînes opératoires that has allowed us to disentangle it. The Chinese tea men who came to Assam in the 1830s worked closely with Bruce to establish a model for growing and manufacturing classic Chinese teas. The model is interesting because it selects from a spectrum of Chinese practices to assemble a tea technology adapted to conditions in Assam. This tea technology was already oriented toward streamlining and the reduction of skilled labor inputs: in other words, it was a version of Chinese tea manufacture that adapted well to subsequent industrialization. The other Chinese contribution was still more significant. Key elements of Chinese tea savoir faire, in particular cutting back and transplanting, were successfully applied in the domestication of the Assam plant through the 1830s and 1840s. Although these techniques were discarded as soon as a routine of growing from seed could be established, the initial domestication of the wild Assam tea tree was nevertheless the foundation of the Assam-tea industry. In other words, to assess the importance of the Chinese contribution to Indian tea, we have to retrieve and give due weight to a poorly documented episode that lasted only a few years.

The use of the chaîne opératoire has helped us to identify and track key moments, linkages, trends, and inconsistencies in this particular case of technology transfer with greater clarity than textually based analyses typically manage. I suggest that the chaîne opératoire might usefully be applied more broadly in the global history of technology, including the historical study of itineraries and entanglements that is the theme of this volume.

Part 3

Material Complexes in Motion

Chapter 7

The Itinerary of Hing/*Awei*/Asafetida across Eurasia, 400–1800

Angela Ki Che Leung and Ming Chen

The focus of this chapter is on the global itinerary of asafetida as a drug, spice, and plant, tracing the processes and contact zones where knowledge around it was generated, transmuted, altered, disseminated, or disregarded from the fifth to the nineteenth century. Asafetida's long global history was inextricably entangled with the movements of peoples, money, rumors, religious activities, medical and culinary practices, and scientific inquiries across regions and continents in different periods. Asafetida was a malleable and mobile material, creating and depositing values and meanings in some—but not all—relational fields along its temporal and spatial trajectories. Its history reveals erratic and open-ended processes of knowledge integration and disintegration centering on the material that did not, however, add up to the formulation of any straightforward definitions of a single, discrete material. These processes demonstrate, instead, how asafetida "reflected an attitude about the vitality of the world" in contact zones where it was ascribed special properties or "cultural logics" as a drug, a food, or a plant.[1]

Asafetida is described in modern European pharmacopeia as the dried latex (gum resin) from the root of several species of the Ferula herb of the umbelliferous family, grown wild in dry, stony, mountainous regions in Central Asia, including regions between Lar and Yazd in today's Iran, in the Qandahar region of southeastern Afghanistan, and in southern Uzbekistan. The product is also widely known in English literature as hing (of Sanskrit or Hindu origin).[2] Having a strong stench, the gum-resin was used for centuries both as a spice and as a drug in Asia and in Europe. In Europe it was compared to and sometimes equated with silphium, which was believed to have been introduced to Europe from North Africa in the fourth century

BC during the conquest of Alexander the Great and was used in ointments by ancient Greek doctors. The material was then believed to have become rare until its "reemergence" as a plant in the sixteenth century. From then on, the plant producing the resin became an object of great interest for European naturalists.[3] By the mid-nineteenth century, the resin was still used in Europe "as a stimulant and antispasmodic in chronic bronchitis, hysteria and tympanitis."[4] Little is known of its global circulation during the entire medieval and premodern period, however, partly because Western literature has paid little attention to the history of this Central Asian product in the rest of the world and especially in East Asia where it had a brilliant career in the premodern period.

Around the time that asafetida "reemerged" as a plant in Europe toward the end of the sixteenth century, Li Shizhen (李時珍, 1518–1593) in China compiled the monumental and globally translated *Bencao gangmu* 本草綱目 (Systematic materia medica 1596), in which asafetida was given lengthy descriptions as a drug coming from Central Asia or India. Under the heading *awei* (阿魏), as the drug was then commonly called in China, Li provided a list of Chinese transcriptions of foreign terms designating what was considered the same material: *ayü* (阿虞), *xingyu* (*xingqu*, 興渠[瞿], *hing-yu*, 形虞), and *haxini* (哈昔尼).[5] Berthold Laufer clarified in 1915 that *ayü* in fact transcribes the Persian term *anguza(d)*, *xingyu* transcribes the Sanskrit term *hiṅgu*, and *haxini* transcribes Ghazni, a city in today's Afghanistan. None of these terms, according to Laufer, was the origin of *awei*. For him *awei* is a phonetically exact transcription of *ankwaṣ(ḍ)*, a word in Tokharian B—a now defunct Indo-European language spoken north of the Tarim Basin (northern Xinjiang) from the sixth to the eighth century.[6] This transcription suggests that the traders who introduced the product to China were probably a Tokharian-speaking people. Kuchen traders active in the Tarim Basin region between present Afghanistan and China in the seventh and eighth centuries were the most likely candidates.[7]

Li Shizhen's rich historical account of awei sums up the resin's written history in China until the late sixteenth century when it was widely known in East Asian materia medica, bearing witness to its great mobility in the global market. The focus of this chapter is on awei's itinerary as a material and on the ways in which knowledge was constructed around it as a resin or as a plant. Three characteristics of the substance are highlighted: its materiality as a drug, spice, and plant; the significance of the resin's defining yet intangible

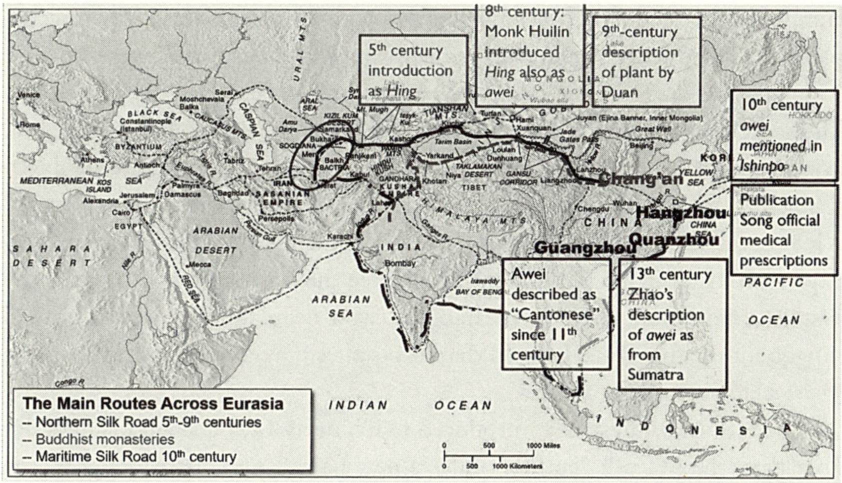

FIGURE 7.1. Map of the main routes across Eurasia, showing the itineraries of *awei*/hing/asafetida (after Hansen, *The Silk Road*, 6–7). The map shows Chang'an as a hub where, between the fifth and ninth centuries, the first Chinese names of asafetida emerged, the first book on materia medica including *awei* was published, the first description of the plant was written down, and it appeared in recipe books. Then in Quanzhou and Hangzhou during the tenth to twelfth centuries *awei* was imported in bulk, publications of official recipe books included *awei*, and the stench of *awei* helped construct new pharmaceutical knowledge and a new interpretation of *qi*.

stench; and its elusive authenticity and identity as a global commodity and as an object of scientific enquiry (see figure 7.1).

Awei as Name, Material, and Knowledge in East Asia, Fifth–Tenth Centuries

Awei left Central Asia and India and began its new journey in China as *xingqu* (a transcription of *hiṅgu*) in one of the first Buddhist vinaya translated from Sanskrit into Chinese in the fifth century, *Shi song lü* 十誦律 (*Sarvāstivāda-Vinaya*, Ten recitations Vinaya). This text was read and translated orally before being transcribed into Chinese by a team including Central Asian Buddhist monks such as Puṇyatara (弗若多羅) from Kasmira and the great Kuchean translator Kumārajīva (鳩摩羅什, 344–413), in the city of Chang'an, capital of the non-Sinitic (of Qiang ethnicity) state of Later Qin (265–450), a great patron of Buddhism.[8] The reading of the vinaya in Sanskrit by Puṇyatara followed by the translation into Chinese was attended by more than three thousand monks and novices at Xiaoyao yuan (逍遙園), a huge complex where Kumārajīva organized his translation projects.[9] This incomplete translated text introduced the term *xingqu* as one of the five resins that a monk could accept from a donor, as a Buddhist code of practice for monks. The same passage also introduced new knowledge on other foods including syrups, greases, root plants, fruits, soups, most of which were transcribed into Chinese phonetically from Sanskrit as they had no existing Chinese names.[10] Xingqu, thus designating a common material in Indian Buddhist culture also expressed its foreignness in China. The three thousand monks attending the historic translation session in Chang'an, who heard of the material called xingqu for the first time, learned that it was a resin, without any idea about its physical appearance or nature.

The resin was physically introduced to China as a tributary trade item not later than the seventh century, under a new name, *awei*, as recorded by the official history of the Sui dynasty (581–618). This book was published in the 620s, and in the chapter on Central Asia (called in Chinese *xiyu*, the western territories) it is reported that the material awei was one of several crops and products of the tributary State of Cao (漕, known in Sanskrit as Jaguda, and Zabul in Arabic), which occupied the territory of today's Afghanistan.[11] By this time, the knowledge of the resin and of the State of Cao as a major producer was already well known in the Eurasian Buddhist world.[12] From this point on, the term *awei* became the most common to designate the resin in Chinese texts of various genres, eclipsing all other coexisting Sanskrit, Persian, Arabic, and Mongolian terms including the first Chinese transcription *xingqu*.

The reception of the material awei as a medical ingredient in China was immediate. It was included in the first Chinese compendium of materia medica, *Xinxiu bencao* 新修本草 (The new compilation of materia medica), published in 657 with imperial endorsement. For the first time in China, the knowledge about awei as a drug was crystallized in writing. It was recorded here as a drug of middle medicinal quality, spicy in taste, without poison but with a repelling smell. It was considered to be an effective vermifuge, useful in dissolving lumps and masses in the abdomen, dissipating bad *qi* (from the body) and protecting the patient against ghosts and bad spirits.[13] It was said to be made from the "juice" of the root of a plant said to look like the indigenous Chinese plant *baizhi* (*Angelica dahurica Benth.et Hook*, another umbelliferous plant), after it is sun-dried and ground into a powder. But an inferior type consisting of segmented roots was also in use. Finally, without specifying the native place of this drug, the compiler of the compendium, Su Jing, highlighted the paradoxical and defining character of awei: having an unusual stench, it was efficacious in getting rid of foul smell.[14]

Awei's unique stench and ascribed efficacy as a vermifuge and its paradoxical effect of stopping foul smell flagged it as an occult drug in China in the eighth and ninth centuries. It was included in medicinal recipes in this period mostly for banishing disease-causing evil spirits and ghosts. Awei was sometimes made into pills that were burnt to fumigate the patient diagnosed to be possessed by bad spirits.[15] Doctors also advised patients to take awei pills to prevent disease transmission by malicious *qi* emitted from corpses.[16] It was used to exterminate vermin such as creeping bugs since the seventh century in recipes treating leprosy, believed to be caused by bugs pullulating inside the body.[17] One of the first doctors to introduce such recipes, Sun Simiao (孫思邈, 581–682), attributed this therapeutic use of awei to an Indian origin.[18] Sun was also one of the first to include awei in composite recipes for dissolving lumps and masses inside the abdomen, establishing a long Chinese tradition of such use in subsequent centuries.[19]

Wang Tao's (王燾) compilation of recipes based on his research in the imperial library, *Waitai miyao fang* 外臺秘要方 (Secret essentials from the imperial collection, ca. 752) was the first medical book to provide a series of recipes using awei for occult purposes such as expelling evil spirits. Together with another newly introduced resin from Central Asia, benzoin (安息香), and sometimes mixed with cow's milk (a substance rarely used in Chinese medicinal recipes), awei was an ingredient in medicines to be taken for the ousting of evil spirits including ghosts and fox spirits that took the form of beautiful women and caused hallucinations such as intercourse with the spirits. Sometimes awei was mixed with toxic elements such as arsenic and various types of animal hair and bones into pills for fumigation to banish bad spirits from a patient or to prevent epidemics in a locality; the pills could also be hung in houses or carried on journeys as a charm to ward off bad spirits.[20] The occult use of awei also appeared in Japanese pharmacopeia not later than the tenth century. *Ishinpō* 醫心方 (Essential medical recipes, 982), one of the earliest and most influential Japanese medical texts by Tanba Yasuyori (丹波康頼, 912–995) based on Chinese medical classics, recorded the use of awei mixed with alcohol in a recipe for the prevention of postmortem contamination (*zhu*).[21]

Buddhist monks working in Chang'an probably played the key role in disseminating knowledge on awei in the early period. Other than their translation projects, several prominent Chinese Buddhist monks traveling between Central Asia and China continued to bring back new knowledge on the resin to their audiences in Chang'an. One of them was Monk Huiri (慧日, 680–748) whose discussion on xingqu shows the unique sensitivity toward the resin as a taboo in Indian culture.[22] Another was Huilin (慧琳737–820), a linguist who had worked in major monasteries in Chang'an.[23] In his major linguistic study, *Yiqie jing yin yi* 一切經音義 (The sound and meaning of the Tripitaka), he explains that "xingqu (*hiṅgu*) is a 'tree juice' that people in the western region (India) put in food. It is what we [the Chinese] now call the drug awei."[24] Huilin here enriched the knowledge on awei by clarifying the distinction between hing as a food in India (with the name *xingqu* transcribing the Sanskrit term) and awei as a medicine in China. His interest in the resin had much to do with his role as a Buddhist monk: the resin was not only a well-known taboo food in Indian Buddhism but also a common drug stored in monasteries along the route he traveled between India and China, via the Kingdom of Khotan, on the southern edge of the Taklamakan Desert in the Tarim Basin. Chinese archaeological findings show that one of the common medicines stored in Buddhist monasteries in the Khotan region was awei.[25] A document listing market prices of awei in the Turfan region in the year 743 illustrates the material's full integration as an accessible commodity by this time in China's medical culture: one *liang* (around 1.3 ounces) of superior quality awei powder was worth eight copper coins, and middle quality and inferior awei cost seven and six coins respectively, which were of the same value as the much-sought-after native ingredient the dried *poria* fungus.[26]

Knowledge of the plant that produced awei, however, was not recorded in writing until the ninth century. Thanks to the curiosity of the literatus Duan Chengshi (段成式, 803–863) who interviewed Central Asian and Persian travelers sojourning in Chang'an, capital of Tang China (618–907), where he spent time as an official and a retiree, the plant was described in writing for the first time. In his influential jottings, Duan indicated the ingredient's geographic origin: Persia and northern India, with the Persian name of the plant *anguzad.* Duan told his readers that the "awei tree could be as tall as 8–9 *zhang* [almost three meters]," with a yellow-greenish bark. "In the third month of the year it develops leaves that look like rat's ears, but no flowers nor fruits. If one breaks the branches a tasty juice will come out that will, after a while, solidify as *awei*."[27] Duan informs his readers that this information was given to him by a Melkite monk from Central Asia, whereas Indian monks told him that the juice of the plant was mixed with rice and beans to make awei.[28] Duan's text has to be appreciated in the context of the cosmopolitan culture of the Tang capital at the eastern terminus of the Silk Road infrastructure at its prime. Chang'an evolved from a religious (Buddhist) center in the fifth century to a political and commercial hub of the Chinese Empire unified under the Tang dynasty from the seventh century, a setting that facilitated the global circulation of knowledge of all kinds, to and from Central Asia, by traders, monks, diplomats, and other sojourners who interacted freely with resident doctors, scholars, and officials.

By the ninth century, then, awei as a medical material was known in writings and was commonly used in China. The material knowledge constructed through it circulated along the Silk Roads, from today's Afghanistan and northern India, via the Tarim Basin to western China, turning up finally at Chang'an at the eastern end, in monasteries, marketplaces, and imperial offices. From this cosmopolitan center, the resin was made known to all East Asian regimes, as *awei* in China, *agi* in Japan, *awi* in Korea, and *a ngui* in Vietnam. However, this knowledge on the material—kept in temples, sold in markets in the form of a powder or segmented dried roots, or in various mixtures—remained fragmented and elusive. No writer on the substance actually witnessed the herb or tree that produced the substance or the production process of the various forms of the material found in marketplaces. The early domination of the Tokharian B term *awei* for the commodity—eclipsing all other transcriptions (Sanskrit, Persian, or Arabic) that differentiate between the plant, its various parts, the resin, and its derivatives—greatly simplified knowledge construction through the material. It was precisely such fragmented and incomplete knowledge about awei that nourished imaginaries on its native place, physical form, materiality, and therapeutic qualities in the later periods that determined its value as a thing.

Awei as a Popular Medicine in East Asia, Tenth–Fifteenth Century

The importance of awei in Chinese materia medica continued to grow after the tenth century, as it now reached East Asia not only by land but also in increasing bulk by sea via Southeast Asia.[29] By 1141, an imperial regulation allowed awei, along with eighty-six other spices and drugs, to be legally traded on the Chinese market.[30] Historians attribute the explosion of medicinal recipes in this period to the great influx of Central and South Asian drugs via

FIGURE 7.2. The *awei* of Canton in materia medica text *Bencao tujing* by Su Song (1061). Public domain.

land and sea. Robert Hartwell, using the data of extant fragments of an early fifteenth-century imperial encyclopedia to measure the medical use of foreign commodities from the seventh to the late thirteenth century, estimated that asafetida was used in 37 percent of recipes treating constipation-related syndromes between the eleventh and the thirteenth centuries (see figure 7.2).[31]

Many such recipes were recorded in official compilations of medicinal recipes, including some state charitable dispensaries, such as the one-hundred-chapter *Taiping shenghui fang* 太平聖惠方 (Recipes of the Imperial Grace during the Great Peace, 992) and the *Taiping huimin heji ju fang* 太平和劑局方 (Recipes of the Charitable Dispensary for Popular Relief during the Great Peace, mid-thirteenth century).[32] Awei's initial occult use in repelling evil spirits and stopping postmortem disease transmission was still common, but its main therapeutic use seems to have become as a dissolvent of blockages, a dissipater of phlegm, bad *qi*, and irregular growths including tumorlike lumps inside the abdomen. Its use was then extended to treating women's reproductive disorders including obstructed menstruation and as an abortifacient. It had acquired a more constructive function by the tenth century when mixed with other ingredients, as shown by the "awei pill" considered to be effective in "boosting the vital *qi* of men, and blood *qi* of women," recorded in the *Taiping shenghui fang* compilation.[33]

The popularity of awei also reflects the global impact of Islamic medicine peaking in the fourteenth century. Similar pharmaceutical uses of the drug can be found in the Arabic tradition, as shown in Ibn Al Baytar's (1197–1248) description of *andjudan* (the Arabic term of the asafetida plant), with a note on the beneficial warming and drying effect of the resin (known as *haltit* in Arabic) on the liver and stomach, and its use in dissolving intumescence produced by heavy foods.[34] A purgative formula from around the twelfth century using asafetida and other ingredients, for instance, is found in the Genizah collection of Egyptian medieval medicine.[35] An early mention of asafetida in English in a late fourteenth-century translation indicated its introduction into Europe.[36] Meanwhile, the first compilation of Islamic medicinal recipes written in China, the *Huihui yaofang* 回回藥方 (Islamic medicinal recipes) published in the fourteenth century, contains recipes based on asafetida. A specialist of this text draws our attention to the similarities, in form and content, between this text and Islamic medical texts such as that by Avicenna (980–1037, known in China as abu Ali).[37] Unlike Chinese medicinal books, this text transcribed *asafetida* not only as *awei* but also in Arabic, Persian, and even Latin terms. The uses of the drug were similar to those recorded in existing imperial compilations of recipes, especially in its use in dissolving phlegm and hard lumps in the abdomen.[38]

The medicinal use of awei was also recorded in other non-Sinitic cultures in East Asia in this period. Some rare texts from the Tangut kingdom controlling parts of western China near today's Xinjiang area between 1038 and 1227 indicated the popular use of awei as a medicine in that short-lived kingdom much influenced by China.[39] The Tibetan medical classic *The Four Tan-*

tras (*rGyud-bzhi*, considered a late eighth century compilation, printed for the first time in 1546) cited the use of awei as a vermifuge and a Wind-dispelling drug, establishing the popular use of the resin in Tibetan medicine. It was in the Tibetan medical tradition that awei acquires an aphrodisiac quality (as in Arabic and Indian cultures) not explicit in other East Asian traditions.[40]

Sources in this period suggest that the product was a highly valued commodity and much appreciated by the imperial court. In the tenth century, the Khotan state offered two valuables to the Song court: white jade and awei. Like jade, awei was part of the lucrative black market trade in this part of Central Asia.[41] One century later, a grandiose tributary mission of the Chola dynasty of southern India in 1077 to the Chinese emperor offered awei together with other valuables such as frankincense, rosewater, cloves, borneol, pearls, and rhino teeth among other precious items.[42] A Chinese scholar official complained in 1045 about the unreasonably high price of awei, which, for him, was only a mediocre and aggressive ingredient whereas milder drugs of finer therapeutic quality were much cheaper.[43] This comment informs us not only about the drug's high value in this period but also the mixed Chinese reception of this foreign thing as a medicinal ingredient. This concern seemed to be growing among experts and would explain the gradual decline of awei's market value in subsequent periods.

This image of awei as a material may also be one of the reasons that it never posed the problem in Chinese Buddhist dietetic practice that it did in India and other non-Sinitic cultures. Believers did not need to avoid awei in meals as it was not an integral part of the Chinese or East Asian culinary tradition. A Chinese Buddhist text of the tenth century stated that of the five spicy herbs forbidden for monks—garlic, chive, green onion, scallion, and xingqu—only xingqu was not a food in China.[44] Despite some sporadic evidence that parts of the awei plant were consumed as a vegetable in western China in the early tenth century, this use did not persist, largely because of its stench.[45] In this regard, the reception of asafetida in China was similar to that in Europe.[46] However, for a brief period in China under Mongolian control from the late thirteenth to the early fourteenth century, awei was used as a condiment and added to various soups and dishes with game and other meat to enhance the taste. An influential book on diet at this time, *Yin shan zhengyao* 飲膳正要 (Essentials of food and drink, 1330) by the Mongol doctor Hu Sihui (忽思慧), was uniquely informative on the question. The spice, called *kasni*, was put in dishes and soups based on mutton, deer, and bear meat to enhance the taste.[47] This practice, however, was at best marginal and ephemeral in the Chinese culinary culture in its long imperial history.

Asafetida in the Construction of Modern Botanical Knowledge, Sixteenth–Nineteenth Centuries

Asafetida's global visibility increased suddenly from the sixteenth century onward, with new transoceanic routes facilitating the circulation of humans and materials and intellectual movements that inspired unprecedented global efforts to retrieve drugs recorded in Dioscorides's classic.[48] Gathering, compiling, and sharing information and materials both heard about and witnessed firsthand, comparing them with those described in classic texts, European travelers and their native collaborators in different parts of the world created new networks of knowledge making.[49] In the process, a new relational field formed, within which asafetida emerged as a plant.

One of the first global travelers interested in asafetida as a plant was the Portuguese Jewish doctor Garcia da Orta (1501–1568). In 1534 he sailed for Portuguese India as chief physician aboard a Portuguese fleet and in 1538 settled in Goa to practice medicine. Based on his Indian experience he authored the *Colóquios dos simples e drogas he cousas medicinais da India*), which reveals his extensive knowledge about Asian drugs and spices. He was one of the first Europeans to point out the different Arabic and Indian names for asafetida resin and for the plant, and he reported especially on the popularity of the resin as a food in India. But he also admitted that, despite his familiarity with the resin, he had never seen the plant, which grew deep inland, and did not know what it looked like. "No people known to me use anything but the gum which is obtained by making cuts in the tree."[50]

It was probably the desire to penetrate this tantalizing myth about asafetida that drove the German doctor Engelbert Kaempfer (1651–1716) to travel all the way to the region of Lar along the Persian Gulf to witness the harvesting of the resin. Kaempfer belonged to a generation of European doctors in the service of European trading companies who traveled far and wide seeking out firsthand information on the natural world.[51] Serving as a physician in the Dutch East India Company, he traveled to Russia, Persia, India, Siam, the East Indies, and Japan during the decade between 1683 and 1693.[52] He published the observations of his travels, *Amoenitatum exoticarum*, in 1712 in which seventeen pages were devoted to observations on asafetida, including detailed drawings of the plant and its parts and an illustration of the harvest-

FIGURE 7.3. *Hingiseh* or asafetida plant, in Kaempfer, *Amoemitatum Exoticarum* (1712). Public domain.

ing he witnessed in 1687 (see figure 7.3).[53] R. Carrubba translated those pages from Latin into English for the first time in 1979. From Kaempfer's account, the reader learns that the asafetida plant was part of the umbelliferous family and grew wild in large meadows in remote arid and stony mountains. The gum was generated in the root and harvested in the spring in four phases—mid-April, late May, early June, and early July—by peasants who had to take a night's climb of some fourteen miles to the asafetida fields. They cut the stems and leaves to expose the top of the root in order to collect the liquid coming up in several slicing stages until all the liquid from the root was harvested. The sap collected in the third stage, called Pispaas, was believed to be the best, with the firmest consistency. Kaempfer also noted that, by his time, overharvesting had already left very few older and larger plants (more than twenty years old) that were capable of producing greater quantities of sap.[54]

Kaempfer's identification of the plant and detailed description of the production of the resin formed the basis for subsequent descriptions and for discussions among European naturalists about the asafetida plant itself. Two important figures in this debate were Hugh Falconer (1808–1865) and his friend John Royle (1799–1858), both naturalists with medical training and global travelers as physicians in the British East India Company. Falconer identified Kaempfer's asafetida plant as *Ferula Narthex Boiss.*, a plant that he not only observed in its natural site but also cultivated with seeds brought back from Persia, a common practice by naturalists of that time. John Royle fully recorded Falconer's account of the plant with detailed drawings of the various plant parts. He raised a new point about the fruit or seeds of the plant, imported in India from Persia and Afghanistan under the name Anjoodan and widely employed by Indian physicians.[55] This implies that, contrary to previous belief, other parts of the asafetida plant besides the resin were also used as medicine or food in India.

European botanists continued to study what they believed to be the asafetida plant that was now cultivated in European botanical gardens with seeds brought back from Asia by various naturalists.[56] John Balfour (1808–1884), professor of botany at the University of Edinburgh, in 1841 provided meticulous descriptions of the plants cultivated in Edinburgh, further differentiating the odor of the main plant (strong garlic odor), the flowers (sweetish), ripe fruits (asafetida odor), cotyledons and early leaves (not fetid), young root (bitterish taste).[57]

Such new knowledge on the asafetida plant soon found its way back to East Asia. Japanese Rangaku (Dutch learning) scholars and doctors Otsuki Bansui (1757–1827) and Otsuki Banri (1785–1837) translated various European materia medica and compiled their translations in *Ran'en tekihō* 蘭畹摘芳 (Extracts of Dutch botanical learning, 1815). This work includes a long section on asafetida, comparing knowledge collected by Kaempfer and other European naturalists with that on awei in Chinese materia medica.[58] The Chinese translation of European materia medica, especially John Royle's work of 1876, was undertaken by British missionaries some seventy years later. The translator, while naming the text "Compendium of Western pharmacopeia," admitted that the asafetida from the plant *Narthex Asafoetida* of Falconer was indeed awei in China.[59] However, such knowledge had little impact on contemporaneous Chinese pharmacology and definitely did not placate a growing Chinese concern about the authenticity of the drug that was found on the market.

The Defining Stench and Elusive Authenticity

Stench

The distinctive and defining stench of asafetida shapes its diverse careers as a food or as a medicine in different cultural contexts along its long spatial and temporal trajectories. The foul smell of the material was often the first thing that Chinese and European writers remarked on in their writings. Chen Cheng (陳誠, ca. 1365–1457), a diplomat sent to Central Asia by the Ming government (1368–1644) around 1415, provided the first Chinese eyewitness description of the plant. He found "in the city of Shahrokia, more than 500 *li* east of Sarmarqand, a stinking herb . . . of about one *chi* in height. Its branches and leaves resemble an umbrella. It thrives in the spring and dies in the autumn. The stench is unbearable. The juice taken from it while it is still alive can be made into a paste that we call *awei*."[60] Similarly, Garcia da Orta wrote in 1563 that "The nastiest smell in the world for me is Assa-Fetida." The Portuguese, he continues, called the resin "the food of devils," whereas the Indians "have become accustomed to it." It was also known as "Devil's Dung" in Europe.[61] John Royle told his readers that the "intolerable aliaceous odour" was what distinguished asafetida.[62]

The East Asian and European aversion to its strong smell explains the material's failure to enter their culinary traditions, differing from its long history as a condiment in India and Persia. We have seen that it was not a taboo food for Chinese Buddhists as it was for the Indians, because it did not tempt the Chinese at all as a food. Lucien Leclerc who translated Ibn Al Baytar's thirteenth-century description of asafetida also commented, in the early twentieth century, that the "Orientals" (meaning Indians and Persians) had different types of *andjodan* for alimentary uses, whereas Europeans did not, as the odor was too strong.[63]

However, it was precisely this stench that defined awei's unique value in Chinese pharmacopeia. Strong odors of Central and Southeast Asian drugs introduced in China deeply influenced the ways in which Chinese experts analyzed and classified pharmaceuticals. Kou Zongshi (寇宗奭), the twelfth-century official responsible for procuring drugs for the Song government, in his influential book *Bencao yanyi* 本草衍義 (Augmented materia medica 1116) began to redefine the "*qi*" quality of drugs as their "odor"—rather than as their "nature," as had previously been the case. Before this work, drugs were divided into five tastes (acid, salty, sweet, bitter, and spicy) and four "*qi*" (cold, hot, warm, cool). But Kou reinterpreted the meaning of "*qi*" as "odor," divided also into four categories: fragrant, stinking, fishy, urine-like. He reinterpreted the four previous categories of "*qi*" as four "natures" (*xing*) of drugs. Awei, together with garlic, salty fish, and sweat-soaked socks were listed as examples of "stinking" matters.[64]

More important, the "stench" of a drug was directly related to its specific therapeutic efficacy.[65] Before the popular fourteenth-century saying on the value of awei pointed directly to its stench ("There is true awei amid much fake; that which stinks and removes stinking is the most precious"), awei's healing power had already been explained in a twelfth-century recipe book, the *Shengji jing* 聖濟經 (Classic of imperial charity, ca. 1111–1118) attributed to the Song emperor Huizong: "People do not know that the stench [of a drug] has a function . . . salty fish is beneficial to the intestinal organ, as its stench was great enough to scour blood stasis."[66] Awei's stench thus explained why it was efficacious in treating indigestion and lumps in the abdomen.

This reanalysis of the "*qi*" of drugs as odor continued to develop in the late imperial period, and awei continued to represent the category of stench in pharmaceutical handbooks.[67] The influential late Ming doctor Miao Xiyong (繆希雍, 1546–1627) further explained the quality of awei: "Fragrance facilitates the natural flow of Blood and *qi*, whereas stench reverses their flow; thus [to minimize the abrasive effect of reversed *qi*], one needs to reinforce the weak stomach and spleen of the patient first before *awei* is taken to dissolve lumps and stagnations."[68] The standard way of processing awei by Chinese apothecaries also revealed the concern of the stench as being an indication of abrasiveness: after being ground into fine powder in a clean bowl, it had to pass over a liquor heater to take up the aroma.[69]

Asafetida's unique stench was also a key to identifying the material in European medical culture. For centuries it was considered a variety of or a substitute for the ancient and highly valued Cyrenaic silphium, a superior resin thought to have become extinct. Nineteenth-century botanists attempted to fix its identity by quoting the Roman botanist Dioscorides (ca. 40–90), and the Islamic physician Avicenna (980–1037) who both thought there were two kinds of resin, one with a strong stench from Persia and another with a lesser smell from Cyrene.[70] However, since the defining stench was intangible and could not be measured and since most East Asian and European users and writers on the material had seen neither the plant nor the processing of the

FIGURE 7.4. Authenticity testing of *awei* in the anonymous text *Buyi Leigong paozhi bianlan* (1591). Public domain.

ingredient, the authenticity of the material remained elusive and became a perennial issue for both the consumer and the botanist.

Rumors and Authentication

The elusiveness of the "true" identity of awei was an issue for the Chinese as soon as it was introduced as xingqu. A fifth-century Buddhist sutra translated from Sanskrit contained this verse: "When one consumes xingqu, one should take the authentic (*zhenshi*, lit. "true, real") product. If one consumes the fake and abandons the authentic, no good will be done. A thousand doctors would not be able to save such an idiotic person."[71] In other words, the problem of hing's authenticity was known to the Indians and, through them, to the Chinese at the very beginning of the thing's journey in East Asia. The problem worsened after the resin became a global commodity in the seventh century under the dominating term *awei*, which did not distinguish the plant and its parts from the resin, reaching a first crisis in the eleventh century when its use in medical recipes was rapidly expanding in East Asia. Materia medica books from then on sometimes illustrated awei with a nondescript treelike plant called "Guangzhou awei," suggesting that Guangzhou, the global port, was now taken to be the native place of the plant; others claimed that the plant also grew in Southeast Asia, southwestern China, and even the Yangzi region.[72] Moreover, sources did not agree on the exact part of the plant that contained the resin, whether it was the stem, the leaves, or the root. The only thing that everyone agreed upon was the stench. The crisis soon generated techniques to verify the drug's authenticity (see figure 7.4). Three methods were described: putting the product in a copper container overnight, and the copper would turn silver white if the thing was authentic; putting the product in a juice of *wudou* grass overnight, and authentic awei would be blood red in the morning; putting the resin on a pomelo tree and authentic awei would dry up the tree quickly.[73]

Uncertainty about awei's authenticity continued to grow, however, despite all the efforts. A thirteenth-century scholar official, Zhao Rukua (趙汝适), talked to Southeast and Central Asian merchants in Quanzhou (Fujian), a major trading port in southeastern China. He obtained the information that awei sold in China was mostly from Persia via the Samboja kingdom (today the southeast portion of the island of Sumatra). He also heard a widespread rumor that awei was in fact goat meat corrupted by the poisonous resin, which further thickened the shroud of mystery around awei (see figure 7.5). Later

FIGURE 7.5. Illustration of *awei* as poisoned goat meat in the 1885 edition of Li Shizhen, *Bencao gangmu*. Public domain.

experts, including Li Shizhen, found the story unfounded, especially when awei was known to be nonpoisonous.[74] How should this strange rumor be explained? Did it have something to do with the unusual packaging of the most highly priced Kandahari Hing observed in India before its exportation: "it is sewn up in goat skins, forming small oblong bales, with the hair outside . . . with an odor recalling that of garlic and oil of caraways"?[75]

Li Shizhen's contemporary, Chen Jiamo warned consumers that awei on the market was often a counterfeit made of garlic. Others suggested ways to distinguish the good from the inferior, the real from the fake: Miao Xiyong pointed out that one should judge the quality of awei by its color. The superior resin should be yellowish, while inferior stuff was black. Chen Shiduo of the early seventeenth century told his readers that true awei would float midway in water, while the fake would simply sink.[76] The discussions on awei continued to focus on the verification of its true identity. Sometime before the late sixteenth century, another popular saying was formulated to highlight the difficulty of obtaining "true" awei, comparing this with the ease of getting true *huangqin* (scutellaria root, *Radix Scutellariae*).[77] Awei had by then become almost synonymous with counterfeit.

Adulterated Commodity

The Chinese concern about awei's authenticity as a commodity was mostly triggered by changing market prices, especially when the ingredient was losing popularity in the East Asian pharmaceutical market. By the late fifteenth century, the official price of the drug was still high, in absolute terms, at two *guan* (about two taels of silver) per catty, but relatively low compared to five for myrrh, and three for ambergris.[78] Garcia da Orta observed in the sixteenth century that the high cost of asafetida was partly caused by the fact that the gum deteriorated quickly and the Indian suppliers manipulated the material to keep up the price.[79] This remark hints at the common practice of adulteration of the raw resin in India before export, later confirmed by Japanese Rangaku authors of the eighteenth-century *Ran'en tekihō*. They informed their readers that the high value of hing in India (where it was used widely as both a medicine and a spice) caused the flooding of fake and adulterated asafetida on the global market.[80] In 1751 the Swedish merchant Peter Osbeck wrote in his travel notes that, in Canton (soon to become the only international trading port in China), "the Chinese get many commodities from several parts of Asia, and in particular parrots, ivory, tortoise shells, and asafetida."[81] Being the only one of the four imports with a questionable identity, asafetida's value was probably maintained as an exotic import.

Nineteenth-century commercial information, including trade statistics from India, shed light on the circulation of the different types of asafetida in the world market that could account for the erratic prices. Such information provides important clues to the circulation of the commodities, which may help us understand, with hindsight, some of the earlier fragmented knowledge about awei in East Asia. Bombay was the world's largest trading hub of asafetida in the nineteenth century, importing all types of resins from Persia and Afghanistan and exporting a part of them to the rest of the world after "manipulation." From 1884 to 1889, 37,297 hundredweight (one hundredweight is about 40–50 kilograms) of hing were imported by sea, while 6,020 hundredweight *hingra* were imported by land from Kabul (*hingra* was the raw material of asafetida exported to Europe and probably also of awei to East Asia). Of the grand total of imports, only 8,586 hundredweight were exported. This export declined further between 1886 and 1890, when the total import held steady at 37,306 hundredweight, with only 2,014 hundredweight exported. These figures also demonstrate that India was the biggest consumer of hing, taking up roughly two-thirds of the total import. It was also the major "manufacturer" of hingra.[82] Finally, it seems that all imported resins were adulterated in various ways before they were sold on the market.[83]

Under the three main categories of imported "raw" resins in Bombay (hing from *F.alliacea*, *Boiss.*; *Kandahari hing*; and hingra from *F. foetida*, *Regel*), there were many categories of adulterated products on the market with great price differences. All the hings, especially the supreme *Kandahari hing*, were more expensive than the hingra that was mostly for export to the West and to East Asia.[84] Good quality hing was worth as much as eighty rupees per hundredweight in the late nineteenth century. The average value was about fifty-five rupees.[85] Hingra from Persia and Afghanistan was significantly cheaper, being valued at only about twenty rupees per hundredweight on the average.[86] The prices were obviously determined largely by the biggest Indian market.

By the 1850s the Chinese were no longer concerned with awei's high prices but were baffled by its unreasonably low prices, which they interpreted as being the outcome of its adulteration. A famous doctor witnessed that, "in the shops, awei was faked with foreign garlic . . . the provinces of Zhejiang and Jiangsu [where the doctor worked] are thousands of miles away from the Western countries [where the drug was produced], yet the price [of this gum] is very low. From this we can tell how much faked resin is being sold."[87] The awei seen by the Chinese doctor was likely the same inferior material as was sold by the Indians to Europe.[88]

Authenticated Plant?

European naturalists traveling widely in different parts of Asia after the sixteenth century searching for the "true" asafetida plant were increasingly frustrated by the difficulties of their quest. Ibn Al Baytar, Garcia da Orta, and John Royle all admitted that there was a great deal of confusion created by the terms in different languages (Sanskrit, Arabic, Latin, Tokharian B, Chinese, and so on) designating the various plants producing the resin, their various parts, and the resin itself.[89] Most European naturalists seemed to agree that there were actually two or more types of asafetida possibly from different plants, one more fetid than the others, also with different colors. John Fryer (ca. 1650–1733), the British travel writer and doctor who served as a surgeon of the East India Company in the 1670s, went so far as to claim that the asafetida produced in Persia and consumed in Europe was not Indian hing.[90] The search of European naturalists for the plant from the seventeenth to the mid-nineteenth century further complicated the picture, because they could not show exactly which ferula plants furnished the resin sold in the European market.[91] None of them had actually seen the production chain. Balfour's claim in 1860 that asafetida was furnished by Ferula Asafoetida of Linnaeus and Ferula Persica of Willdenow was never substantiated.[92] The resin had had a global circulation since the sixteenth century, but few consumers had seen asafetida in its natural habitat or witnessed its production and export processes.

For European botanists interested in the identity of asafetida, the problem underwent a dramatic turn after the mid-nineteenth century, when fuller records of Indian trade became accessible, and European trader naturalists began to interact more directly in the global network of indigenous merchants, artists, gardeners, and pharmacologists to direct the collection of botanical data to be sent back for analysis in Europe.[93] The French pharmacologist Nicolas Guibourt wrote in 1850 that the Indian asafetida he obtained from a Parisian pharmacologist was very different from what one could find in the European market.[94] By the 1890s it seems a consensus had been reached among European naturalists that the mystery of the identity of the commercial asafetida in Europe was solved. In 1891 the British pharmacologist William Dymock (1834–1892) wrote that, although silphium of Cyrene could no longer be obtained, the gum resin that was sold on the European market—for a long time believed to be Asian (Indian) hing—was in fact something different. He believed that the asafetida in European commerce was indeed not Indian hing (from *Ferula alliacea, Boiss.*) but was what Indians called hingra (from *Ferula foetida, Regel*). The former, with a stronger stench, was from a smaller plant that grew on the hills of Khorasan (modern-day Iran), whereas the latter, as witnessed by Kaempfer in the late seventeenth century, was from a tall plant that grew in western Afghanistan. There were several key figures in the process of identifying the resin, a process that took place in the 1870s and 1880s between India and Europe. Dymock was the British military surgeon in Bombay; Ardeshir Mehrban was a Persian merchant who procured the plant of hing for Dymock; Daniel Hanbury (1825–1875), a British pharmacologist and botanist, studied the samples sent by Dymock to London; the Swiss botanist Pierre-Edmond Boissier (1810–1885) confirmed the identity of the plant producing hing as *F. allicacea*; and James Aitchison (1836–1898) was the Scottish botanist who identified the plant producing commercial asafetida (hingra) in Europe as *F. foetida*.[95] However, Dymock and George Watt (1850–1930), a British botanist and reporter on economic products with the Government of India, who both claimed that the mystery was solved, remained ambivalent themselves as to the exact types of hing and asafetida hingra that were available in the Indian and European markets. They identified asafetida as "certain species of Ferula yield[ing] either *Hing* or *Hingra*, or both these drugs. . . . [D]ifferent systems of extraction and manipulation, or diversified conditions of climate and soil, produce both *Hing* and *Hingra*."[96] In other words, knowledge constructed by nineteenth-century European merchants and botanists on the Ferula plants and resin justified rather than fully clarified the confusion.

Looking back over the economic and medical itineraries of awei in China from the perspective of this nineteenth-century information, we may want to conjecture that the supply and price of awei have been much determined by the Indian market since the fifth century. It would seem that relatively unadulterated and expensive asafetida from Central Asia could be found in China at the beginning of its importation in the seventh century until perhaps the fourteenth century. This was the period when awei was widely used in many medicinal recipes and, for a short time, in culinary practices. With the global circulation of awei as a commodity, or commodities, from the sixteenth century, a discourse of unverifiable authenticity developed around the material, while its use and value as a drug had already been declining in East Asia.

Conclusion

The long and looping global itinerary of asafetida as a drug/spice/plant begins as an interaction of humans with certain biological, social, and cultural needs with materials of various kinds in multiple temporal and spatial contexts. Asafetida's unusual stench becomes linked to an exceptional transformative power on the human body and spirit and has led to its enduring significance in religious and medical realms.[97] In major hubs such as Chang'an, the material known as hing (xingqu) or awei emerged in a written body of knowledge in culinary and medical contexts that incorporated different cosmic logics in Arabic, South Asian, and East Asian cultures. It was a desired spice, even an

FIGURE 7.6. The search for silphium goes on today. A modern imagined representation of silphium. *Horticulture* (F&W Media, 2010).

aphrodisiac, and was at the same time taboo for Buddhist monks in South Asia. It was a key occult drug used for expelling evil spirits and vermin in all Asian cultures. In East Asia, in particular, awei redefined the explanatory paradigm of healing and medical theory, its unique stench again playing a key role. But the same stench disqualified it as an East Asian and European food.

This changing materiality did not disrupt its popularity in South Asian markets but, in East Asia, caused a gradual decline after the fourteenth century. Being a traveling material with an unknown raw state that went through uncontrollable production processes before it reached East Asia, awei was increasingly brought into doubt as a valued commodity, as demand for it increased from a global market. Confusing information about its provenance and authenticity was coupled with changing principles of drug use in East Asia. In centers such as Canton, Hangzhou, and Quanzhou, conflicting information was particularly abundant. Meanwhile, drugs perceived to possess drastic transformative powers lost their appeal and gave way to milder, often local ingredients. Asafetida's unique stench, moreover, amplified its perceived abrasive nature.

Europeans picked up the global interest in and demand for asafetida, however, as East Asians were turning their backs on it, albeit in a completely different context. The opening of new transoceanic routes after the sixteenth century, combined with a search for ancient knowledge, ignited a search for materials mentioned in Dioscorides's *Materia Medica*. To compare asafetida with the classical material known as silphium—and from there to unravel the differences between various Ferula plants—became a tantalizing project for traveling diplomats, doctors, merchants, local traders and gardeners, and natural historians in major European trading companies, medical faculties, and their botanical gardens, and in the nineteenth century in newly created laboratories. Silphium remains an object of research even today (see figure 7.6). As seeds and plants were exchanged, acclimatized to new regions, and the chemical contents of all types of asafetida were investigated, a whole new set of knowledge accumulated around the material complex "asafetida." The "true" appearance of asafetida, its plant, and its manufacture remained more elusive than ever, however, even as knowledge about it was codified anew in a network of global hubs of trade and scholarship, including Bombay, London, Paris, Geneva, Edinburgh, Edo, and Canton.

Chapter 8

Smoke and Silkworms

Itineraries of Material Complexes across Eurasia

Pamela H. Smith, Joslyn DeVinney, Sasha Grafit, and Xiaomeng Liu

A remarkable sixteenth-century French compilation of mostly practical recipes for various art and technological processes contains much evidence of the movements of materials: both short-span itineraries within Europe—including silkworms and the blue dyestuff woad between southern France and Spain, dyes and pigments from Italy, amber from the Baltic, metals from Germany—as well as long-span pathways of dyestuffs such as turmeric and stick lac from South Asia, cochineal from Central America, and the tree resin, dragon's blood from the Canary Islands and North Africa, techniques of damascening armor from the Near East, and "damasking" cloth by resist dyeing it with "Moresque" templates, likely derived from the Ottoman Empire. Among all this evidence of the movement of materials and techniques across Eurasia, there are two unusual and puzzling recipes which are the focus of this essay, one labeled "Medicine of the orientals against all maladies" and another with the heading "The Work done in Algiers."[1]

The anonymous manuscript in which these recipes appear is a 170-folio first-person account of processes carried out in a workshop, together with recipes and observations collected perhaps on visits to other workshops. Most of the manuscript is written in the same hand, although a scribe seems to have been involved in taking down some parts. The anonymous author is an experienced practitioner but does not appear to have been part of an identifiable trade association. He knows some Latin, although far from perfectly. Perhaps he is the son of a craftsman, trained in a workshop, with grammar school or even some university training, who then went to work for a rich merchant of Toulouse (where the manuscript seems to have been compiled), or for a noble family, such as the Béthunes, whose library contained the manuscript after its probable composition in the 1580s and before the Béthunes donated their books and manuscripts to the king's library (now the BnF) in the seventeenth century.[2] The manuscript's 170 folios contain a collection of mainly technical recipes in no apparent order for objects that might have formed the contents of a type of collection much in fashion in the sixteenth century called a *Kunstkammer* (Chamber of Art). Ms. Fr. 640's recipes would have pleased a collector with such a *Kunstkammer* because this kind of collection sought to show off the imitation of nature's artifice by means of the artifice of the human hand. The manuscript contains instructions for the draughtsman, recipes for pigment making, metal casting and metal coloring, imitation gem production, arms and armor making, grafting trees, surveying land, a practice of taxidermy to manufacture monstrous composite animals, making papier mâché masks, and much more. The author may have been based in Languedoc, as he mentions towns of this area, particularly Toulouse.[3] He may have had ambitions to publish a "book of secrets," such as that published under Alessio Piemontese's name for the first time in 1555, which then circulated widely throughout Europe. Although they sound esoteric, books of secrets generally comprised collections of mostly straightforward recipes for various art, medical, and utilitarian ends.

"Medicine of the orientals against all maladies"

Of over forty medicinal recipes in the manuscript, "Medicine of the orientals against all maladies" on folio 77r is the only entry that refers to "orientals," the only one that advises taking in the smoke of a dried herb as a treatment, and the only example of the use of a pipe.[4] Investigating the remedy in this recipe and the entry's careful drawing of the pipe opens up a wider history of pipes and smoking that suggests the global movement of medical and ritual practices across Eurasia and beyond over a very long time. "Medicine of the orientals against all maladies" is a neatly written recipe, second from the top of fol. 77r between recipes "Against redness of the face" and for "Fatty earth" (figure 8.1).[5]

The recipe begins with the command to "dry rosemary in the month of May."[6] Next, "fill this bowl with powder made of it [rosemary], and put a lit charcoal on top."[7] After filling the bowl of the clay pipe with the crushed rosemary and charcoal, then "receive the smoke by a quite tightened mouth, and a part will come out by your nose. But if you want to purge the head also pinch the nose. Against colds, rheums, and other maladies."[8] It is noteworthy

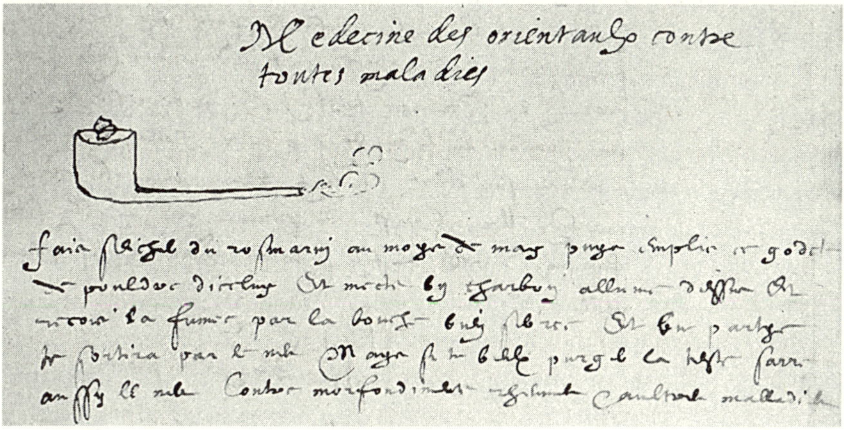

FIGURE 8.1. Sketch of the pipe for taking in the smoke of rosemary, in the entry entitled "Medicine of the orientals against all maladies." Fol. 77r, Ms. Fr. 640, "Recueil de recettes et secrets concernant l'art du mouleur, de l'artificier et du peintre." Département des manuscrits, Bibliothèque nationale de France. Public domain.

that these instructions direct the reader to bring the smoke into the head (not inhale it into the lungs).[9] Drawing rosemary smoke into the head to remedy colds and rheums fits squarely within general humoral theories of balance—as well as imbalance—in specific organs caused by "accumulations of crude or otherwise morbid, harmful matter," or "fluxes." One of these fluxes, a "catarrh" or a "rheume" was attributed to "watery, mucous or corrupt matter that had accumulated in the head."[10] Thus, the instruction to take smoke into the head aligns with the common sixteenth-century notion that treatments should target the specific location of the ailment. Moreover, as rheumes and catarrhs were "cold" ailments, and rosemary was considered a hot and dry herb, it stands to reason that it was used against head colds, as described in folio 77r.[11] The use of aromatic smoke as a remedy for colds or fluxes is related to the concept of infected air, in which airs could be corrected or protected by good smells, especially of aromatic herbs such as rosemary.[12] This concept led to practices of fumigating bodies and living spaces by taking in the aroma or smoke of strong-smelling substances.[13]

Rosemary

Rosemary occurs as an ingredient in another medical recipe in the manuscript, folio 102v, which instructs how to make an "infusion of anthos or rosemary" intended "for the elderly" by distilling the oil of rosemary, a common method for extracting the plant's aromatic oils and other substances perceived to contain medicinal virtues.[14] Rosemary was viewed at this time as beneficial in aging and enhancing memory, but the medicinal virtues of rosemary possess a long and global history.[15] In ancient Greece, rosemary was worn as a wreath upon the head for better memory during examinations and was also placed under pillows or beds to prevent nightmares.[16] Rosemary appeared in the writings of Dioscorides (40–90 CE) and Galen (130–210 CE) and, by the latter Middle Ages, possessed rich symbolic meanings in Europe.[17] Christian legends reported that the shrub received its virtuous blue flowers from the mantle of Mary and would not grow taller "than Christ's height on earth," ceasing at the age of thirty-three to grow higher, increasing only in breadth.[18] Sixteenth-century references to rosemary frequently mention its use by the Romans to treat jaundice after drinking.[19] The name "rosemary" comes from the Latin compound of *ros* and *marinus* (*rosmarinus*) or "sea dew," which alludes to its native habitat along the Mediterranean coastal region.[20] In his widely circulated *Historia Stirpium* from 1542, Leonhart Fuchs (1501–1566) refers to the works of Dioscorides and Galen when outlining rosemary's properties, and he mentions the herb's Greek name, *libanotis*, which was "called *rosmarinus* by the Romans."[21]

By the sixteenth century, rosemary's reputation as a kind of medical panacea was widely acknowledged in European herbals and medicinal treatises. The fourteenth-century English friar Henry Daniel (c. 1315–c. 1385) wrote an entire treatise on the herb, describing the numerous medicinal virtues of rosemary and listing a variety of medical conditions it could be helpful for, from gout, colds, and hair thinning to nightmares, memory loss, and aging skin.[22] The first known English herbal of 1525 calls for boiling the "hot and dry" herb and drinking the water, "for it is much worth against all evils in the body."[23] Rosemary was seen as especially efficacious for problems of the brain. Shakespeare's widely quoted "there's rosemary, that's for remembrance" is also found in the herbal of Rembert Dodoens (1517–1585): "the Arrabians and their successours Physitions, do say that Rosemarie comforteth the brayne, the memory, and the inward senses, and that it restoreth speech."[24] Like Fuchs, Dodoens also cites Dioscorides and Galen about its property, when boiled, to work against jaundice. Pietro Andrea Mattioli (1501–1577), in the 1579 French translation of his commentaries on Dioscorides, provides nearly three pages on the herb.[25]

Rosemary's virtues stemmed from its nature as a "hot and dry" herb. Fuchs mentions rosemary as a hot condiment herb, and the herb appears in other contemporary herbals in culinary recipes for the balancing of the body.[26] This explains why rosemary is used to treat a cold ailment in the recipe on folio 77r, albeit as a fumigant—also a common usage. Friar Daniel's manuscript suggests that if "þou hawe gret cold in þin hed take þe smoke of his bark in þin nase and þou shalt felyn helpe."[27] Fuchs states that "the later physicians say that rosemary as a fumigant eases a cough and dripping [noses], and what is especially noteworthy, when burned it renders the house safe during pestilence by its fumes diluting the danger of impure air." He notes that "today [rosemary] is planted everywhere in gardens and pots. However, in Narbonne, in France, it grows so abundantly that the inhabitants burn no other wood."[28] Dodoens states that "rosmary groweth naturally, and plentifully, in divers places of Spayne and France, as in Provence and Languedoc. They plante it in this countrie in gardens, and mayntayne it with great diligence."[29] The herbal writer John Gerard (1545–1613) also noted that "rosemarie there is such plentie thereof in Languedocke, that the inhabitants burne scarce any other fuell."[30] Clearly, rosemary grew abundantly in France during the sixteenth century, and botanical knowledge about the herb circulated in southern France in texts.[31] The availability of rosemary and knowledge about its medicinal characteristics were not the only reasons for its popularity as a fumigant. Sixteenth-century beliefs about the relationship between smell and medicine were also important.

Smell

Aroma has played a role in medical thinking since antiquity. In *Airs, Waters, and Places* and *Nutriment*, Hippocrates mentions "bad smells" in the context of disease diagnosis and prevention.[32] Until the late seventeenth century, physicians expressed a belief in a porous "breast- or nipple-like" plate separating the nasal and cerebral cavities. In *The Olfactory Organ*, Galen argued that the nose "was no more than a passage which carried smells up to the true olfactory organ, the brain itself."[33] Thus, with no barrier of protection, the brain could be affected by both "good" and "bad" smells. In addition to the physiognomy of sensory reception, the view that smells were material substances also contributed to their significance in narratives of health. In his work on smell and health, Richard Palmer lists instances of smells as part of treatment for "cold complexions of the brain, for various kinds of headache, for catarrh, for melancholy, and for nervous conditions."[34]

The intersection of medicine, aromatics, and perfumery present in Ms. Fr. 640 thus reflects common links drawn between scent and health in the sixteenth century. The development of distilling and the extraction of essential or volatile oils meant that the use of volatile scents—whether in scented candles, rosewater, or aromatic herbs and flowers—was commonly employed to perfume clothes, bedding, soaps, spirits, and liqueurs.[35] Because it was widely available, rosemary was extensively employed during periods of plague, both in its essential oil form and also burned as a fumigant, to act as a prophylactic or remedy.[36] The connections between olfaction, aroma, fumigation, and health were common in early modern Europe, but the burning of aromatics for both bodily and spiritual health goes far back into the human past. It is recorded in Greek and Roman practices, for example, where it was part of Dionysian/Bacchic rites and is embodied during Christian ceremonies in the burning of aromatic resins such as frankincense and myrrh as materials that through their smell and smoke materialized ideas of spirit and directed thoughts to the heavenly sphere.[37] As one recent author put it, "smells were concretely evident when sights, sounds, and tastes were not; olfactory experience was tangibly perceived, although it did not involve the body's limbs as did touch. These qualities made smell a singularly effective means of discerning divine presence or absence, demonic activity, or moral condition. Such significations expressed the cosmological orientation of the ancient Mediterranean world, whereby olfactory codes served to structure human experience in relation to divine order."[38]

The Pipe?

Rosemary use was thus common throughout the Mediterranean and especially in southern France where Ms. Fr. 640 was compiled. But what about the pipe? How common was it to smoke rosemary in a pipe? By the mid-seventeenth century, Nicholas Culpeper (1616–1654) echoes earlier texts in recommending rosemary for a variety of maladies but also provides examples of other materials to use as fumigants aside from rosemary, including rue, bitumen, and crushed amber.[39] Significantly, in *The English physician* (1653), he advises "to burn the Herb [rosemary] in houses and chambers" to "correcteth the air in them" and says "the dried Leavs shred smal and taken in a Pipe like as Tobacco is taken, helpeth those that have any Cough or Phtisick, or Consumption."[40] While Culpeper's text is witness to an increased use of pipes for smoking in the later seventeenth century, the Ms. Fr. 640 recipe seems to be a rare usage of the pipe in the late sixteenth century. Historians writing on the history of pipes in a global context typically assume that pipe smoking is associated solely with tobacco, taking for granted that the smoke is inhaled into the lungs, not held in the head as recommended by Ms. Fr. 640.[41] For example, in their introduction to *Smoke: A Global History of Smoking*, Sander L. Gilman and Zhou Xun begin their history with Columbus's encounter with Native Americans smoking dried tobacco leaves. Widely cultivated in the Americas by both indigenous peoples and colonists by the sixteenth century, tobacco had been used in ritual and medicinal practices for millennia since its first cultivation in the Americas between the sixth and fourth centuries BCE.[42] By the sixteenth century, tobacco twigs were being burned by North American Indians to treat chest conditions, perhaps indicating that the smoke was inhaled into the lungs.[43] Thomas Harriot (1560–1621), however, in his 1588 text about his travels to Virginia discusses the benefits of using tobacco and speaks of taking it into the head: "the leaves thereof being dried and brought into powder: they take the fume or smoke thereof by sucking it through pipes made of claie into their stomacke and heade; from whence it purgeth superfluous fleame and other grosse humors, openeth all the passages of the body: by which meanes the use thereof, not only preserveth the body from obstructions; but also if any be, so that they have not beene of too long continuance, in short time breaketh them: wherby their bodies are notably preserved in health."[44]

The spread of tobacco smoking apparently led to the diffusion of the clay pipe as a smoking implement around the world, although tobacco was also consumed in cigars, by sniffing, and by chewing.[45] According to Gilman and Xun, what surprised sixteenth-century Europeans was the use of the pipe to deliver the smoke instead of fumigating in "rooms full of smoke." The English were perhaps the first to manufacture white-ball clay pipes toward the end of the sixteenth century.[46] A 1573 entry in a chronicle notes an unusual practice of "taking-in of the smoke of the Indian herbe called 'Tobaco' by an instrument formed like a little ladell."[47] Meanwhile, the French in Florida had been introduced to tobacco smoking with clay pipes earlier in the sixteenth century. The chronicler John Sparke reported on John Hawkin's voyage to the West Indies in 1565, writing, "The Floridians when they travell, have a kind of herbe dried, who with a cane and an earthen cup in the end, with fire, and the dried herbs put together, doe sucke thorow the cane the smoke thereof."[48] It is probably impossible to precisely date and trace the introduction of clay pipes into Europe, but it seems that this type of pipe was new in the period of Ms. Fr. 640, and the author-practitioner's inclusion of a sketch of the pipe may have come from his recent encounter with it.[49]

Medicine of orientals

Finally, an obvious question about the author-practitioner's heading "Medicine of orientals" is to ask what "oriental" meant in sixteenth-century France. Although the term generally was used to refer to "Easterner," the reason for the use of the term on folio 77r is impossible to ascertain from the body of the recipe, especially as the author-practitioner does not seem to associate "oriental" with "exotic" but instead uses "oriental" simply to denote the origin of materials.[50] It is especially puzzling because rosemary was widely grown in France at the time but was not recorded in sixteenth-century Eastern—specifically Chinese—medicine as being smoked in pipes. After its introduction from *Daqin* (the premodern Chinese name for the Roman Empire) around 200 CE as an ornamental plant in the royal garden, rosemary is recorded for the first time in the eighth-century *Bencao Shiyi* 本草拾遺 (A supplement to materia medica) as preventing and expelling harmful *qi*, and "when burned, it exorcises evil spirits." The early tenth-century *Haiyao bencao* 海藥本草 (Foreign materia medica) also suggested its use as a mosquito repellent when mixed with notopterygium root and burned.[51] By the sixteenth century, Li Shizhen's 李時珍 *Bencao gangmu* 本草綱目 (Compendium of materia medica) quoted both these uses for rosemary but evaluated it as an aromatic only used for sachets.[52]

The term "orientals" in Ms. Fr. 640 is perhaps instead related to the diffusion of pipe-smoking practices from another part of the "East." Tobacco smoking had spread to Asia and Africa by the sixteenth century, and to South Asia, where the Mughals "mixed tobacco with other more familiar leaves, spices and sandalwood" and "drank" it (meaning "inhaled") by means of a water pipe or hookah.[53] The use of water pipes in Persia, India, and China would have been known to sixteenth-century Europeans via trade and travel.[54] During the early usage of clay pipes in Europe, therefore, dry pipes and Eastern water pipes may have been conflated, leading to the use of "oriental" in Ms. Fr. 640.[55]

While the question of why "orientals" was used in this manuscript is thus likely to remain unanswered, it is clear that the recipe "Medicine of the orientals against all maladies" is a product of the global itineraries of materials, practices, and objects in the early modern world. The use of dried rosemary in a clay pipe, although perhaps not as common as the use of tobacco, is not surprising given the widespread availability and use of rosemary in France as an aromatic fumigant and as a medicine in a variety of forms. The introduction of the clay pipe to Europe as a smoking implement is hard to date precisely, but it appears to have been a relatively new item in the mid- to late sixteenth century and was perhaps still associated with the spiritual meanings it possessed for Native Americans or possibly conflated with fumigation practices from the East that, like the burning of aromatics in Europe, could have both spiritual and medicinal significance. The author-practitioner's inclusion of a careful drawing of the pipe in the recipe on folio 77r probably indicates his unfamiliarity with the pipe, but whether he had personal experience using the aromatic smoke of rosemary in a clay pipe as a medicine or he learned about this practice from other sources, perhaps even copying it from another text, will—like many questions to do with long span routes of materials, objects, and practices—remain obscure. Even more tantalizing questions arise with regard to the entry on folios 52r–v of Ms. Fr. 640, in a recipe entitled "The Work done in Algiers."

"The Work done in Algiers"

"The Work done in Algiers" on folios 52r–v of the manuscript is an odd mix of gold-seeking alchemical aims—almost the sole such "alchemical" entry in the manuscript—combined with quite practical considerations and ordinary ingredients. In many ways, apart from its alchemical aim of making a powder that transforms antimony into gold, this recipe is of a piece with the rest of the manuscript in its attention to practical details, first-person perspective, and materials common to a workshop. Its practical tone is worth quoting in full:

> The Work done in Algiers
>
> Take a colt of three or four years and feed it on barley and straw cut in the manner one feeds horses in Spain, and water it with good fountain or river water. I do not know if it would be good sometimes to water it occasionally with water of sulfurous baths, and to sometimes give it fenugreek or other hot foods, for the intention of the worker is to use the heat of his dung, and the climate here is cooler than that of Algiers. Keep it in a warm and close place and make sure that none of its manure and urine should be lost, of which you will make a heap or two in order that while one cools the other will be at the appropriate heat to

> continue. Also take a large glass flask as thick as you can, and one finger thick if it can be so made, and of the capacity of one pitcher or earthen jug. Around the feast of St John put into it a dozen and a half chicken eggs, that is to say, the yolk without the egg white and the germ. Others say sixty yolks. And with this dozen and a half of egg yolks put in half an ounce (others say sixty eggs, half a pound) of female silkworm eggs. And after having thoroughly luted [coated with protective clay] the flask (I do not know if there needs to be air for the generation), put it and bury it in the heat of the dung up to the neck, and leave it there until several worms are engendered. And then remove the flask and do not bury it in the dung any longer, but only keep it placed on the hot layer of dung until all the worms will have eaten and consumed one another, bustling and stirring, and only one remains. When this is the case you need to feed it at regular intervals, day and night, with the aid of two men, who by intervals will take care of it, and you will feed it with an egg yolk covered with a gold leaf or with a liquid yolk with the gold leaf incorporated, and take care that it does not want for such food (some say one egg yolk per hour, others three, but the thing itself will demonstrate the practice). Nourished in this way it will achieve its growth in seven weeks or two months and will become like a snake, one span and four fingers long, and one pound in weight, and as the wings begin to arise on it, one will need to make it die, doing so with a charcoal fire in a ring around the bottle one span distant from it, and then stopper and lute the bottle well in order that it does not exhale. [52v] Or to be safer, retire from there until the fire completely died down and all is cooled, for its exhalation would be dangerous. And for the occasion when you feed it with pincers, wash your mouth with good vinegar and take some preservative and plug yourself up well. Once it is dead, put it in a linen cloth or a canvas of silk and fold it and hang it from the ceiling where the air and sun dry it. Once it is quite dry, pulverize it in a mortar and keep this powder carefully, because one dram of this on 3 pounds of molten antimony reduces it to finer gold than the other one. But it does not have as much weight. For this work you also need to choose the oldest antimony that you can, which has often been melted and finely hammered into sheets or other works, and purify it before by melting and throwing it into some honey and vinegar. The term of the work is nine months from the feast of Saint John until the 25th of April.

This recipe appears among instructions for making ink with which to print plates, for the proper mode of harvesting linseed for oil, and for keeping pigments fresh and their colors bright. A folio further on, there appear purely practical instructions for raising silkworms (the Spanish silkworm eggs, it is noted, producing one and a half times as much as the ones from Languedoc), as if the Work of Algiers, too, is just another process for silkworm growth. Indeed the two entries do appear to be connected, if not for producing silk: "The Work done in Algiers" on 52r–v and "Silkworms," a page away on fol. 53v–54r, both involve growing silkworms from eggs, and the dates mentioned in "Algiers" for starting the process of growing silkworms in the flask are aligned with the end of one complete lifecycle in "Silkworms"—worms, pupation, moths, mating, and laying of eggs—in order to begin the cycle again by hatching the worms from eggs. The author-practitioner is familiar with early modern sericulture practices, and his detailed account of raising silkworms seems to be the result of firsthand experience.

Compared to other contemporary works on sericulture, however, the author-practitioner's account is lacking in one crucial particular—it contains not nearly the same detail on deriving silk from the cocoons but instead provides much more information on the entire life cycle that produces eggs. This is a marked difference from other literature on silkworms. Of course, it may be a coincidence that a recipe for using silkworm eggs to create a gold-producing powder, "as they do in Algiers," occurs one folio earlier (fol. 52r), but there are other similarities that could indicate that the purpose of "Silkworms" is not to extract silk threads from cocoons but, rather, to prepare freshly laid silkworm eggs for "The Work done in Algiers."

Holy Days

The overlap of timing seems especially important: The "Algiers" entry concludes, "the completion time for such work is nine months from Saint John's day [June 24th, near the summer solstice] until 25th April." The corresponding timing in "Silkworms" states: "They [the hatched moths] finish spinning and laying eggs in three weeks and around Saint John's Day. And at that time one keeps their eggs and seed until Holy Week." The occurrence of Holy Week (the week immediately preceding Easter) shifts each year, as does Easter (always the first Sunday after the first full moon occurring on or after the spring equinox), but April 25th is the latest possible date upon which Easter can fall.[56]

The correspondence of these dates in the two entries betrays a deeper connection to a cycle of growth and rebirth observed both in nature and in the Christian liturgical year. John the Baptist's Day, occurring close to the summer solstice (the longest day in the year), followed by the significant period of nine months (the length of human gestation) to the feast of the Resurrection (occurring near the spring equinox), a time of rebirth of the natural world and the turning point of the solar year—all add clear symbolic dimensions to these entries on silkworms. Moreover, the author-practitioner seems to be thinking of resurrection when he writes about the silkworms shedding their skins on folio 52r that "from their birth until the time when they make their cocoons and their prisons, sleep and rest 4 times, and each time remain 4 or five days resting without eating, as if they were dying for rebirth another time." This reference to resurrection is also present in a poem on the nature of silkworms by "Bishop Vida," who is briefly cited in the margin of folio 53v. The following excerpt from Vida's *The silkworm: a poem in two books* is representative:

> Imprison'd dark within the silky ball
> Compleat their labours, and resign their breath
> Exhausted, sinking in the shades of death.
> Go wond'rous race, thrice happy artists go
> Meet fate content, to you the sisters owe
> A second birth; kind Venus shall requite
> Your toils, and raise you to the realms of light[57]

This "second birth" is reminiscent of the author-practitioner's words about the silkworms, "dying for rebirth another time." A later poem by seventeenth-century Christophe Isnard also connects silkworms with the resurrection of a phoenix:[58]

> This celebrated artisan, of an immortal life,
> Comparable to no other being in its fate
> Leaves and takes up life again, as soon as it emerges,
> And the end of its course, is followed by another.
> If for this the Phoenix, is the envy of others,
> This worm would have occasion to raise itself with greater strength,
> For being beyond the reach of death
> Its form is ravished [that is, carried away] from its body four times.[59]

Silkworms, then, were not just prized for their economic value but could also be employed for their symbolic power.

Alchemy

The "Algiers" recipe involves hatching silkworm eggs in an enclosed flask where they feed on egg yolks and, eventually, consume each other. When the sole surviving silkworm grows to resemble a serpent, it is burned and produces a toxic vapor. Its ash is to be used to transmute metals into gold. A common preoccupation across the practices and the knowledge systems of the major religions and medical texts across Eurasia was the relationship between body and spirit. Practices and substances in the material world were understood to body forth the processes and significances of the spiritual realm. Nowhere was this more prominent than in the laboratory practices and the texts of alchemy. Alchemy is often misunderstood and caricatured as a vain search for gold, but as any reader of alchemical texts soon comes to see, it is also a complex of esoteric and exoteric practices and beliefs—simultaneously pure pious meditation and practical bodily techniques, often to produce the most quotidian of goods such as dyes, acids, and mineral compounds. It resulted not only in many useful substances but also much theorizing about the natural world. Through its regularized processes of transformation by fire, acids, and (what we would call today) chemical reactions, alchemy gave much scope to its practitioner for bringing about the transformation of substances and the generation of new materials. It can be understood as an early modern "science of transformation." One of its enduring mysteries was how to transform materials from one state to another, from solid to liquid, and finally, to volatile spirit. It is not surprising, then, that ordinary artisans and religious reformers alike regarded alchemy and its transformation of materials—especially metals—to be a demonstration of the connection between material and spiritual worlds.

This recalls the resurrection of the silkworm and the process in the flask during the "work of Algiers." The toxic fumes generated by burning the surviving silkworm in the flask is significant in light of another recipe in Ms. Fr. 640 that also produces a potent agent, a "horrible poison," by enclosing lowly creatures—in this case, snails—within a flask. This process (spanning folios 55r–v) involves allowing the snails to putrefy into a poison that will kill a person when spread on a board on which he steps, or on his stirrups. Although snails rather than silkworms are involved in this particular recipe, it contains several parallels to "Algiers": the snails are sealed in a glass bottle, fed boiled egg yolks, and kept in the bottle buried in horse manure until the substance turns into a powerfully toxic substance. Like "Algiers," the preparation of this poison also happens around midsummer, "in the month of June and July."[60]

These recipes for closing silkworms and snails in flasks to transform them into a powerful material appear related to those in a twelfth-century metalworking account by the pseudonymous Theophilus in which basilisk powder is the key ingredient for making "Spanish Gold."[61] Here, basilisks are raised in sealed brass vessels. They develop from chicks hatched from eggs laid by cocks, growing serpent tails in the process, before being finally burned into ash from which "red gold" can be produced.[62] As Pamela Smith has treated elsewhere, there are similar accounts in another metalworking text, the *Rechter Gebrauch der Alchimei* (1531), which outlines a process of sealing lizards in a vessel to create gold: "In a pot wide at the bottom and narrow at the top, with a cover that has air holes in it, place nine lizards in the milk, put the cover on, and bury in damp earth. Make sure the lizards have air so that they do not die. Let it stand. . . . The lizards will have eaten the brass from hunger, and their strong poison will have compelled the brass to transform itself to gold."[63]

There are many similarities between the recipes in Ms. Fr. 640 for the work of Algiers, the "horrible poison," and the processes of creating gold and gold-producing agents in the texts of Theophilus and the anonymous book of metal and alchemical recipes, the *Rechter Gebrauch*: enclosing low creatures in a flask, feeding them foodstuffs not normally eaten by them (eggs for silkworms and snails; brass, dirt, and milk for lizards), and transforming the animals into a highly valuable or inordinately poisonous substance through a process of putrefaction and regeneration. On folio 98r, the author-practitioner also notes the relationship between metals, lizards, and gold: "Leadsmiths say that making a lizard die in the melted tin makes the tinning become very golden," adding more prosaically, "Or else putting in sal ammoniac."

These recipes for producing agents of transformation from various types of low creatures may be connected with their physical properties: snakes and lizards shed their skins; lizards can regenerate their tails when detached; lizards, snakes, and frogs emerge from the ground fully grown after freezing winters; and insects, salamanders, frogs, toads, and other such creatures were regarded as generating spontaneously from putrefying materials.[64] These associations with generation and regeneration make sense of practices such as dried frogs being used in amulets for childbirth.[65] We have already seen that the author-practitioner and his contemporaries made the silkworm a metaphor for spiritual resurrection, but no earlier metalworking texts known to the authors of this chapter make use of the silkworm as a powerful alchemical agent to produce gold.

Silkworms and *Gu*

Instead, the author-practitioner's use of silkworms has an intriguing similarity to a body of knowledge from China that dates back to 1400 BCE, the concept of *Gu*.[66] *Gu* refers to procedures associated with certain—usually minority—cultures in the southwestern regions of China. These procedures had the aim of producing wealth or poison. They worked by sealing creatures such as poisonous snakes and insects in a vessel, where they consume each other until there is but one survivor, which is called the *ku*.[67] The traditional Chinese character for *Gu* is succinctly captured in the strokes of its written form. *Gu*, in fact, is formed from two radicals, *Chong* and *Min:*

蟲 (*chong*) + 皿 (*min*) = 蠱 (gu)

Chong is both an all-inclusive term in premodern Chinese culture designating a broad range of legendary creatures and mundane animals and narrowly defined as a category that refers to small residual creatures including worms, insects, snakes, and small reptiles.[68] *Min* literally means vessel or receptacle. The *Gu* character is formed by the character for *Chong* in its upper register while *Min* constitutes the lower register. Thus, the character itself embodies the placing of worm bugs in a vessel for the process of *Gu*.[69]

Around the twelfth century, *Gu* began to be associated with another term: *Jincan* (金蠶), which translates as golden (金) silkworm (蠶). The concept of the golden silkworm predates its associations with *Gu* as "'the silkworm that generates silk thread, literally spitting out money.' *Jincan* 金蠶 (golden silkworm) from Xu Xuan's 徐鉉 (916–991) *Jishen lu* 稽神錄 (Records of the investigation of spirits) tells of a man finding (and discarding) a gold-colored silkworm larva inside of a perfectly round stone."[70] Twelfth-century Chinese scholars identified several morphological varieties of *Gu* based on the surviving creature from the vessel, of which the golden silkworm (*jïncan*) was the most powerful.[71] The twelfth-century author Cai Tao 蔡绦, a scholar and official in the Northern Song dynasty, was the first to record the practice, noting that the golden silkworm originated from Sichuan and spread to provinces including Hunan, Hubei, Fujian, and Canton.[72] From that point on, a variety of texts began to identify and classify the *Jincan*, including medical texts that provided antidotes for its harmful properties. One text from the Southern Song dynasty, most likely from the first half of the thirteenth century, contains a concise and informative account of the practice: "Southern people rear golden silkworms. The silkworm has a golden color and feeds on silk textiles from Sichuan. Its excrement is used to poison people by adding it into food. Whoever eats this will die. The silkworm can bring fortune to its owner, and make him rich. It cannot be killed by water and blade. [If one wants to get rid of it, one must] place the silkworm into a gold or silver vessel and leave it on the road. Whoever picks it up will become its new owner. This is called 'marrying out the golden silkworm.'"[73] Various sources discuss the duality inherent in the *Gu* practice: it is capable of bringing both fortune and failure, of being both a poison and a benefit, similar to processes in European texts.

Another similarity of the "Algiers" process and *Gu* practice can be found in the timing of the processes: recall that in "Algiers," the silkworm eggs are to be fed egg yolks in a flask that was buried under manure around the Feast of Saint John, the date associated with the summer Solstice in the Gregorian calendar. The traditional Chinese *Gu* preparations occurred on the "fifth day of the fifth month."[74] This was during the Duanwu 端午 Festival, associated with the summer solstice in the lunar calendar.[75]

In "Knowledge in Motion," Pamela Smith explored the material complex of ideas and practices around vermilion making, which involved the color red, blood, gold, and lizards. This complex, integrated into Arabic and Latinate alchemical discourse, is especially noteworthy in the use of lizards in processes that aimed to transform and ennoble substances. She follows this complex from China through the Islamicate world to Europe.[76] The intriguing recipe in Ms. Fr. 640, "The Work done in Algiers," seems to indicate yet another trace of this complex and to suggest the very long span movement of a material complex—made up of materials, practices, and ideas and centered on the transformation of substances—from China, through the Islamicate world of North Africa and Spain, to Toulouse in southern France, where it crystallized in writing in the recipe compilation of a curious craftsman.

Conclusion

Both processes examined in this essay contain echoes of the various material, ritual, and textual resources with which human beings in different places and times explored the relationship of body to spirit and investigated the great mysteries of generation, transformation, and regeneration. In the case of the Native American pipe, this new object was incorporated into a field of materials and practices involved in spiritual and physical prophylaxis and healing.

In the case of the silkworm, a thing variously valued and meaningful in different spheres of human interaction—agrarian, economic, metaphorical, and religious—emerged as part of a knowledge system that sought to explore and explain material transformation.

Chapter 9

Itineraries of Images

Agents of Integration in the Buddhist Cosmopolis

Tansen Sen

Tracking the movements of objects over long distances, and sometimes over the *longue durée*, reveals the inadequacies of limiting studies of "connectivities" or "border crossings" to a specific region or subregion. The broader geographical space within which such connections or interactions took place must be taken into account in order to comprehend the complex and accretive ways in which objects and ideas moved. In order to follow such itineraries, we should think beyond models of linear diffusion and examine the ways in which objects continued to move, even when the people who first carried them ceased to travel. These itineraries were recurrent movements in which several different groups of people participated in sometimes subtly and sometimes drastically transforming itinerant objects. Objects and ideas on some occasions filtered back transformed, often unrecognizable, to their nodes of origin. Moreover, circulating objects generated knowledge across the routes they traversed, triggered the movement of people, and shaped the material cultures of societies at multiple locations.

The spread of Buddhism initiated a flow of texts, ritual objects, images, artistic traditions, and languages over long distances. It also facilitated the congregation of people from different regions and varied traditions at urban centers, oasis states, and port cities. These movements and congregations led to the creation of new ideas, modifications of artistic traditions, and a broader dispersal of concepts and designs. State patronage, the proselytizing zeal of large monastic institutions, pilgrimage activities, and the spiritual needs of itinerant traders were the underlying features that nurtured the successful transmissions of Buddhism across most of the Asian continent. A noteworthy aspect of these Buddhist transmissions was the relative ease with which doctrines and other aspects of the religion underwent transformations so as to fit the preconceptions and needs of new groups of people. As a result Buddhist texts, artistic traditions, and the languages used to proselytize the doctrine differed from region to region. Many of these regions also created their own doctrinal emphasis and their own pilgrimage and transmission centers. Thus while Islam promoted universalism through a single sacred text and pilgrimage site, Buddhism offered multiple options with regard to both. The spread of Buddhism was not a simple unidirectional diffusion of ideas, beliefs, and objects. Rather, it was a multidirectional process within a cosmopolitan space and faith, which continuously evolved and expanded as it gathered in multiple groups of people, polities, languages, ritual objects, and interpretations.[1]

Unlike Sheldon Pollock's Sanskrit cosmopolis, which encompassed South and Southeast Asia,[2] the world of the Buddhists—which included monks, lay believers, translators, and artists, as well as "opportunist" rulers and traders—was significantly larger, more diverse, and involved more discourses, dialogues, and debates among people from varied regions. This diversity made Buddhism one of the most multifaceted religions, with numerous different ways to practice devotion and express faith, with a wide array of culture or subculture specific images, iconographies, and cosmologies and generative of enormous corpuses of texts in several different languages. The focus of this essay is on ways in which the movement of Buddhist images helped integrate widely separated regions of Asia and created a sense of common identity among the Buddhist followers living in different cultural zones despite significant transformations and reinterpretations in the material objects themselves.[3] It is argued that the itineraries of images were crucial in creating and sustaining linkages within the Buddhist cosmopolis for over a millennium and half.[4] While the connections between South Asia and China are at the center of this focus, discussion of other parts of Asia is included to demonstrate the intra-Asian linkages associated with the movement of Buddhist images.[5]

Buddhist Images on the Move

It is doubtful, as was contended by European scholars in the nineteenth century, that the earliest anthropomorphic images of the Buddha emerged only under Hellenistic influences.[6] It is clear, however, that Greek artistic traditions and beliefs had discernible influence on the style and motifs found in early Buddhist art, especially those that emerged in the Gandhāra region (present-day Afghanistan/Pakistan). The location and date of the earliest

Buddha image remains under debate, with Robert DeCaroli noting that the date—which ranges from the first–second century BCE to the first century CE—is "far less controversial than their location. In part, this is because location is closely tied to the issues of identity, cultural diffusion, and political control that have run as an undercurrent through this debate."[7] DeCaroli points out that the early images of the Buddha were of a "transregional" style common to artists in both the Mathurā (in present-day Uttar Pradesh state in northern India) and Gandhāra regions.

Art forms representing Buddhist teachings and alluding to the life of the Buddha also began appearing between the first–second century BCE and the first century CE. These "aniconic" depictions, which DeCaroli defines as "a loose custom or preference by which the representation of certain categories of figures were to be avoided and in which anthropomorphic figural art held a potency beyond that of mere symbols," emerged at some of the key Buddhist sites, including in Sāñcī (in the present-day Madhya Pradesh state of India).[8] Images of donor figures depicted on inscriptions at several monastic institutions in the Gangetic region were the third type of motif that developed during this period. Gregory Schopen has argued that monks and nuns were instrumental in the creation of the so-called cult of images that popularized these donor images and subsequently propagated the figural representations of the Buddha within a monastic context.[9]

The Kuṣāṇa reign in Central Asia and northern India during the first two centuries CE contributed to the flourishing of several art forms, including figural representations of the rulers in several media such as coins and stones. DeCaroli argues that the Kuṣāṇas "introduced the new social customs that allowed for this type of artwork to be used in a wider range of contexts."[10] State patronage not only augmented the production of the portraiture and figural images of the rulers but also influenced the artistic traditions of Brahmanism, Jainism, and Buddhism. A similar trend took place in the Deccan region of India, where the Sātavāhana rulers (first century BCE to second century CE), almost at the same time as the Kuṣāṇas, patronized the production of art forms leading to a wider impact on the local artistic traditions.[11] During the reigns of the Kuṣāṇas and the Sātavāhanas the first images associated with Buddhism began to spread beyond South Asia through overland and maritime routes.

Buddhist images—including small portable statues of the Buddha, Buddhist divinities, and replicas of renowned temples (especially the Mahābodhi

Temple in Bodhgayā); talismans and votive tablets; and paintings that narrated the previous lives of the Buddha—often played a critical role in the early spread of Buddhist doctrines, as they seem to have moved over long distances even before the transmission of the doctrines and the exposition of relevant philosophical ideas.[12] Images were often the main media through which most lay believers encountered, understood, and practiced Buddhism. These things were carried by merchants, missionaries, pilgrims, and other itinerant travelers. During the first and second centuries CE, members of the Sogdian and Parthian merchant communities, originating in regions of present-day Tajikistan, Afghanistan, and Iran, were the foremost transmitters of Buddhist images to China. The Kuṣāṇa Empire greatly facilitated the movement of these merchant groups between South Asia and Han China, and with them Buddhist images and doctrines.

The role of merchant groups is hardly acknowledged in Chinese texts concerning the initial transmission of Buddhism. Rather, the most famous Chinese legend associated with the introduction of Buddhism into the region revolves around the Han Emperor Ming's 漢明帝 (r. 58–75) dream of a golden man, with a halo around his head, who flew into the imperial palace. On his inquiring about the possible meaning, his officials explained that the dream was probably about the Buddha, who was venerated in the "Western Regions." Emperor Ming soon dispatched envoys to bring to Han China people who could explain the teachings of this "sage of the West." The Han envoys returned in the year 67 CE with the first two "Indian" monks named Jiaye Modeng 迦葉摩騰 (Kāśyapa Mātaṇga?) and Zhu Falan 竺法蘭 (Dharmaratna?). These two monks are credited for initiating the translation of Buddhist texts at the imperially funded "first" Buddhist monastery in East Asia known as the Baima si 白馬寺 (White Horse Monastery). This story of an official transmission of Buddhism was composed two to three centuries after Buddhist doctrines and images first entered Han China, with the objective of legitimizing the transmission of the foreign religion by inventing a role for the Chinese emperor in the process.[13]

Archaeological evidence suggests a more random transmission of Buddhism and one that seems to have been more integrated with local beliefs from an early stage. The images found engraved on Mount Kongwang 孔望山 in the coastal area of Jiangsu province are examples of this phenomenon. Dating from the second and third centuries, these images include various representations of the Buddha, in standing, seated and *parinirvāṇa* postures (see

FIGURE 9.1. The Buddha and donor figures, Mount Kongwang (second–third century CE). Reprinted from Sen, *Buddhism, Diplomacy, and Trade.*

figure 9.1). They are interspersed with traditional Chinese motifs and images of human figures wearing costumes that suggest their Central Asian origins.[14] Another early image of the Buddha, in seated posture, is found in a cave tomb at Ma Hao 麻浩 in Sichuan province and also dates from the late second or early third century.[15] The use of Buddhist images in Chinese tombs suggests the incorporation of Buddhist imagery in the local mortuary tradition. It is not clear, however, if the patron of the tomb was aware of the identity of the figure portrayed in the image or the teachings associated with Buddhism. Wu Hung has argued that these early images should not be termed "Buddhist" given the context in which they were used. Wu Hung writes that, although at the initial stages several elements from Indian Buddhist art filtered into China, none of them had

> either an inherently Buddhist content, or a Buddhist religious function. Rather, as novel forms, they served to enrich the representations of Chinese indigenous

> cults and traditional ideas. It would be misleading to identify these works as early Chinese Buddhist art and take them as the true embodiment of the original Buddhist meaning. In fact, these works cannot even be seen as reflecting a fusion of Buddhism and the Chinese tradition. They only reflect a random borrowing of Buddhist elements by Han popular art. In my opinion, this was the dominant situation when Buddhist art was first introduced into China. In this tenuous way, nevertheless, Buddhist art gradually gained a foothold in a vast and unfamiliar land.[16]

Despite such ambiguity in the use of images in the popular realm, it is evident that by the third century Buddhist objects were being employed for their supernatural and miraculous powers, aspects that would in the next few centuries become more widespread with the translation of Buddhist texts and the visits of Chinese monks to South Asia. These early images serve as examples of the ways in which objects in motion function as agents in spreading ideas—no matter how transformed or incomplete the concepts may seem to have become—and linking far-flung regions of the world. These dynamics continued in the making and venerating of Buddhist images advocated in some of the translated texts and by the members of the monastic institutions that subsequently reached China. Images, too, especially paintings narrating the previous lives of the Buddha and representations of Buddhist "hell" and "paradise" were employed to spread the basic teachings of Buddhism, particularly those related to the ideas of karma and retribution, which were initially unfamiliar to Chinese worshipers. Some Buddhist sculptures also acquired supernatural qualities because of their perceived ability to move and have humanlike feelings, forming important narrative tropes in Chinese Buddhist miracle tales.[17]

The Buddhist images reaching China during the first millennium CE originated from different parts of South Asia, each with unique stylistic features and motifs. Gandhāran, Gupta, Pāla, and the artistic renditions from Nāgārjunikoṇḍa in south India were some of the major traditions that entered China. Styles that developed in Central Asia also had significant impact on the making of Buddhist images in China. These images spread through different routes and means: South Asian artisans traveled to China expressly to make Buddhist statues and paintings. Indian painters named Shijiafotuo 釋迦佛陀 (Śākyabuddha?), Juecheng 覺稱 (Buddhakīrti?), and Tongzhi 童智 (Kumārabodhi) during the Northern Wei period (386–534) are known to have made paintings that adorned several local Buddhist monasteries.[18] Similarly, a Sri Lankan monk artist in China during the Tang period made clay images of Buddhist divinities installed at monasteries in Luoyang.[19] On other occasions, Chinese monks and diplomats visiting South Asia returned with Buddhist images and other artifacts. The diplomatic mission sent to South Asia in 645 and led by the Tang diplomat Wang Xuance 王玄策 (fl. seventh century) included a Chinese artisan named Song Fazhi 宋法智 (fl. seventh century) who drew several images of Buddhist figures that were subsequently replicated in China.[20] Buddhist drawings, rubbings, and replicas were also brought to China from India by Xuanzang 玄奘 (600?–664) and displayed prominently at the Tang capital.[21]

John Kieschnick has pointed out the importance of Buddhist images in the practice of Buddhism in China. "Images never ceased to be a central feature of Chinese Buddhist devotion." Kieschnick specifically details the impact of Buddhist images on Chinese ideas of sacred power as well as on material culture and social relationships, and at the same time, they served to instigate debate about the legitimacy of foreign icons. Through the circulation of these images, "Buddhist symbolism," as Kieschnick notes, "became increasingly widespread" in China.[22] This included the swastika, the dharma wheel, and the lotus. The Chinese also invented their own images and modified those entering China from South Asia and elsewhere. This was often done in the process of local practitioners' adapting of new images, as well as providing prospective converts familiar contexts to engage new practices. This combination of foreign and local helped create a dual identity for the Buddhists outside South Asia. While the foreign images helped the followers associate with the broader Buddhist world, the local innovations created a notion of the sacredness of their own land. This duality was crucial in creating and integrating the Buddhist cosmopolis.[23]

Itinerant Images and the Diversity of the Buddhist Cosmopolis

Conceiving of itineraries as staged, circulatory, converging, and disjunctive helps us better understand the movement of Buddhist images across the Asian continent. These facets of itineraries have been discussed in detail in the first two chapters of this volume. Indeed, the spread of Buddhist images from South Asia to China, their transformation en route to or in China, and the subsequent transmission to other regions of East Asia and elsewhere reveal multiple bumps and blocks, disjunctures and differences—as well as the "effort" of the local as discussed by Arjun Appadurai.[24] Such movements and itineraries created unique linkages and modes of circulation among different societies, their transformations and local reinterpretations suggest the convergence of distinct ideas and values, and the further spread or the reverse transmission of these transformed images denotes the circulatory itineraries of images.[25] A wide range of imagination, knowledge, and perceptions resulted from this process: As people living in China and Japan encountered Buddhism and new customs and practices through these images, they added fresh content and motifs, creating distinct forms and "forms of circulations."[26]

New types of images emerged based on the understanding and interpretations of texts that were rendered into local languages. The impact of the *Lotus Sutra* on the production of art forms in East Asia, especially Japan, is perhaps the most notable. The transmission of the text to Japan, in its Chinese translation, led not only to the creation of images related to the teachings contained in the work but also to the production of art forms that portrayed the lifestyle of the aristocratic class who were the main sponsors of the artists illustrating the text.[27] Images also ceased to travel when cultural norms of a region prevented the spread of certain iconographic styles. This is evident from the abrupt cessation of iconographies of bare-chested female figures drawn in Indic style found at the Kizil caves in the Kucha region of present-day Xinjiang province. Chinese aesthetics with regard to the depiction of female bodies in art form seems to have blocked the eastward spread of such representations.[28]

Other than a few exceptional periods such as during the Northern Qi (550–577) when attempts were made to closely copy Indic styles, Buddhist images produced in China mostly followed local aesthetics and traditions, serving both religious and secular purposes.[29] Although, for example, paintings on the walls of caves such as those in Dunhuang incorporated local motifs and styles to propagate the doctrine or express the faith of the followers, Buddhist statues allegedly created in the likeness of Empress Wu Zetian 武則天 (r. 690–705), who usurped the Tang throne in the seventh century, were built to promote political legitimization.[30] A result of efforts to produce localized forms of Buddhist art meant that, by the seventh century, China emerged as a leading producer and exporter of Buddhist images. The art produced during this period is, as Dorothy Wong notes, sometimes referred to as the "Tang International Buddhist Art Style," which was "a synthesis in both style and

FIGURE 9.2. Star maṇḍala from Japan (late seventeenth century). Gift of the Friends of the Philadelphia Museum of Art (1978-45-2). Philadelphia Museum of Art.

content of fresh influences from India and the existing tradition of Buddhist art in China that had developed since the fifth century."[31]

Wong points out that this "Tang International Buddhist Art Style" was preceded by the "first" international Buddhist art style beginning "with the Gupta (fourth–sixth centuries) style achieved in the fifth century in centers such as Mathurā and Sārnānth, which then became the classic idiom copied throughout Asia." Wong also suggests the emergence of a "third" international Buddhist style, the "Pāla International Style," derived from "the Gupta style prevailed from Nepal to Tibet, Myanmar, Thailand, and Indonesia from the eighth through the 13th century and beyond."[32] Although these three "international" styles of Buddhist art appeared during different periods, they often mixed, intermingled, moved, and converged in different directions and sites.

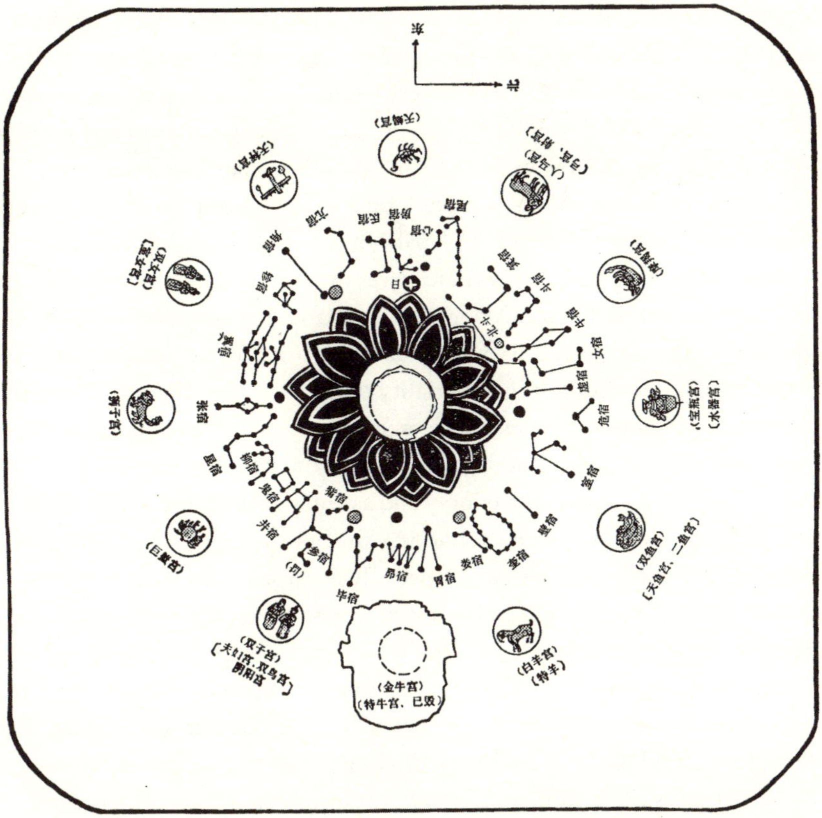

FIGURE 9.3. Astronomical diagram from the Xuanhua Tombs (twelfth century). Reprinted from Sen, "Astronomical Tomb Paintings." After Hebei sheng wenwu guanlichu (Hebei Cultural Relics Administrative Department) and Hebei sheng bowuguan (Hebei Provincial Museum), "Hebei Xuanhua Liao bihuamu fajue jianbao" (A brief report on the excavation of a Liao painted tomb from Xuanhua, Hebei), *Wenwu* 8 (1975): 31–39, 44.

Many of the images associated with Tantric teachings popular in East Asia between the seventh and tenth centuries, for example, were influenced by the Pāla style. However, the clergy in Tang China and Japan modified and created their own rendition of these images, which mixed local ideas, beliefs, and motifs. The mandalic representation of the cosmos, wrathful Tantric deities, and even the mantras (*dharanis*) written in Indic script that emerged from such mixtures were powerful images employed in rituals, ceremonies, and mortuary practices in several regions of East and Southeast Asia. In fact, as Hiram Woodward has demonstrated, there were similarities and mutual influences between Tantric images produced in Tang-Song China, Tibet, Cambodia, and Java.[33] In the case of Java, Woodward suggests the possibility of a local monk named Bianhong 辯弘 in Chinese sources contributing to the design of the upper terraces of Borobudur based on the knowledge he acquired studying under the Tantric masters in Tang China during the late eighth century.[34]

Tantric ideas emitting from Tang China were also responsible for the creation of new varieties of images in Japan. In the seventh–eighth centuries, the Japanese developed unique ways of representing the Buddhist cosmos through paintings known as the Star Maṇḍalas (in Japanese, *Hoshi mandara* 星曼陀羅) (see figure 9.2). Employed in Tantric Buddhist rituals, maṇḍalas are viewed as connecting devotees to the materiality of the universe. The Star Maṇḍalas, with representations of celestial objects frequently in anthropomorphic forms, originated from Tantric Buddhist texts containing synthesized Hellenistic and Brahmanical ideas on astronomy and astrology. These texts were rendered into Chinese by Indian monks working in Tang China and rapidly gained acceptance throughout East Asia especially because of the state patronage of Tantric Buddhism. The diagrams are manifestations of the convergence of different ideas and traditions, some of which spread with Buddhism and others that arrived or developed locally in China and Japan independently. These Japanese diagrams entered the streams of Tantric circulations in East Asia, were transmitted back to the clergy in China, and had significant impact on the lay believers of Buddhism. During the eleventh and twelfth centuries, for instance, images derived from these diagrams found their way into the tombs belonging to Han Chinese residents of the Khitan (Liao) Empire.[35] These tomb paintings (see figure 9.3) are also evidence of the enduring feature of East Asian Buddhist practices, starting from the Han tombs at Ma Hao, that mixed the teachings of Buddhism with Sinitic mortuary traditions.

There were other images produced in China, and not associated with Tantric Buddhism, that traveled across several regions of the Buddhist world. These included the representations of the Chan/Zen monk Bodhidharma, the bodhisattvas Dizang 地藏 and Guanyin 觀音, and the so-called laughing Buddha named Budai (in Japanese, Hotei) 布袋. The creation of these images and their associated cults indicate the local character of Buddhism and the importance of China as a center for disseminating Buddhist doctrines and iconographies. The image of the famous Buddhist divinity Mañjuśrī (Wenshushili 文殊師利) riding a lion may have originated in China and spread to all parts of the Buddhist world (see figure 9.4), indeed, even monks from

FIGURE 9.4. The Buddhist divinity Mañjuśrī. (Dunhuang, ninth century). © Trustees of the British Museum.

South Asia believed that this "bodhisattva of wisdom" lived on the Chinese mountain Wutai and traveled there expressly to pay obeisance.[36] Paintings of Mañjuśrī and maps of Mount Wutai 五臺山 were produced for local consumption and for export to places such as Tibet, Japan, Nepal, and India. The widespread acceptance of Mount Wutai as the abode of Mañjuśrī validated the integration of China into the Buddhist realm and legitimized its status as a center for the transmission of Buddhist ideas and images.[37]

Similar to the story of the Han Emperor Ming's dream, the official introduction of Buddhism to Japan is also associated with an account that involves the local ruler and an image. In 538, a bronze image of the Buddha and other artifacts were gifted to reigning Japanese Emperor Kinmei 欽明天皇 (r. 539–571) by diplomats sent by the ruler of the Korean polity called Paekche. The bronze image, which most likely originated in either China or Korea, soon became a significant element of the legend as it was said to have miraculously reappeared after being discarded in a river.[38] The story gave the presence of Buddhism in Japan a founding genealogy—and one associated with the ruler—while it incorporated yet another region into the Buddhist realm. With Buddhism established in Japan, an effervescing and multidirectional interaction of Buddhist objects, images, and ideas ensued in East Asia. Ideas and objects locally produced in the region not only circulated between China, Korea, and Japan but also influenced the seminomadic societies and polities in Central and northwestern Asia, as well as parts of Southeast Asia. The emergence of an East Asian sphere did not terminate Buddhist circulation to and from South Asia. Monks from Korea visited India to procure religious texts, images, and relics, and Buddhist missionaries from South Asian monastic institutions such as Nālandā continued to travel to China and Korea as late as the fourteenth century.[39]

Unlike their Chinese and Korean counterparts, Japanese monks did not visit Buddhist sites in South Asia until the sixteenth century. However, their longings for the Buddhist holy land can be discerned from a series of images they created to visually represent the geography, peoples, and even the Chinese monks visiting the holy land in South Asia. By the fourteenth century, if not earlier, the Japanese were making maps of India based on the seventh-century travel records of the Chinese monk Xuanzang (see figure 1.8a–b in chapter 1, this volume). They also created visual representations of famous South Asian monks, including those who were active in China. The most intriguing of these are the images of the Tang monk's meetings with

FIGURE 9.5. Xuanzang at Nālandā (fourteenth century). After Rambelli, "The Idea of India," and reprinted from Sen, *Buddhism across Asia*.

the South Asian ruler Harṣa and his teachers at Nālandā (figure 9.5).[40] All of these representations were imagined, often mediated through Chinese texts, and expressed the longing for and belonging to a Buddhist world that were felt by Japanese worshipers. Like the Star Maṇḍalas, these are examples of images on the move, albeit not physically but, rather, as part of the imagination of distant, yet interconnected, locales in the Buddhist cosmopolis.

Almost at the same time as these images were circulating in East Asia, Mongol Iran was intimately in contact with various parts of South Asia and China. The Ilkhanate official and prolific author Rashīd al-Dīn in *Jami al-Tavarikh* (World history) composed in the fourteenth century included a section on the biography of the Buddha. He based this "Life and Teachings of the Buddha" on information provided by a Kashmiri monk named Kamalaśrī and two Chinese informants. One of the existing manuscripts of this work, dated 1314, contains three images from the life of the Buddha: the scene of temptation (see figure 9.7d below), where the evil Mara attempts to prevent

FIGURE 9.6. The birth of the Buddha, illustrated version from *Majma al-tavarikh* (fifteenth century). The Walters Art Museum, Baltimore, Maryland. Public domain.

the Buddha from attaining enlightenment; the Jetavana monastery, where the Buddha preached; and the holy site Kuśīnagara, where the Buddha attained nirvana.[41] Two centuries later, paintings of the Buddha appeared in Hafiz-i Abru's *Majma al-tavarikh* (Chronicles). Much of the information on the Buddha's life in this later text was copied from Rashīd al-Dīn's work, but the illustrations, the birth of the Buddha (figure 9.6), Buddha's meeting with a Brahmin, and his nirvana are different in content and style when compared to the illustrated manuscript of *Jami al-Tavarikh.* They are more similar to the Japanese representations of an imagined India, yet another example of the accretive and looping itineraries of objects, images, and ideas in the multicultural, interconnected, and imagined cosmopolis of Buddhism.

Conclusion

The circulation of Buddhist images continued even after the colonization of Asia by European powers. In fact, in the late eighteenth and nineteenth centuries, Buddhist images from China were introduced into India by Chinese immigrants. These immigrants, numbering around sixteen hundred in 1901, had settled primarily in Calcutta (now Kolkata), the capital of British India, and they built several Chinese temples and shrines, including some dedicated to the Guanyin. They also brought with them images of Ruan Ziyu 阮子鬱 (1079–1102) and Liang Cineng 梁慈能 (1098–1116), two localized Chinese Buddhist divinities whose temples originated in the Sihui region of Guangdong province. The images of these two buddhas had spread to present-day Malaysia, then from there they reached Calcutta. Members of the Chinese immigrant community also built temples in several other holy Buddhist sites in India and had statues of the Buddha and other divinities, such as the Guanyin, imported from China or Myanmar. These Chinese Buddhist temples and shrines indicate how durable and persistent the circulations of Buddhist images were.[42]

In the colonial period Buddhist images and artifacts also traveled not only as part of religious practice but also as part of the attempt to collect, catalog, and curate antiquities from the colonized world. Like other dimensions of the larger colonial enterprise intending to demonstrate imperial power and European cultural superiority over the ancient (and then weakened) Asian civilizations, these activities included the excavation of historical sites, the removal of objects from their original sites—purportedly to preserve them as well as for their display at museums and exhibitions around the world.[43] The British, for example, removed a significant number of Buddhist statutes, inscriptions, and

FIGURE 9.7A. The assault of Mara (with aniconic representation of the Buddha). Amaravati (second century). Wikimedia, Creative Commons, public domain. https://commons.wikimedia.org/wiki/File:MaraAssault.jpg.

FIGURE 9.7B. Dunhuang rendition of the assault of Mara (tenth century). MG 17655, Guimet Museum. Wikimedia, Creative Commons, public domain. https://en.wikipedia.org/wiki/Mara_(demon)#/media/File:Dunhuang_Mara_Budda.jpg.

FIGURE 9.7C. Modern rendition of the assault of Mara story in Burmese style in a Buddhist temple in Penang, Malaysia. Wikimedia, Creative Commons. Public domain. https://commons.wikimedia.org/wiki/File:Mara_demon_nat_and_Buddha.jpg.

paintings from Central and Southeast Asia and moved them to their empire in India or transported them to galleries in England. The French, Germans, and all other imperial powers, including the Japanese, employed similar methods with comparable objectives. A large number of objects were also smuggled from Asian sites and sold at auction houses in Europe and the United States. Some of these colonial collections and stolen objects are now being returned to the countries of origin, sustaining, in certain ways, the circulatory journeys of Buddhist images and artifacts.[44]

So Buddhist images followed complicated itineraries, moving initially from South Asia to China, where they underwent local transformations and reentered new paths of circulation. Iranians, Chinese, Southeast Asians, and several other groups of people contributed their own local aesthetics and meanings to Buddhist iconography during this process of circulation. The four final images (figures 9.7a–d) of the evil Mara's encounters with the Buddha prior to the Buddha's enlightenment, each coming from a different cultural sphere, illustrate the resultant local character and diversity of artistic representations in the Buddhist world. They also show the connectivities among the four regions through the employment of similar motifs. It was argued that in order to understand and trace the itineraries of Buddhist images, a broad canvas of geographical space and temporal framework is needed. The use of such spatial and temporal analysis reveals the contribution of itinerant images in integrating the Buddhist cosmopolis that encompassed most of Asia.

FIGURE 9.7D. Persian rendition of the assault of Mara (fourteenth century). Wikimedia, Creative Commons. Public domain. https://commons.wikimedia.org/wiki/File:Jami_al-Tawarikh_001.jpg.

Anthropologists have demonstrated that the demand, desire, control, and value of goods may lead to commodities taking on "life cycles" of their own.[45] Others have emphasized the "agency of objects" in creating new contexts and courses of action.[46] The itineraries, or "life cycles" of Buddhist images included transmission, modification, transformation, further movement, and sometimes return to original sites of creation in new forms. The "life cycles" also pertained to the changes in the medium of expression, from texts to paintings or vice versa, for instance, as well as the contexts within which the images were placed—cave temples, monastic institutions, homes, royal palaces, or museums. As Buddhist images moved from one place to another and took on new shapes, meanings, and contexts, they triggered imaginations, fostered knowledge, and created fresh linkages within and across several societies. Through such processes, far-flung regions, where Buddhism was interpreted and practiced in diverse ways, became integrated, thus linking the Buddhist cosmopolis not only through faith in the teachings of the Buddha and the pilgrimages and the missionary work of the Buddhist monks but also through the itineraries and life cycles of images that sustained these linkages.

Chapter 10

Itineraries of Inkstones in Early Modern China

Dorothy Ko

What kinds of objects, practices, and knowledge are most amenable to continental transmissions across Eurasia? In this chapter I approach this question from the negative, by examining the circulation of knowledge about the making and use of a class of objects that is ubiquitous in East Asia but virtually unknown elsewhere—the inkstone.

The Inkstone as Material-Emotional Complex and Four Propositions

An inkstone (or inkslab; ink palette), a piece of polished stone about the size and thickness of an outstretched hand, is at once an instrument for grinding ink for writers and painters, a collectible object of art, a token of exchange between friends or sovereign states, and an inscriptional surface on which texts and images are carved and reproduced by way of ink rubbings (figure 10.1). As such, for over a millennium, the inkstone is entangled with the production of elite masculinity, texts, and the culture of *wen* 文 (culture, literature, civility) in China, Korea, and Japan.[1] More than an object or artifact, the inkstone is a conduit of such dynamic consternation of embodied skills and emotional investments that we may call it a "material-emotional complex."[2]

Curiously, this iconic object in East Asia has by and large escaped notice by Europeans travelers, tall-tale tellers, and curio collectors. In early modern Europe, the fascination with the Chinese writing system took the form of a fixation with the shapes and strokes—the alleged "pictorial" quality—of the written characters themselves, and not with the implement behind their making.[3] One reason for this general neglect might have been the incommensurability of writing systems, the very incommensurability that helped ignite

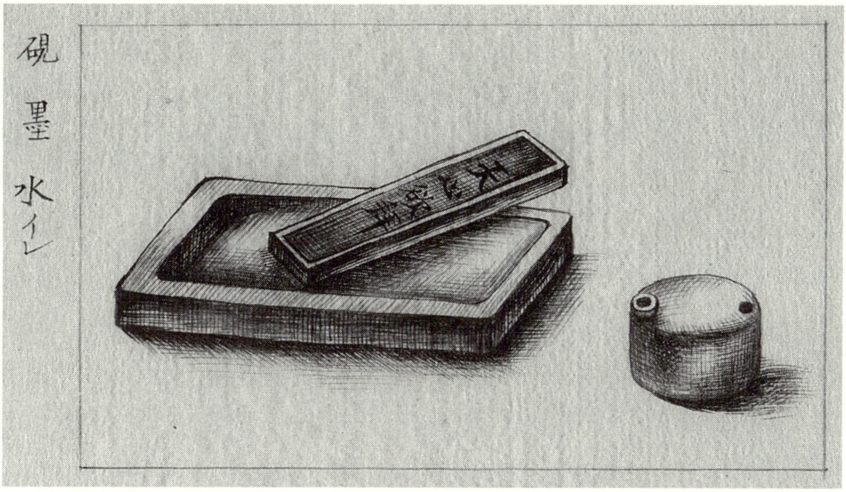

FIGURE 10.1. Drawing of an inkstone, inkstick, and water-dropper (Japan, c. 1878). Ink drawing on paper, 5.5 x 9.2 cm (image), 9.2 x 11.5 cm (sheet). Library of Congress Prints and Photographs Division, Washington, DC.

the fad of ideographia. In contrast, the material and cultural meanings of the inkstone were readily translatable in Korea, Japan, and Vietnam, where the male scholarly elite shared the Chinese writing system (forming what some Japanese scholars have called a Pan-Asian "cultural sphere of Chinese writing").[4] According to the research of Sixiang Wang, inkstones were routine gifts exchanged between envoys from the Ming (1368–1644) and Qing (1644–1911) courts and the Chosŏn (1392–1910) court in Korea.[5] Future research may illuminate their salience in the tributary relations that China formed with Japan and Vietnam.

In Japan the inkstone was an emblem of aristocratic pedigree, treasured as much for its material and design as for its efficacy in producing ink. In a poignant short tale recorded in the twelfth century, "Breaking the Inkstone," a courtier included a beautiful lacquered one decorated with gold and silver in his daughter's dowry, hoping to attract the emperor's attention. When a servant broke it by accident, he set off a series of tragic events. The popular tale, a celebration of loyalty, male bondage, and religious renunciation, circulated in multiple textual versions and pictorial formats in medieval Japan, cementing the enduring association of the inkstone with literary culture and civility.[6] Today, Japan remains the most significant repository of collections of inkstones outside China, which are virtually nonexistent in Europe and North America.

Writing systems aside, what other factors account for the mobility of inkstones in regional transmissions but their "stickiness" (to borrow Harold Cook's term) in transcontinental crossings? A more complete answer—one that may have relevance beyond the case of inkstones or any single class of objects—has to begin with an understanding of the modern/nonmodern difference in the propensity of things, bodies, and space. Before he can convey the "cultures of accommodation" and fluid itineraries of Jewish cross-cultural traders in the twelfth century, for example, Amitav Ghosh has to warn his readers to let go of "the universal, irresistible metaphysic of modern meaning" based on binary thinking and a linear, progressive time line. Traveling between Egypt, India, East Africa, Syria, Morocco, and Spain, these brokers and financiers "bear witness to a pattern of movement so far-reaching that they make the journeys of later medieval travelers, such as Marco Polo and Ibn Battuta, seem unremarkable in comparison." Even more remarkable is their expansiveness. Making the Middle East and even India their home, these traders could be simultaneously Jewish and Muslim before a partitioning of the world of meaning made these positions irreconcilable.[7]

This simultaneity of possibilities is the very precondition for "entanglement," according to the feminist philosopher Karen Barad. She writes, "to be entangled is not simply to be intertwined with another, as in the joining of separate entities, but to lack an independent, self-contained existence." In this world of fluid beings and becomings, "individuals [be they humans or things] do not preexist their interactions; rather, *individuals emerge through and as part of their entangled intra-relating*" or "intraacting" (emphasis added.) She proposes a relational framework of "agential realism" through which agency and knowledge making can both be reimagined, as a matter of intra-action. Another concept relevant to the theme of entangled itineraries of the present volume is "enfolding" which as Barad explains is the "ongoing materialization of the world in its intraactive becoming." Like individuals, space as an observable phenomenon emerges only in this reiterative process of intra-action, as does time.[8]

Inspired by these empirical and theoretical works, I formulate four propositions in considering itineraries of materials: (1) The framework of global-local is imprecise, because the local is not one place or point but many. (2) Scale is important: continental itineraries such as those between Europe and Asia are "big loop" transmissions wrought of a series of "medium" and "small loop" intermediary transmissions.[9] (3) "Itineraries" are not the abstract and

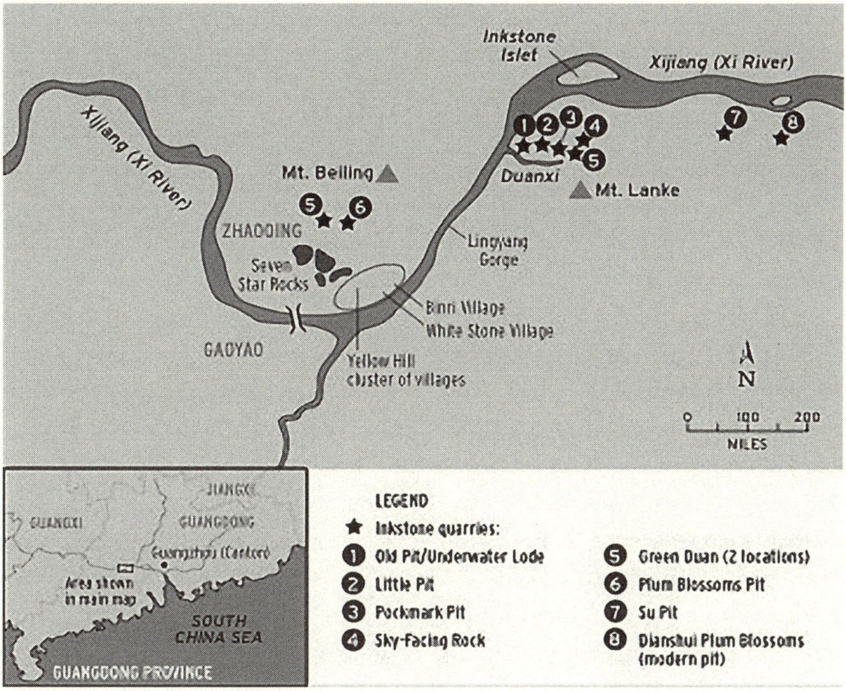

FIGURE 10.2. Map of the Duan area, Zhaoqing, Guangdong, China.

rational lines of Euclidean geometry (which has no body nor texture) but, rather, nonlinear threads of entanglement and traces of bodily gestures and movement.[10] (4) Instead of the inert space of modern Cartesian fiction, the nonmodern space through which material complexes travel is active, enfolded, or wrinkled.

In her contribution to this volume, Bray suggests that nontextual codifications—for example, the graphic-based "technography" of *chaîne opératoire* diagrams—are better suited to the task of mapping complex itineraries over wrinkled spaces than linear texts alone. As a first step toward charting the mobile and sticky trajectories of the inkstone in China, East Asia, and Eurasia, I describe a simplified "operational sequence" (chaîne opératoire) of the quarrying and making of inkstones.

Itineraries of Inkstone in the Qing Empire

The making of inkstones involves specialized knowledge that was distributed widely but unevenly through the Qing empire, clustering around a handful of desirable quarries that yielded stones suitable in hardness, smoothness, and porosity for grinding ink.[11] These are the "wrinkled" crevices in which knowledge accrued. Since the thirteenth century at the latest, Chinese scholars and painters have deemed that the best materials for inkstone making come from only two mountain ranges on the opposite shores of the Xijiang River in Guangdong province on the far southern edge of the empire (figure 10.2). I call this geographical enfolding of land and water the Duan area (centering on the present-day city of Zhaoqing) and I call the stones Duan stones.[12]

Hundreds of households in a cluster of villages collectively called Yellow Hill (Huanggang 黃崗) have for centuries made a living exclusively from mining and carving the stones from Duan quarries nearby. The stoneworkers developed a rich ritual tradition surrounding a patron saint, the Strongman Wuding, and various tutelary deities. From one generation to the next, knowledge was transmitted by way of apprenticeship, aided by ink rubbings of the actual stones made. Reversing the conventional division of labor, many families have entrusted the more respected tasks of carving and polishing trade inkstones to daughters and wives, leaving the men with cutting boulders, fabricating boxes, and handling retail or wholesale businesses. However, the master carvers—those who made one-off inkstones intended for the collector's market—are primarily male. Many of these vertically integrated family enterprises, traceable at least to the sixteenth century, are still operating today.[13]

From the hands of the members of stoneworking families, raw stone blocks and finished products in standardized formats and massive quantities, wrapped in protective sheaths of algae fished from the Xijiang River that flowed in front of the village, found their way to every corner of the empire through new and established trade routes.[14] There was a stationery shop in every town and an inkstone on every student's desk. With an estimated literacy rate as high as 10 percent of China's 450 million people in the nineteenth century, the making of quotidian inkstones was a sizable business.

At the high end of the market, the raw stone blocks landed in the ateliers of some of the empire's best-known artisans and even scholars, who carved them into inkstones that were and are today deemed singular objects of art, their value determined by a balance between the mineral features on the stone, the name of the owner, and the fame of the carver (an eighteenth-century inkstone can fetch between one thousand and fifty thousand US dollars, or more, in auction houses in China today). The choicest raw blocks were often sent as tribute to the imperial warehouse in the capital, where they were carved into inkstones by teams of artisans drafted from Guangdong for the emperor's own use, or as gifts to be bestowed on his favorite scholar officials. Thereby the inkstone—as a material-emotional complex—returned to society and initiated new circuits of sentimental and commercial transactions. The itinerary of inkstones thus crisscrossed the public/private, core/periphery, court/noncourt, and commodity/noncommodity divides of the empire in small and big loops.

The Genesis of Textual Knowledge

Specialized knowledge about using, washing, and repairing inkstones was as mobile as the stones themselves but circulated only within literate society. In accordance with the desirability of the inkstone as a collectible object, a tradition of connoisseurship treatises in print took shape in the tenth century. Ranking stones and identifying their origins by their surface mineral features figured as the main concerns of the writers and readers, lending the entries a dry, catalogue flavor that became characteristic of the genre. Although not the earliest publication, the *Inkstone Chronicle* (*Yanshi* 硯史) by Song calligrapher and painter Mi Fu (1051–1107 CE) is seminal in this regard. The entry on stones cut from the lower part of a certain mountain (Mount Xiayan 下巖) in Duan reads: "The stone is fine-grained, and when struck it gives a clear sound. Its Mynah's eyes are round, surrounded by a ring of azure cloud that often sparkles brightly."[15] Mynah's eye, dots of high concentration of iron in mica that often appear in concentric circles of varying layers and colors—each pattern named after a different bird; each eye further divided into "alive" and "dead"—is one of about thirty prized mineral features associated with Duan stones.[16]

Readers of these treatises learned to discuss the shades of color, texture, and the profusion of surface mineral features (with an elaborate nomenclature to describe each), to infer from these material qualities the origins of the stones, and to keep abreast of a shifting hierarchy of ranking as new quarries were opened and new mineral features named. By the eighteenth century, this body of literature (thirty-one titles in my hard drive) had become highly self-referential, wrought of quotations of quotations, and frequently reanthologized. To gain novel insights and advantage over competitors, collectors occasionally made the arduous trek to the quarries to interview the stoneworkers and to acquire stones. There, a curious form of enfolding occurred, whereby the knowledge of the scholar and the experience of the stoneworker became intertwined.

From the times of Mi Fu, specialized textual knowledge about the geological nature of Duan stones and their material qualities is made by people occupying two competing subject positions: the out-of-province scholars who wielded the writing brush and the Yellow Hill stoneworkers who wielded hammers and chisels, which they fabricated themselves from quenched iron. The latter "spoke" when Mi Fu inaugurated the convention of scholars citing interviews with Duan stoneworkers to boost their authority claims. In this symbiotic manner, the stoneworker's craft knowledge not only informed but was also validated by the writings of the scholars.

Although it is tempting to construe the scholars' quotation of the craftsmen as yet another case of the word or mind colonizing the hand, quite the opposite is true when one reads between the lines. Although mostly illiterate, the stoneworkers' vicarious agency and authority came through in surprising clarity, their voices often still audible through the writings of the scholars, the ventriloquists. For centuries, these craftsmen, as the native informants to visiting scholars, manipulated knowledge about Duan inkstones and the criteria of their assessment. The prevalent valuation schemes that connoisseurs promulgated in their treatises were by and large *theirs*. They affected the vicissitudes of fads for particular hues and mineral features by making new stones available to the buyer. Their power to name new quarries, stones, and mineral features also generated new tastes in the marketplace.

Two developments in the eighteenth century disrupted this cozy codependence between the out-of-province scholar and the stoneworker: the rise of technocratic culture and evidential scholarship in the empire. New kinds of writers—private assistants to bureaucrats—began to make their way to the quarries in conspicuous numbers. Although outside the formal bureaucracy and paid out of the official's private funds, these itinerant technocrats managed hydraulic, mining, and fiscal projects that constituted the core of statecraft concerns. They possessed what Chandra Mukerji calls "logistical power" and were leaders of a "logistical and epistemic culture of working."[17] Their reports from the Duan quarries increasingly stake authority claims on their own personal observation, experience, and privileged access to the quarries in the late eighteenth and early nineteenth centuries.

A second and related development is the currency of philological or evidential scholarship that became dominant among reform-minded scholars in the empire in the second half of the eighteenth century. Eschewing abstract moral speculation, they embraced a "scientific" method encapsulated in the slogan "seeking the truth from concrete facts" (*shishi qiushi* 實事求是).[18] Devoted to precise empirical research, scholars sought to ground their knowledge of history and art in direct encounters with ancient artifacts in situ instead of hewing to received traditions of textual transmission. Many journeyed to the far reaches of the empire in search of ancient steles carved onto stone tablets or cliffs as models of ancient scripts and sources of texts unsullied by printed transmission. One of the major centers of evidential or philological research in the empire was the Ocean of Learning Academy in Guangzhou, about one hundred kilometers to the east of the Duan quarries. Thus two out of four modes of practical knowledge transmission described by Dagmar Schäfer—Agora and Scholarly Arts—converged in the Duan area.[19]

Eyewitnessing and Its Reality Effect

In this milieu, a third subject position in the discourse of inkstone knowledge came into being in the early nineteenth century: the local expert. This new authority figure had to stake his claims by rebuking both the out-of-province scholars and the stoneworkers, thus opening a fissure in the existing knowledge system. Instead of relying on familiarity with the textual tradition or interviews with native informants, the authority claim of the local expert rests on a new mode of knowledge making that resonated with that of the itinerant technocratic assistants and evidential scholars: eyewitnessing. A new way to perceive, describe, classify, and evaluate Duan stones came into being.

Emblematic of this new mode of knowing is a genre of maps of the Duan quarries, carved on the very backs of inkstones or incised on wood blocks and printed as illustrations of books, depicting the underground tunnels as if the viewer had X-ray eyes. This visual format of codifying information in inkstone connoisseurship has no precedent before the early nineteenth century. One graphic example is the map on the back of an undated Duan inkstone, probably from that era (figure 10.3). It offers a somewhat naturalistic depiction of the topographical features of Mount Lanke (unmarked on the map), terminating in a named summit in the upper right corner. Two labels mark the (accurate) locations of the Little Pit (Kengzai) and Pock Mark (Mazi) quarries nesting in the folds of the mountain. What appears as shading on the ink rubbing is the result of low relief carving and skilful rubbing. Also visible in this veristic depiction are the human activities and the sites of cultural memories that animate the place: among them a fisherman sailing down-

FIGURE 10.3. Map of (the Caves) Inside the Old Pit (*Laokeng neitu*). Ink rubbing of the back of a Duan inkstone in the collection of the Tianjin Museum. Chen Yu, *Duanyan minsu kao*, 90.

stream in a boat with a woven canopy typical of the region and two shrines central to seafaring and quarrying rituals.[20]

On the same map the viewer encounters a second, distinctly unnaturalistic visual strategy that anticipates the penetrative gaze of X-ray radiography: to the left of a shrine guarding the entrance to a quarry is a tunnel that to modern eyes resembles a digestive tract with seven lobes, each named. These are the caves, cut more and more deeply into the belly of the earth as stoneworkers sought to follow the veins of the most desirable stone in the empire, the finest of the fine Duan stone from the most celebrated quarry, the Underwater Lode (Shuiyan, literally "water rock").[21] The arching tree in the lower left corner of the inkstone, the tall mask of a hidden boat, and the forward leaning body of a man rowing a scull, all direct the viewer's gaze to the series of connected caves. Its frontal presentation constructs the viewpoint as being directly across the Xijiang River, which in reality corresponds to no human settlement; the vantage point is that of a bounty hunter on the opposite shore bargaining for a ferry, his eyes fixated on the holy grail of inkstone making.

The general recognition of the Underwater Lode as the most desirable among the Duan caves dated to at least the year 1600, when a celebrated mining operation yielded quality stones.[22] The feverish pursuit of stones from this one quarry reached new heights in the nineteenth century. As more stones were retrieved from deeper recesses in the submerged tunnel, more people in the Central Plain became cognizant of the name, generating desire for the stone and knowledge about it among a new group of collectors. The aforementioned map on the back of an inkstone signals a new epistemic priority because of this "tunneling effect": a desire to see through the bedrock of Mount Lanke so as to discern the relative positions of the caves dug during successive operations. The structure of knowledge has shifted from an ordering of the seventy-some quarries in the Duan area to an ordering of the stones gathered from the *same* shaft of the submerged quarry but in different years.

This fixation on—indeed the fetishization of—the inside of the Underwater Lode found graphic expression in a map carved on the back of another Duan inkstone (figure 10.4). The billowing waves, the semicircular boats dwarfed by them, and the foliated river banks direct the viewer's gaze to the outcrop of rock in an indescribable shape that juts into the river. The Underwater Lode, turned inside out and replete with schematic lobes clearly labeled, appears more hollow than solid despite its larger-than-life size. An inscription on the front of the inkstone reads: "Made on a winter's day, the year Bingxu of Daoguang [1826]."

The maker or inscriber of both inkstones called their maps *tu* 圖, which Francesca Bray suggests is best rendered as "technical images."[23] To summarize arguments made thus far: The appearance of *tu* maps, a new mode of visual communication about Duan quarries in the first half of the nineteenth century, serves to reorder the hierarchy of Duan stones. The emergent hierarchy at once positions the Underwater Lode as being superior to other Duan quarries while inaugurating a system of internal differentiation also within it. More than displaying inert information, *tu* also works to incite action: in this case the quest of literarily grounded knowledge about the inkstone on one's

FIGURE 10.4. Map of Duanxi (Duanxi tu). Carved on the back of a Duan inkstone. Collection of the Metropolitan Museum of Art, New York.

desk, tracing its origins back to a particular cave on the southern edge of the empire.

Seen in this light, part of the forces that propelled the appearance of maps of the Underwater Lode can be sought in the currency of the philological method of precise and impartial research. The fad of in situ observation encouraged more outsiders from the Central Plain to journey to Guangdong and more officials stationed in or passing by the Duan area to write about the geography and geology they observed with their own eyes.[24] The exorbitant prices commanded by the fine-grained stones, limited in supply, and the jostling in an overcrowded career path for scholars, whose wager was that an expertise in inkstones might open doors to new patronage networks, also explain the zeal in the search for truth deep in the subterranean tunnels in an increasingly microscopic vision.

The related visual trope of X-ray penetration is analogous to the rhetorical devices used by treatise writers of "I saw it with my own eyes" or "I heard it from the native or old stoneworker" in the production of a reality effect: all are authority claims that are intrinsically paradoxical. Although both the visual and verbal tropes give the illusion of authentic on-the-ground knowledge, they also position the viewer or writer as an outsider who could not have crawled inside the shaft, no wider than ninety millimeters in diameter and submerged under water except during mining season. Even if he had been able to gain entrance during a mining operation, the writer would not have been able to see what he most desired to find under the dim illumination of burning wick dipped in lard. The problem was not one of visibility but epistemology. The challenge of the connoisseur was to infer from the surface mineral features and color of the stone on his desk the exact location of its origins in the shaft (which grew to over one hundred meters by the nineteenth century). But the shaft, now emptied of its lode and the bedrock that encased it, yields no clue to that correlation. All that one could have seen would have been subterranean reservoirs and piles of waste rock littering the floor. The would-be eyewitness would have found himself crawling naked in a tunnel, which is by definition empty and perhaps in imminent danger of collapse. The hypothetical situation is a near comical refutation of the philological creed of "seeking the truth from concrete facts" that are gathered from personal observation.

But of course this inconvenient reality was never made explicit in the discourse. The knowledge useful to the inkstone connoisseur is constructed ex post facto and shored up by faith in the authority claims of someone *else*. It is the opposite of what is usually meant by authentic, in situ, local, or indigenous knowledge.

He Chuanyao, the "Local" Expert

Without crawling into the shaft, the expert who shifted the ground of knowledge about the Underwater Lode was He Chuanyao (fl. 1820s–1830s), author of the slim treatise *Discerning Inkstones from the Treasure Inkstone Studio* (*Baoyantang yanbian* 寶硯堂硯辨), completed in 1827 and published eight years later (hereafter *Discerning Inkstones*). An aspirant scholar, He Chuanyao was a famous man in Gaoyao, a town directly upriver from the Underwater Lode, although Gaoyao was a very small place. His home, on the northern side of town, housed a collection of inkstones, heirlooms passed down from his father. Other local collectors praised it as impressive, but no one outside Gaoyao, let alone the province, seems to have heard of it.[25] Nevertheless, He Chuanyao was the rare "local" expert who lived a stone's throw from the quarries and managed to garner enough logistical and financial support from visiting outsiders to have his treatise published.

In its content and style of presentation, *Discerning Inkstones* ruptures the existing discourse of inkstones in three ways. To begin with, it is the first treatise to include a map of the shaft of the Underwater Lode, showing the relative locations and parameters of the four main caves—the East Cave, Main Cave, Minor West Cave, and Major West Cave (figure 10.5). The map on the back of the first inkstone discussed above is likely to have been modeled on the diagram in He's book.[26] One may think of this illustrated treatise, with its choreographed interplay of texts and diagrams (as well as the dancing words in diagrams), as a precursor of "technography," a recent method of studying technology in its concrete ethnographic and geographic settings that combines graphic and textual styles of knowledge.

Reflecting the growing commercial interests in exploiting the quarries, a second innovation introduced by He, in concert with his mapmaker Huang Peifang 黃培芳, is to measure space in terms of manpower and to use it as a technique of visual codification. The total length of the shaft (and the distance between any two points in the tunnel) is calibrated in terms of the number of stoneworkers needed to line up in the tunnel one-man across. Also interesting is how the stoneworker's body serves as a unit of measurement: the height of the ceiling is indicated by the words "sitting man" or "standing man" (figure 10.5).[27]

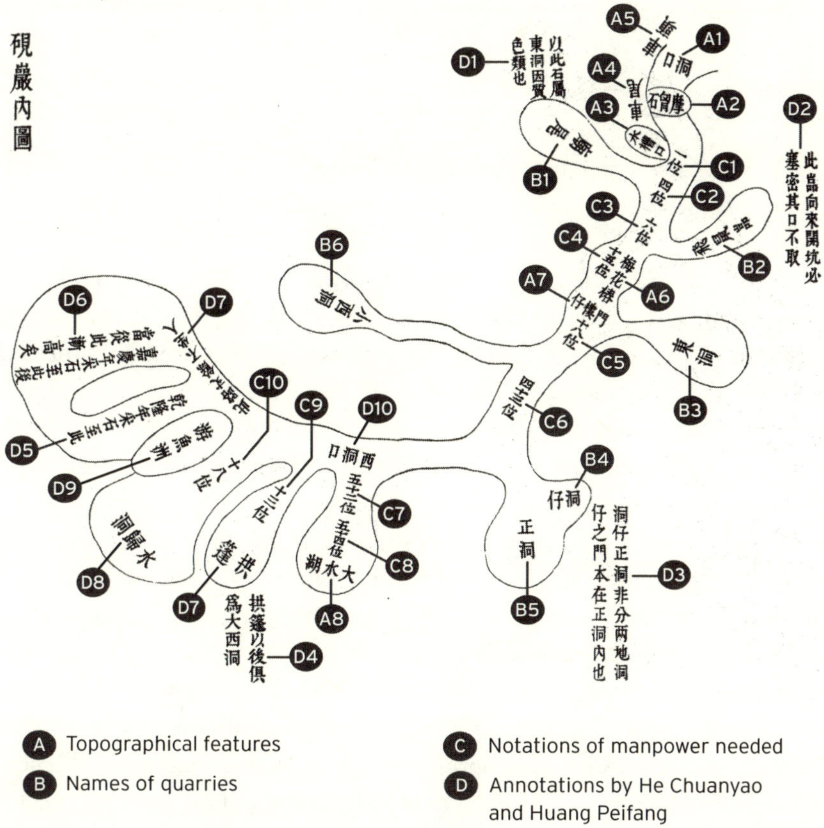

FIGURE 10.5. Map of Inside the Inkstone Quarry (that is, Underwater Lode), made by mapmaker Huang Peifang. He Chuanyao, *Baoyantang yanbian.*

He Chuanyao made a third, perhaps more far-reaching, intervention in reorganizing the ranking of the quarries. In his book He summarily dismissed the valuation prevalent among out-of-province scholars at the time, that the most superior stone hails from the Main Cave, followed by those from the East Cave and the two West Caves. Embedded in the structure of He's table of contents (entitled "A ranking in descending order of the Four Caves") is his revised hierarchy:

I. Major West Cave (including Canopy Cave)
   Appended: [Seven] quarries that resemble it: Old Su, Sky-Facing, Dragon Craw, Old Lode, White Stripe, Little Pit, Pock Mark

II. Main Cave (including Wee Cave)
    Appended: [Eight] quarries that resemble it . . .
III. Minor West Cave
    Appended: [Seven] quarries that resemble it . . .
IV. East Cave (including Temple Tail Cave)
    Appended: [Four] quarries that resemble it . . .[28]

He Chuanyao simplified existing geography by subsuming three previously independent caves of the Underwater Lode (Canopy, Wee, and Temple Tail) under his nomenclature of the Four Caves. His organization elevates the Major West Cave, the most recent to open, to the pinnacle of the hierarchy. In subsuming all worthy quarries to the Four Caves of the Underwater Lode, He Chuanyao established the supremacy of the Lode by making it the norm against which all other stones are to be assessed.

Understandably, the entry on the supreme Major West Cave is the first and the longest in the book. A few excerpts convey He's rhetorical strategies: "[Stones from] the Major West Cave come in five strata. The top stratum is greenish-purple in color, with a coarse and thick nature [*zhi* 質]. . . . Next comes the second stratum. Its 'Sky blue' [a mineral mark] appears as the color of the sky after the morning star has faded but the sunrays have yet to appear" (16). He's lyrical description appeals to shared sensory experiences. Another characteristic of He's prose is that, instead of justifying his valuation, he simply listed the unique characteristics of each of the five strata, as well as two sublayers within the third stratum and three sublayers within the fourth stratum. Of the bottom sublayer of the fourth stratum, he wrote: "There are green 'Five-colored nails' and white 'Five-colored nails' [another mineral mark], both so hard that they repel the knife. None of the stones from the 'miscellaneous quarries' have these features. Nowadays people use them to tell the stones apart, but they are in fact defects" (19). The entire raison d'être of his knowledge system is the connoisseur's mandate of authentication. He comes to the aid of the uninitiated by making the distinctions between the authentic and the fake unequivocal and discernible. His "appendix" might as well read, "Beware of forgeries!"

On the identification of the origins of stones and deciphering the authentic from the fake, He Chuanyao exudes a confidence rarely seen in other treatises. Instead of making authority claims that out-of-town authors staked on "natives" or "old stoneworkers," He simply stated what he knew in a deadpan

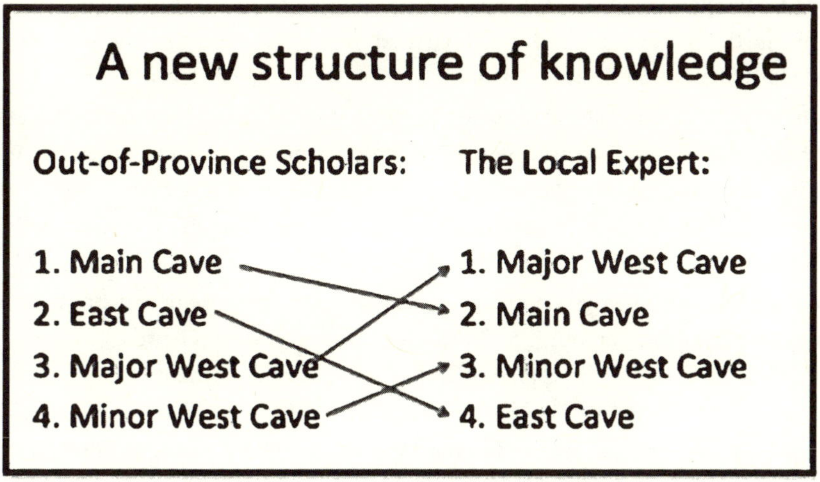

Figure 10.6. He Chuanyao's new hierarchy of ranking quarries within the Underwater Lode. *Baoyantang yanbian* (Discerning inkstones from the Treasure Inkstone Studio).

tone of certainty. Adverbs such as "always," "inevitably," "never seen," and "occasionally" give the impression that he has seen it all and can size up any one stone in the context of a total system of comprehensive knowledge.

He Chuanyao grounded his authority on his own judgment and sensory experiences. He "examined [surface details] under direct sunlight" (18, 22, 27), "tapped [with the knuckles]" to discern if the echo resembles that of metal, wood, or ceramic tiles (17, 28, 34), "turned the stone sideways" to make out the direction of the grain of the slate (35), "scratched away" surface features with his fingernails to see how deeply embedded (22, 37), or "tested with a knife blade" to determine if a particular kind of joint in the stone is sandy or rocky in nature (28). Each test, however intrusive, is necessary because it yields a sign about the exact origins of the stone, in terms not only of a certain quarry but of the exact location of the chamber in the shaft. When all the signs are added up, the truth would reveal itself. He Chuanyao sums up his method: "Discerning [even slight] differences from [what appears] to be similar; discerning the [underlying] similarities from apparent differences" (40). It sounds as if he had taken a page directly from the method of the philologists.[29]

To sum up, in his radical revision that places the Major West Cave at the top of the hierarchy of the Underwater Lode (and by extension all Duan stones and, hence, all stones in the empire; see figure 10.6), the "local expert"

He Chuanyao reshuffled the hierarchical judgment of inkstone connoisseurship in three ways. He redrew ("clarified") the parameters of the four caves of the Underwater Lode in a more precise manner; he devised a new classification scheme that encompasses twenty-four of the Duan quarries; and he introduced a universal discourse of "color" (the hue, mineral features, and other surface qualities pleasing to the eye) and "*zhi* 質" (nature, materiality, the quality that makes an inkstone conducive to grinding ink without hurting the brush) by which all stones are to be described and evaluated. The *zhi* of a stone is nothing mysterious; the experienced connoisseur can know it by closely examining, tapping, or scratching the specimen on hand.

He's ranking of the quarries and standards of judgment, unique in his times, circulated widely in the empire after a scholar with more respectable academic credentials, fellow Guangdong native Wu Lanxiu (1789–1839; a *juren*-provincial graduate 1808), incorporated them verbatim in his own treatise completed around the same time, *Chronicle of Duanxi Inkstones* (*Duanxi yanshi* 端溪硯史). A classicist at the Ocean of Learning Academy in nearby Guangzhou, one of the regional hubs of philological research in the empire, Wu offered comprehensive and impartial coverage of all salient systems of knowledge about Duan inkstones. One reader of this popular book singled out He Chuanyao's opinion as the most "original, nuanced and accurate" whereas the rest of the book is "nothing special." Another reader opined that He's ability to discourse on the minutiae of the real and the fake surpasses even that of experienced "old stoneworkers."[30] It stands as the canonical view among scholars and connoisseurs today.

He Chuanyao's efforts, influential as they are in changing the prevalent ranking of Duan stones among his readers, did not obliterate the natural advantage enjoyed by the stoneworkers. His lasting contribution was to produce for himself a new form of brush-based authority, thereby conjuring a new subject position—the "local" expert—as an alternative native informant to the stoneworker. When he accused the stoneworkers of profiting from misnaming stones and ranking inferior stones highly, he inadvertently revealed the considerable power that the stoneworkers exercised over the consumers. In the final analysis, the contest reveals tensions and fissures in overlapping but contending knowledge systems involving agents occupying three subject positions on two shifting fronts: (1) between the out-of-province visitors and the local scholars, both of whom wrote, and (2) between the scholars who wrote, foreign or local, and the largely illiterate stoneworkers.[31]

Concluding Thoughts

Similar to the itineraries of the inkstone albeit at a different scale, global movements of asafetida, tea, and other materials mentioned in the present volume were conducted not only through uneven terrains but also through spaces that were not yet Cartesian and far from inert. Itineraries in nonmodern worlds took shape by negotiating with spaces that are not objectively "out there," but foldable, collapsible, and stretchable entanglements (in Barad's usage) created in the very process of movement.[32] The ascendance of the "local expert" He Chuanyao, however, hints at a different configuration of space and knowledge in the early nineteenth century. In their paradoxical accent on both "subjective" eyewitnessing and the penetrative mode of visual narration (which presupposes "objective" distance), not to mention rationalized space/time for commercial exploits, He's maps announce a transition to early modernity. His textual discourses, meanwhile, hark back to the millennial tradition of connoisseurship literature established by such Song paragons as Mi Fu.

To return to the question posed at the beginning of this chapter, why are the itineraries of inkstones so mobile in regional loops but sticky in transcontinental loops? Itineraries of material-emotional complexes became sticky when they hit a bump or slip into folds in wrinkled spaces. These obstacles are places where dense meanings and socialities accrue and, as such, are as deserving of research as unimpeded movement. In the case of the inkstone, it is the density of the nexus of knowing practices sedimented (or "iteratively enfolded," to use Barad's term) over a millennium in the Duan area, one of the myriad small places in the empire, that renders the knowledge sticky.

The scholars' consistent reliance on the stoneworkers' insights points to a second reason for this stickiness. Knowledge about inkstones is ultimately embodied in the hands, eyes, and minds of the prospectors and miners in the Duan villages. As such, it is resistant to the "scientific" method of the evidential or philological scholars whose métier is the written word. For all his erudition, the scholar's predilection for investigating local phenomena in situ is doomed to fail. Empirical studies cannot smooth the wrinkles and bumps in the knowledge field because what they most desire to know can never be found materially where the stones originated. The challenge of the connoisseur—his "science," if we may—was to infer from the surface mineral features and color of the stone on his desk the exact location of its origins in the shaft.

But there is nothing to be observed in the now empty tunnel even if he were to descend into it. In this sense, the only "authentic" knowledge possible is the irreproducible experience of the miners.

To put it differently, when space is active instead of inert, the "local" is not a place—any one place with knowable parameters—that yields authentic knowledge. The "local" is but a situational construct that recedes and expands depending on one's attentiveness and interest. There is no "local" in the Duan area; it is a meaningful category only to someone coming from the outside. Does the "local" refer to the province of Guangdong, separated from the Central Plain by mountain ridges, traditionally the land of exile? Does it refer to He Chuanyao's native town of Gaoyao, the Yellow Hill villages, or the shaft of the Underwater Lode? The "local" is wrought of elusive geography that became solidified or recognizable only when it was made manifest by dueling authority claims.

Although textualization has been crucial in facilitating the movement of a host of materials analyzed elsewhere in this book, the inkstone case exposes the limitations of texts as a vehicle in transmitting knowledge and meaning.[33] Part of the reason the Duan stoneworkers retained considerable discursive authority vis-à-vis the out-of-province scholars is that the truth about an inkstone (even to a scholar connoisseur) is inherent in the palm-sized stone itself, each with its unique shape, weight, texture, and constellation of mineral features. It is a matter of no small irony that the most valuable knowledge about the writing implement "travels" not by way of texts but, in and of itself, in its irreducible singularity and materiality. The physical and epistemic density of the inkstone did not preclude mobility—it was rather agile in traversing society and court—but conditioned its uneven motion, producing bumpy spaces and gaps in its wake.

In ending, it may be productive to ponder the relative merits of actual objects—three-dimensional models, if we may—as a third way of encoding information and transmitting dense knowledge beside texts and diagrams. Consider the "myriad-treasures cabinet" (*duobaoge* 多寶格), a desktop storage device made of hard or soft wood and comprised of disparate drawers, trays, and niches fitted into a cuboid or rectangular structure like a jigsaw puzzle (see figure 10.7a–c). For those in the know, hidden compartments would swing open upon the pressing of a concealed button or latch, drawers can be slid out and covers of boxes removed, revealing spaces within spaces, and treasures upon treasures.

FIGURE 10.7 (a). A myriad-treasures cabinet, at once an instrument of concealed storage and of intimate display. Collection of the National Palace Museum, Taipei.

These cabinets might have derived from portable stationery boxes for traveling scholars and rationalized storage for curio collectors in the Ming dynasty called "containers of tens and hundreds of things" (*baishijian xia* 百十件匣). They became the prerogative of Qing emperors from Kangxi (r. 1661–1722) on, fabricated by artisans in the imperial workshop to the emperor's specifications. As the emperor's playthings, both the materials and contraption of the cabinets became more fanciful in the Qianlong reign (1736–1796). The historian Pamela Crossley suggests that they embody Qianlong's new ruling ideology of "universal sovereignty" on the heels of the demise of the Zunghar Mongols and the retreat of the Muscovite Russians, the two other contending powers with Eurasian ambitions, in the 1760s. The management of this multiethnic empire requires a new ideology that envisions the empire as an assemblage of simultaneous locations: Out of one, many.[34]

FIGURE 10.7 (b–c): "Out of one, many." Myriad-treasures cabinet with thirty curio items, *Zitan*-wood, L25 cm x W25cm x H21cm. Collection of the National Palace Museum, Taipei.

The multicompartment cabinet is likewise an assemblage of simultaneous locations and enfolded dimensions. As such it embodies in its material format the notion of active and convoluted space, the pivotal role of movement initiated by private knowledge and desires, as well as multiple and competing locales. Perhaps this dynamic vision may serve as one of the models for imagining the material and epistemic worlds that were made by the entangled itineraries in materials, recipes, practices, and knowledge.

Part 4

Convergences and the Emergence of New Objects of Knowledge

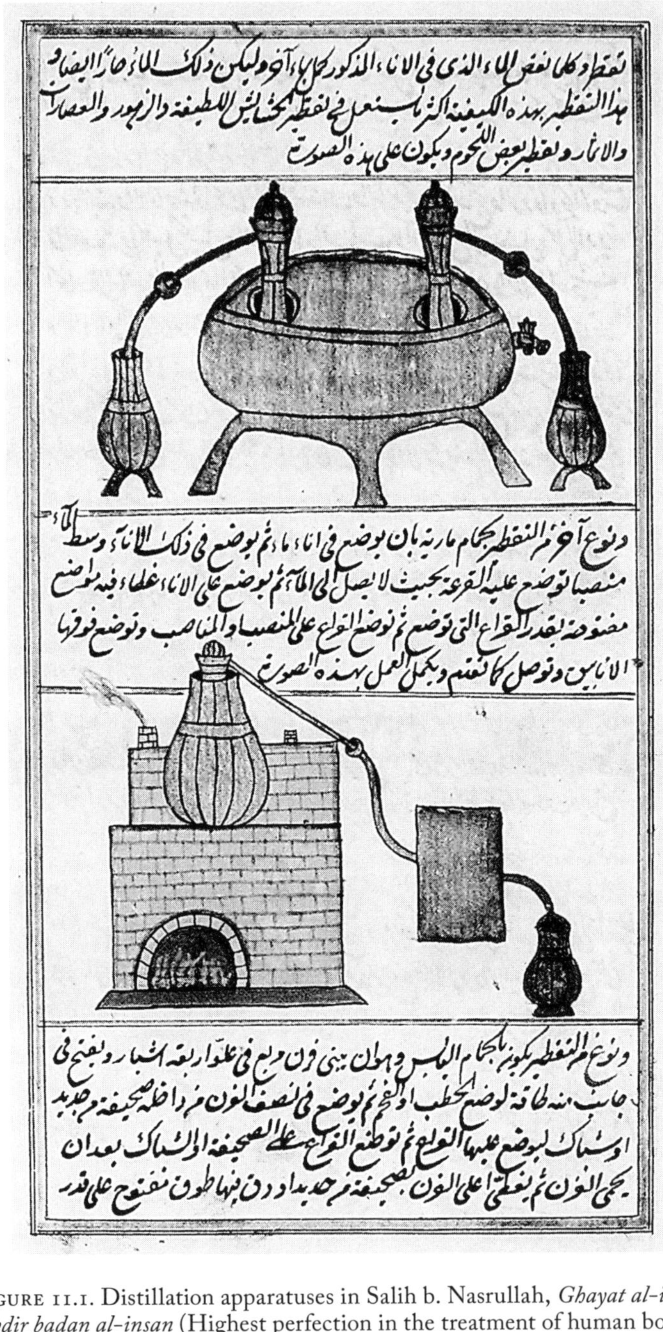

FIGURE 11.1. Distillation apparatuses in Salih b. Nasrullah, *Ghayat al-itqan fi tabdir badan al-insan* (Highest perfection in the treatment of human body), copy dated 1722. Fol. 26a, MS Ayasofya 3682, Süleymaniye Library, Istanbul.

Chapter 11

Convergences in and around Bursa

Sufism, Alchemy, Iatrochemistry in Turkey, 1500–1750

Feza Günergun

Ottoman texts rarely depicted chemical utensils and apparatus prior to the late seventeenth century when Ottoman physicians became acquainted with the European iatrochemical books heavily illustrated with laboratory equipment. These early images of laboratory equipment appear in *Ghayat al-itqan fi tadbir badan al-insan* (figure 11.1), a pharmacological treatise in Arabic composed after European formularies by Salih b. Nasrullah ibn Sallum (d. 1669), the chief court physician and his colleagues in Istanbul. These colored, painting-like illustrations contrast with the simple drawings of the Turkish iatrochemical texts composed by the dervish physicians of Bursa. Two examples are *Künuz-i hayat el-insan kanun-ı etibba-i feylesofan* (figure 11.2) of Ömer bin Sinan el-Izniki (d. early eighteenth century) of Iznik (Nicea), near Bursa, and *Cevher el-ferid fi tıbb el-cedid* (figure 11.3) by Ömer Şifai el-Burusavi (d. 1742), head physician to the *darüşşifa* (hospital) of Bursa.

Simple or detailed, such illustrated manuscripts, together with the alchemical texts previously composed by Sufis, raise many questions about practical alchemy in the Ottoman Empire, for historians have previously found no evidence of such practice. However, these texts show familiarity with practical alchemical processes and techniques: there are alchemical connections in the names of the apparatus depicted, such as *beyzi-i felasife* (the ovoid of philosophers); *alet-i hermes* (Hermes's instrument), and the processes, including *teaffin-i hukema* (philosophers' putrefaction), one step in the production of the philosopher's stone that aimed to remove the impurities from matter.[1] How did these texts emerge? And why were they produced by Sufi physicians native to or resident in Bursa? A search for answers to these questions necessitates a deeper investigation into the unexpected itineraries of alchemical texts across

FIGURE 11.2. Chemical utensils in Ömer b. Sinan el-Izniki, *Künuz-i hayat el-insan kanun-ı etibba-i feylesofan* (Treasures of human life: Canon of the philosophical physicians; 1695). Fols. 344b–45a, MS T7083, Istanbul University Library, Istanbul.

Eurasia into Anatolia, at first from the east before the seventeenth century and subsequently from the west. This chapter focuses on the itineraries and convergences of people and texts that made Bursa a node of interaction and brought about the crystallization of a field of alchemical knowledge in Anatolia, which, in turn, prepared the ground for an encounter with European iatrochemistry and the subsequent incorporation of practical alchemical tools and techniques into this field of knowledge.

This itinerary of alchemical ideas and texts begins with the movement of peoples in Asia Minor, a part of the Byzantine Empire. Following the defeat of the Byzantine army by the Seljuks at Manzikert in eastern Anatolia in 1071, nomadic Turkomans from Khorasan, Transoxiana, and Iran began to dominate and inhabit its fertile plains. Sufi mystics wandered along with the Turkoman tribes, believing such settlement was their mission. In the thirteenth century Anatolian towns were devastated by a Mongol invasion that forced Turkoman tribes to migrate further westward. The Mongols' retreat from Anatolia toward the end of the fourteenth century led to an increase in

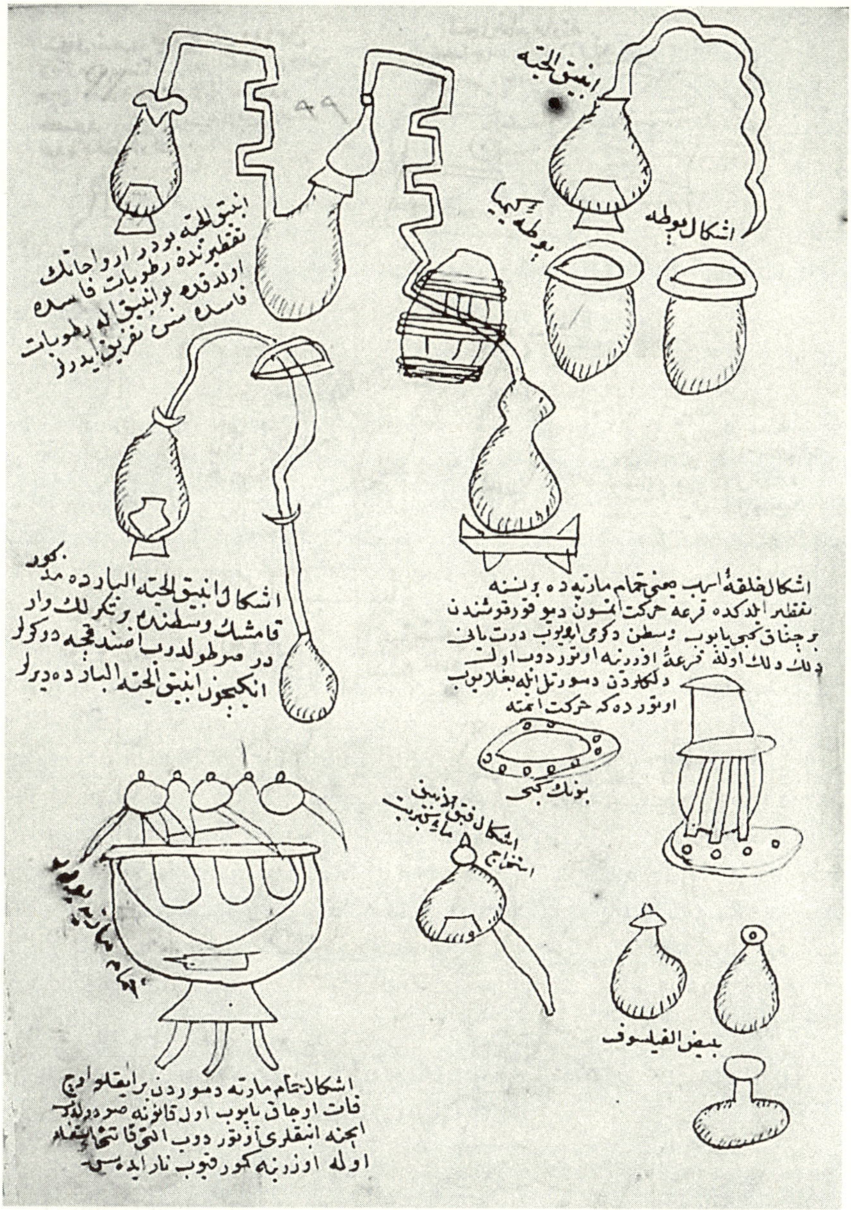

FIGURE 11.3. Chemical apparatus in Ömer Şifai, *Cevher el-ferid fi tıbb el-cedid* The unique gem of the new medicine] (1698). Fol. 99a, MS Hamidiye 1020, Istanbul University Library, Istanbul.

the Turkish population of Asia Minor and paved the way for the establishment of *beylik*s (principalities). With the gradual Islamization of Anatolia, Sufi mystics (locally known as dervishes) originally from Central Asia established themselves in the peninsula.[2] While most settled in and around central Anatolian towns, many moved to the northwestern part of Asia Minor historically known as Bithynia, and they settled on the Byzantine border together with the Turkomans. Following the fall of Bursa (Prousa ad Olympium) and Iznik (Nicea) to the Turks in the early fourteenth century and the flourishing of these cities as Ottoman capitals, the area attracted more Sufis, together with the *ulema*, the religious scholars. Although the seat of the sultan moved to Edirne in 1365 and then to Istanbul in 1453, Bursa remained an "imperial city" and preserved its prominence as a commercial, industrial, and intellectual center. The orthodox and heterodox Sufi doctrines, which originated in a geography stretching from Central Asia, Iran, through the Middle East and reaching Iberia, found their way to Bursa. The Sufi *tekke*s (lodges) embraced these teachings while madrasas were mostly concerned with orthodox teaching such as Muslim canonical jurisprudence, theology.[3] Alchemy, regarded as unorthodox lore, was thus mostly nurtured by the Sufi mystics. The texts they produced, besides carrying the alchemical knowledge formulated by Jabir ibn Hayyan (ninth century), al-Razi (ninth–tenth century) and al-Jildaki (al-Jaldaki, fourteenth century) constituted a body of knowledge that set the stage in the seventeenth century for the adoption of iatrochemical therapies introduced from Europe. This chapter follows the movements of these ideas, focusing on (1) the relational field formed by Bursa's commercial and intellectual position that attracted the Sufi orders, (2) the introduction of alchemical texts to Anatolia and their further production by Sufi mystics, and finally (3) the translation and compilation of iatrochemical texts into Turkish by dervish physicians in and around Bursa.

Bursa and Iznik as Centers of Commerce, Crafts, and Learning

Bursa was a major city at the western end of the Silk Road in the fifteenth and sixteenth centuries, sitting at the crossroads of trade routes connecting Central Asia, India, and the Near East with the Balkans and Europe. Spices, coffee, dyes, textiles, hides, and timber were the major traded goods of the region renowned for its silk industry.[4] Although raw silk of Tabriz was spun, dyed, woven, and exported to Europe, fabrics from Venice and Florence were also marketed in the city. Bursa supplied the sultan's household and Istanbul's rich clientele with textiles and a large variety of goods.[5] Iznik, situated northeast of Bursa, excelled in pottery and tile making. The area abounded in raw materials (white sand, clay, timber, and water) necessary for the development of a ceramic industry. The city flourished especially from the mid-fifteenth century on, when workshops began to manufacture tableware and pottery for the palace in Istanbul, as well as tile revetments for mosques and shrines that Ottoman sultans and dignitaries alike competed in building. Iznik, however, declined as a center of artisanry from the seventeenth century because of the diversion of trade routes, persistent malaria epidemics, and the relocation of workshops to the imperial capital.[6]

Both Iznik and Bursa were endowed with madrasas from 1330s on.[7] Madrasa graduates traveling to Cairo, Damascus, Baghdad, Tabriz, Herat, and Samarqand to enhance their knowledge were instrumental in transmitting Eastern lore to Anatolia. The political unrest that followed the astronomer sultan Ulugh Beg's death in 1449 paved the way for the flow of scholars from Khorasan and Transoxiana. The annexation of Cairo and Baghdad to Ottoman lands in the sixteenth and seventeenth centuries fostered the mobility of scholars with their learning and texts to Anatolian towns including Bursa and Iznik.[8]

The Burgeoning of Sufi Orders in Bursa

Anatolia was already home to a number of various Sufi orders—mostly seen as heterodox Muslims—in the thirteenth century, after which new Sufi traditions emerged. Early mystics of Anatolia, the Abdalan-ı rum expanded Sufi traditions such as the Qalandari, Wafai, and Babai that began in the East (Khorasan, Transoxiana, and Iraq). They established their lodges (also known as *zaviye*), in rural Asia Minor and engaged in farming. Sufis also participated in military raids against Byzantium and in the siege of Bursa. The allocation of land parcels or the endowment of incomes by Ottoman sultans following the conquest of Bursa in 1326 encouraged Sufis to settle in cities. For example, the Turkoman Geyikli Baba (the dervish with a deer) was among the first sheikhs to settle in Bursa. The sultan Orhan endowed the Wafaiyya lodge with the income of a village. Sufis resident in towns entered guilds as a part of the prevailing social organization known as the *futuwwa*.[9]

The fall of the Byzantine Empire in the mid-fifteenth century and the rise of the Ottoman state as a new power transformed Bursa into a stable and fertile environment and attracted more Sufis from Khorasan, Iraq, Bukhara, and Baku, who then established their Anatolian lodges with the financial support of their local recruits. Among these orders were the Zayniyya, the Qadiriyya and its branch Ashrafiyya, the Khalwatiyya, and the Naqshbandiyya. By the sixteenth century, Sufi orders that had formerly emerged in Anatolia such as the Mawlawiyya, Bektashiyya, and the Jalwatiyya had also established their lodges in Bursa. The fate of these orders depended on the political and economic conditions of the era. The Zayniyya order was known as a community of luminaries because they were considered able to balance the exoteric and esoteric meanings of the Quran. Thanks to the support of sultan Mehmed II (r. 1451–1481), they came to dominate the area in the fifteenth century. The Khalwati, who were critical of unorthodox sects, opened up to a dozen of *tekke*s in Bursa under the patronage of sultan Bayezid II (r. 1481–1512). The Ashrafiyya order, established in Iznik in the fifteenth century to complement the teachings of the Qadiriyya with (heterodox) Sufi elements and eclipsed in the seventeenth century, gave birth to the Rumiyya order.[10]

From the fifteenth century onward Bursa grew into a multiethnic city with its many religious practitioners inscribed in its topography: certain neighborhoods came to bear the names of Muslim religious scholars, sheikhs, or Sufi lodges. Muslims and non-Muslims (Jews, Armenians, Greeks, and Copts) resided in separate neighborhoods or dwelled together. The wave of Celali rebellions that struck Bursa in the early seventeenth century could have affected the population negatively, including the Sufis, yet the contemporary Ottoman traveler Evliya Çelebi records over forty Sufi lodges in Bursa, among them seventeen Khalwati, nine Qadiri, three Naqshbandi, and one from each of the Zayni, Gulshahi, Saidi, Rıfai, Badavi, Jalwati, Bektashi, and Mawlawi orders. Besides being spaces for initiation into Sufism, lodges offered sociability and service for the disciples of the sheikh, functioning as places to study, copy, and compose texts and commentaries.[11] Bearing in mind the corpus of alchemical manuscripts compiled or copied by Sufi mystics, we might expect that lodges would be the very places where knowledge of alchemy circulated. Surprisingly, however, among the 320 volumes (62 titles) belonging to the nine dervish lodges of Bursa, no books of alchemy are recorded.[12] As alchemy was considered an unorthodox art, one can surmise that books related to alchemy were not present in the lodge, which was a pious endowment. Sufis must have kept alchemical texts among their private books.

Early Sufi Poetry as Alchemy in Anatolia

Sufi allegorical poems, imbued with alchemical knowledge, were composed in Anatolia from the fourteenth century onward. An early example is *Risale-i Aşık Paşa*, by the Sufi poet Aşık Ali Pasha (1271–1332).[13] The poem clearly makes an analogy with the spiritual path and alchemical processes. Among other alchemical operations, it recounts the preparation of whitening and reddening waters, added to mercury to make the soul like silver or gold—that is, honourable. Was Aşık Pasha initiated to alchemical texts within the Wafaiyya community in Amasya or in Mamluk Egypt where he took refuge and then was imprisoned for many years before being allowed to return to Anatolia in 1332? As the poem is undated, it is hard to know. The fact that his ancestors were from Khorasan and that Amasya was under Ilkhanid (Mongol) rule suggests a transmission from Khorasan, possibly through Persia, which has always been a route for the entry of knowledge to Anatolia from eastern regions. The Cairene Sufi Sayyid Hüseyin Ahlati (d. 1397), the author of an alchemical text *Kimiya*, was originally from Ahlat in present-day Turkey, and his disciples are known to have been active in western Anatolia. In the late fifteenth century, a treatise in Persian dealing with alchemy, medicine, and astrology was dedicated to sultan Bayezid II by Azizullah b. Ataullah Hindi (that is, the Indian, fourteenth century).[14]

Spiritual Paths and Alchemy in Bursa in Late Fifteenth Century

Sufi texts alluding to alchemy appeared in Bursa from late fifteenth century onward, if not earlier. The *Kaside-i Şeyh Safi* (The ode of Sheikh Safi) is a poem in Turkish of about forty rhyming couplets, composed by Safiyüddin Efendi (d.1513), the sheikh of the Zayniyya order in Bursa between 1494 and 1513.[15] The poem addresses the *talibîn* (aspirants) desirous of entering the spiritual path. As in the earlier Anatolian poems, an analogy is drawn between the alchemical processes and the transformations the aspirant must undergo in order to become a *mürşid*, a spiritual guide of the order. The text abounds with operative terms such as *hâk* (pulverizing), *tahmir* (reddening), *tebyiz* (whitening), *tesvid* (blackening), *tasid* (sublimation), *teklis* (calcination), *taktir* (distillation), *tathir* (purifying), *tedbir* (treating), *hall* (dissolving), and *akd* (congealing), along with names such as *kibrit* (sulfur), *zibak* (mercury), *ukab* (sal ammoniac), *pulâd* (steel), *dühn* (oil), *şeb* (alum), *barud* (saltpeter), and other chemical substances. These are intermingled with Sufi terms related to the soul and the self. The term "elixir" is not used, but the aspirant is given nine glasses of a certain "water." There are no recipes for making silver and gold; instead we have descriptions, with alchemical terms and analogies (e.g., birds with green wings) of the long processes of transformation awaiting the aspirant. Sufi texts comparing the process of becoming a "perfect man" with the transformation of base metals into silver and gold, with alchemical recipes, continued to be composed and copied in and around Bursa through the seventeenth and eighteenth centuries.

The Transmission of Alchemical Knowledge to Bursa from Egypt and Iberia in the Sixteenth Century

An early attempt to introduce the work of al-Jildaki (d. 1342), the Khorasan alchemist who flourished in Damascus and Cairo, came from Mohammad b. Omar al-Antaki (d. 1531, Bursa), alias Mullah Arab.[16] Mullah Arab was born in Antioch (Antakya, in southern Turkey) to a family originally from Transoxania. Educated in various towns of Iran and the Middle East, he entered the service of the Mamluk Sultan al-Ashraf Qaytbay (r. 1468–1495/6) in Cairo in the capacity of preacher and mufti. Upon the death of Qaytbay, he migrated to Bursa. Mullah Arab was praised by Bayezid II for his erudition and sermons. His knowledge of alchemy may also have captured the curiosity of the sultan. Mullah Arab was a member of the Hanafi School of jurisprudence and wrote on the life of Prophet Mohammad. Following his arrival in Anatolia, he epitomized two notable works on alchemy and occult sciences—namely, al-Jildaki's *al-Burhan fi asrar ilm al-mizan* (Demonstration of the secrets of the science of the balance) and al-Qurtubi's (d. 964) *Rutbat al-hakim* (Scale of the sage, mid-tenth century).

Mullah Arab epitomized *al-Burhan* in 1512 and entitled it *Talhis al-Burhan* (Epitome of *al-Burhan*). *Al-Burhan* was an encyclopaedic work on a variety of subjects such as the theory of the balance regarded as essential for the success of alchemical operations, with chapters on theology, natural history, physics, and the Balinas (Apollonius of Tyana) commentaries of Jabir ibn Hayyan. However, the larger part of Mullah Arab's epitome portrays the relationships between the planets, the zodiacal signs, and the metals, conceived as the causes determining the "balance" of natures, giving each metal its characteristics in its formation from sulfur and mercury. Mullah Arab's choice of subjects, and his professed aim of making al-Jildaki's text accessible to nonexperts, points to an audience eager to learn about the theory of balance as the work of the Creator.[17] This text, together with the few later copies of *al-Burhan* that are extant in Istanbul libraries, testifies to the transmission of alchemical knowledge from Mamluk Egypt to Ottoman Anatolia.

The *Rutbat al-hakim* (mid-tenth century) by the Andalusian esoteric scholar Maslama b. Qasim al-Qurtubi was one of the most important and earliest texts on occult sciences in the Iberian peninsula.[18] Maslama compiled the *Rutbat* by making use of the *Rasa'il Ikhwan al-Safa*, the epistles of the Brethren of Purity based in Bassorah and Baghdad in the tenth century.[19] By referring mostly to the alchemical texts of Jabir ibn Hayyan and al-Razi, Maslama discussed the nature of the elixir and went on to relate his observations on the oxidation of mercury, emphasizing Jabir's new way of writing practical books with clear demonstrations.[20] The epitome of the *Rutbat al-hakim* made by Mullah Arab is undated, but the author's death in 1531 in Bursa is the terminus ante quem.[21] Sephardic Jews had migrated to Ottoman lands following their expulsion from Spain in the late fifteenth century and were allowed to settle in Istanbul, Bursa, Adrianople, or Salonika.[22] *Rutbat* may have been among the books that scholars and physicians brought from Andalus to the Ottoman Empire. Mullah Arab may have acquired a copy of *Rutbat al-hakim* during his Bursa years, or he may have acquired a copy in the Balkans during the military campaigns he participated in as a preacher. Whatever the case, we know that alchemical knowledge transmitted from the Middle East to Andalus in the tenth century was conveyed to Asia Minor in the early sixteenth century through an Andalusian treatise.

Did Mullah Arab prepare the epitomes for the Sufi mystics of Bursa, or did they approach the Mullah to benefit from his alchemical knowledge? We do not have any clues indicating a possible connection. The extant copies of each epitome are in a library in Kütahya, a town that was once the capital of the Germianids bordering Bithynia. Did Mullah Arab produce both texts for his personal interest, and were these brought to Kütahya by an unknown Sufi follower of esoteric sciences? Yet another question is whether the theory of balance and the mercury-sulfur theory were introduced to Asia Minor through these epitomes? A definite anwer can only be given following a thorough examination of alchemical texts of Jabir and al-Jildaki kept in Turkish libraries.

Al-Jildaki's Anatolian Follower Ali Çelebi el-Izniki

The alchemy of al-Jildaki became widely known in Anatolia from the mid-sixteenth century onward thanks to Ali Çelebi el-Izniki (d. 1609), a Sufi from Iznik (Nicea).[23] Ali el-Izniki is recognized as a descendant of Eşrefzâde /

Eşrefoğlu Abdullah Rumi (d. 1469), who was the founder of the Ashrafiyya order (a branch of the Qadiriyya) established in Iznik and the author of a Turkish commentary to the alchemical poem *Kaside-i Sırr-i Ta-Ha.*[24] According to his own words, Ali el-Izniki traveled as far as North Africa, Arabia, Persia, India, and China and served scholars and sheikhs as a "slave" to learn their arts before returning to Anatolia.[25] He also claimed to have learned the *ilm el-hacer el-mükerrem* (literally, the science of the philosopher's stone) from an Anatolian master who was a disciple of sheikh Ali al-Marjushi al-ama. Al-Marjushi is believed to have flourished in Cairo, and he visited Anatolia during the reign of Bayezid II. He was executed during the reign of Suleiman the Magnificent (1520–1566).[26] These statements—disputable but indicative—put Egypt among the nodes from which alchemical knowledge was introduced to Asia Minor.

Ali el-Izniki undoubtly had recourse to alchemical texts circulating in the Islamicate world. In the preface of *Mecmuatü'l-mücerrebat* (a "laboratory" manual in Turkish), he stated that he collected and studied the books written by both ancient and contemporary authors.[27] He mostly quoted from the following authors: Jabir ibn Hayyan, Ibn Umail al-Tamimi (tenth century), Abu al-Isba b. Tamman al-Iraqi (thirteenth century), and al-Jildaki. In the preface of *Sırr al-Rabbani fi ilm al-mizan* (Divine secrets of the science of the balance) Ali Çelebi clearly mentioned that he used Jabir's *Kitab al-khawass al-kabir* (The great book of properties), the *Kitab al-burhan fi asrar ilm al-mizan* (Book demonstrating the secrets of the science of the balance) that was epitomized by Mullah Arab in 1512, and *al-Taghrib fi asrar al-tarkib* (The elimination of the secrets of the synthesis).[28]

Ali el-Izniki titled one of his works as *Durar al-anwar fi asrar al-ahjar* (Pearls enlightening the secrets of the stones) after reading al-Jildaki's *Kitab anwar al-durar fi idah al-hajar* (Book on the light of the pearls elucidating the stone). The similarity in both titles and the contents relating to the properties and the production of elixirs is remarkable.[29] Ali el-Izniki also produced a number of *müntehab*s (selections), *muhtasar*s (abridged texts), and *şerh*s (commentaries) of al-Jildaki's works.

Why was the influence of the Sufi alchemist al-Jildaki on Ali el-Izniki so great? In the Vienna copy of *Durar al-anwar,* Ali el-Izniki noted that, despite having studied alchemy for so long, he could not grasp a word of it because he had considered only the exoteric meaning of the terms. Upon discovering al-Jildaki's books he was able to unlock the secrets of the art. Besides observing Jabir's authority and conception of alchemical experiments, al-Jildaki was a representative of the mystical and allegorical trend in alchemy.[30] He wrote commentaries in defiance of alchemists who confined themselves to a literal interpretation. A Sufi himself, Ali el-Izniki was probably attracted by the mysticism and the use of allegory embedded in al-Jildaki, the "great compiler of alchemical texts," that incorporated lengthy quotations from earlier alchemical texts.[31] This wealth of information would have attracted Ali el-Izniki, who also found the recipes of former master alchemists such as Jabir and al-Razi, accompanied by al-Jildaki's comments.

Ali Çelebi el-Izniki's Recipes in *Mecmuatü'l-mücerrebat*

Al-Jildaki emphasized the mystical and allegorical aspects of alchemy, but he was also a skillful practical alchemist. Did el-Izniki follow in his footsteps and introduce the "practical alchemy" of Jabir and al-Razi and their recipes to Anatolia? As the texts produced by el-Izniki—and those attributed to him—have not been thoroughly investigated, it is hard to provide conclusive textual evidence. The lack of graphic proof, artefacts, and apparatus for chemical experiments relative to the period under study prevents us from making reliable assessments. In all likelihood, el-Izniki did not have a "laboratory" as al-Razi or Jabir is believed to have had. El-Izniki does not employ terms that indicate a workshop.[32] However, the linguistic features of *Mecmuatü'l-mücerrebat*, a treatise he wrote in Turkish, and some expressions he used in the preface suggest that he worked with a master, without specifying the whereabouts.

Ali el-Izniki implied that he produced or conducted alchemical operations "under the eyes" (supervision) of the masters. The preface also states that he recorded in *rumuz u işarat* (symbols and signs) on sheets of papers the *kemalat* (knowledge) he acquired by trial and labor over the years, lest it be forgotten. When approaching the end of his life, he decided to write a book using these notes to disclose in a clear manner—without signs and symbols—the names of the *ecza* (drugs), their *tertib* (composition), *tathir* (purification), *taktir* (distillation), *vezn* (quantity), the *ateşin miktarı* (intensity of fire), and the *cevherin hall ü akdi* (the dissolving and combination/fusion of metallic ores) so that one would not need to consult a master to perform the operations or recipes. For this purpose, he dubbed his notes *Mecmuatü'l-mücerrebat*, a collection of tried and attested operations or recipes. In Ottoman literature, the term *mücerreb* (*probatum est*) is mostly used by physicians to denote that the formulated prescription (remedy, drug, or therapy) had been employed and proved efficient.

A few times, el-Izniki used the term *gayet mücerrebdir*, which literally means thoroughly tested. Although in most of the descriptions he used the optative or imperative mood (*döşeye*, *saklıya*, *edesin*, *soğudasın*, *kızdurasın*), he also used the first-person singular: "I put a piece of iron in this water, I took it off, one-half was melted."[33] The latter sentence indicates that he may have performed some operations in person.

The penning of this practical guidebook in Turkish for performing chemical operations may be a consequence of his practical activities, because the compilations he made from al-Jildaki's works and a number of other treatises were all in Arabic. At the time, many authors, although their native language was Turkish, preferred to compile in Arabic, because epitomizing or commenting a text in Arabic was probably more convenient for them. Producing Turkish texts would compel el-Izniki to coin Turkish technical terms. Thus, most of the alchemical texts were written in Arabic, the teaching language of the madrasa, although alchemy was not included in its curriculum. Other reasons for choosing Arabic might be the Sufi authors' expectation that madrasa students would be interested in reading alchemical texts. The possible rivalry between dervishes and madrasa scholars may also have led the Sufis to produce in Arabic, the language of the Quran, the sciences, and the madrasa.

One of the seventeenth-century copies of *Mecmuatü'l-mücerrebat* is followed by an Arabic text of twenty-five folios bearing the title *al-Razi's kitab al-asrar*, al-Razi's early tenth-century manual.[34] In the subsequent folios, the Arabic text suddenly turns into Turkish, with lists of recipes. Further comparative studies are needed to ascertain whether this Arabic-Turkish text is derived from al-Razi. The recipes in the margins are evidence that practical alchemy was alive in Anatolia in the seventeenth century. A comparison made between the titles of the eight chapters of *Mecmuatü'l-mücerrebat* and the titles of the seven chapters of section 3 (On procedures) and the appendix (On parities) of *Kitab al-asrar* shows a correspondence in the structure of the two texts.[35] Recipes given in the texts are not the same, however, although the same constituents are involved and similar products are obtained. This gives the impression that el-Izniki modeled his book after al-Razi's *Kitab al-asrar* and inserted different recipes. El-Izniki's recipes are few and brief, and they lack the precision of those of *Kitab al-asrar*. For instance, to produce red ferric oxide, el-Izniki proposes to roast the iron in the open air, where al-Razi opts for the action of humidity on iron filings.

Ali el-Izniki ended his recipes with expressions such as *halis kamer ola* (in case you project this on tin or lead, pure silver will be obtained), or *bakır büradesini kamer eyler* (turns copper filings to silver).[36] No mention is made of the *hacer-i mükerrem* (philosopher's stone), which was supposed to transform base metals to gold and silver. The term "elixir" is mentioned a few times but without medical connotation. It is not clear if he expects to obtain real silver or gold, to transmute mercury and base metals into gold, or to transform them into a goldlike metal by coloring (i.e., aurifaction versus aurification). Jabir, in his *Kitab al-hudud* (Book of definitions), does distinguish between these two processes: the "science of the elixir" is defined as a science that both transforms and colors the low metals to become noble ones.[37]

Was Ali Çelebi el-Izniki a Practical or an Allegorical Sufi Alchemist?

Ali Çelebi el-Izniki was a prolific author. Twenty-eight alchemical works are listed in the epilogue of his *Durar al-anwar fi asrar al-ahjar*, and more than fifty alchemical texts are attributed to his name.[38] His works were copied extensively during his lifetime and up into the nineteenth century, revealing a continuing demand for alchemical texts in Ottoman society. His most popular works deal with the theory of the balance of seven metals, the generation of the metals under the influence of the conjunctions of seven planets, and the properties and preparation of the philosopher's stone, regarded as the subject matter of the "science of the elixir." Aiming to elucidate the secrets of the transformation of base metals to gold, his books are imbued with mysticism, and alchemy is seen as representative of the divine act of creation. The copies of the *Mecmuat el-mücerrebat* modeled after al-Razi's *Kitab al-asrar* as a laboratory manual are rare. Apparently, this text has no mystical overtones, but the descriptions of the procedures may have been regarded by Sufi mystics as describing the spiritual path to Gnosis. In light of current research, it is difficult to claim that a movement of practical alchemy existed in and around Bursa in the sixteenth century if we rely on the few copies of Ali Çelebi's *Mücerrebat*.

Ottoman Sufi Mystics' Involvement in Alchemy

Why did Ottoman Sufi mystics concern themselves with alchemy and copy, translate, and compile alchemical texts? It has been argued that the late sixteenth-century monetary crisis contributed to Sufis' interest in alchemical texts. The rise in prices and the devaluation that followed seriously damaged the livelihood of those depending on salaries up to mid-seventeenth century.[39] How did Sufi mystics make their living in Ottoman lands? Did financial difficulties compel them to have recourse to alchemy? This is a complex issue because the income of the mystics differed according to era, region, and the order to which they belonged.

The early mystics who settled in Anatolia around the thirteenth and fourteenth centuries made their living by cultivating the lands they occupied or that were endowed to them by authorities and from their livestock.[40] After the collapse of the Seljuk rule in the mid-thirteenth century, the emirs (local leaders), aimed to mobilize military support from the locals and began to build and endow lodges for the dervishes who had a significant influence over the local population. Endowments were a means of insuring that assets would remain within one family to provide a lucrative source beyond state control. An alliance existed between dervish groups and merchants: in return for spiritual services (worship, recitation of the Quran), the Sufis would receive remunerations or donations to be recorded as revenues.[41]

Early dervishes of Anatolia exhibited skills in constructing lodges, building and operating watermills, digging wells, channeling water into canals, irrigating lands, setting up ovens, and tanning leather. The popular saying "Bektashi's hoe, Mawlawi's nail" reflects their professional inclination. In order to meet the lodge's expenses they engaged in various crafts and were bound to *Ahilik teşkilatı* (craftsmen's corporations). *Akhis* (craftsmen) followed the *futuwwa* (ethics) in order to attain moral perfection just as Sufi novices followed the *tariqa*, the spiritual practices to reach the ultimate goal. The Jalwatiyya sheikh Üftade Efendi (1495–1580) is said to have made his living in Bursa by sericulture (or another craft related to the silk industry) and button making.[42]

Dervish lodges of the pre-Ottoman and Ottoman periods were *waqfs* (pious foundations). The founder endowed the building of the lodge and provided funds or revenues in order to meet its expenses in conformity with the *waqfiyya* (deed). Founders were men and women of wealth from all social strata. Sultan Bayezid I (r. 1389–1402), according to the deed, had endowed a lodge to the Ishaqi mystics in Bursa, two villages, a large domain with orchards, gardens, pastures, and rivers.[43] All dervish lodges were not so richly endowed, but they were given means to maintain their activities. Modest *tekke*s were maintained through resources provided by sheikhs and disciples:

Koyun [Sheep] Baba, a legendary fifteenth-century dervish, declined the endowment offered by sultan Mehmed II and continued sheepherding.[44] The sheikh of the lodge received remunerations from court members, ulema, state officials, members of the craftsmen's corporation and soldiers for his blessings, recommendations, therapies, and interpretation of dreams, among other services.[45] The community supported the *tekke* members largely with gifts in-kind.

Over the course of time, some Sufi lodges were transformed into institutions of public utility, and the sheikhs became official employees receiving regular salaries. Around 1655–1658 the sheikh of Bursa Mawlawi lodge received twenty-seven *akçe*s (aspers) a day.[46] The medrese teachers who taught in the forty-nine medreses in Bursa in the fourteenth to sixteenth centuries received salaries that ranged between twenty and fifty akçes a day, as stipulated in the foundation deed and according to their proficiency.[47] The average daily salary of the *müezzin* was five akçes, and that of an imam was ten akçes.[48] The daily salary of the Fatih Mosque's timekeeper (*muvakkit*) was ten akçes.[49] As for the salaries of the health personnel, an endowment deed of the fourteenth-century Yıldırım Darüşşifa (hospital) in Bursa stipulated the daily payment of forty akçes (38.4 g. silver) to the first physician and twenty-five akçes (25.6 g. silver) to another physician as well as a ration of wheat and rice per annum.[50] When compared to the above-mentioned salaries the sheiks' salaries were higher than those of the staff of the religious mosque.

During the economic crisis between the mid-sixteenth and the mid-seventeenth centuries, transformation of base metals into silver and gold may have become more interesting, but the alchemical texts studied so far do not hint at such Sufi enterprises. They might have been carried out clandestinely, however. Forgery was widespread in the sixteenth-century Ottoman Empire and was carried out by professional forgers, blacksmiths, locksmiths, and moneychangers (goldsmiths), but mainly by the employees of the state's mint who were equipped with the technical expertise.[51] The practice of adulterating silver and gold yielding concrete results such as coins with lower contents of valuable metals was easier to achieve.

The motivation of Ottoman Sufis to deal in alchemy seems to be similar to that of other Sufi mystics of medieval Islam. Islamic mysticism demonstrated the path to become an *insan-ı kamil* (a perfect man), a gnostic, and to attain union with God. To achieve this, the aspirant, under the guidance of a sheikh, was supposed to pass through a series of *maqam*s (stages or stations) of spiritual development in which he strove to purify his soul and remove worldly concerns. At the end of this *tariqa* (long journey) he attained the *marifat* (spiritual truth) and the *haqiqa* (divine reality). Since spiritual purification, self-denial, and illumination were essential in attaining union with God, Sufi mystics saw the transmutational processes of alchemical purification as a metaphor. Similarly, the tribulations of the dervishes in reaching the ultimate goal were likened to the various alchemical operations (i.e., calcination by fire) carried out to attain transmutation. Christian mystics were attracted to alchemy for similar reasons: by the end of the twelfth century, when Sufi mystics from the East started to settle in Asia Minor, Latin translations of Arabic alchemical works were available in Europe and attracted a considerable number of friars and monks. It was argued that the popularity of alchemy in religious orders in thirteenth- and fourteenth-century Europe went beyond the desire of transmuting base metals and was conceived as a spiritual process, the journey to spiritual perfection. Penalties issued by the church aimed to prevent the practice of alchemy for personal salvation. Franciscans, for instance, believed in their ability to distill a substance—the elixir—that could purify, transmute, and ennoble the human body and the soul at the same time.[52]

The Blending of Alchemy with Sufism

The analogy between the alchemical processes and the stages of spiritual development is not explicitly expressed in the Ottoman alchemical texts of the sixteenth and seventeenth centuries studied so far. An eighteenth-century treatise, however, bears witness to how the "science of the elixir" was correlated with the spiritual transformation of a mystic aspiring to become a perfect man. This is the *Risale-i ferahiyye* (Treatise of contentment) by the Sufi physician Mehmed Fahreddin, son of the physician Ahmed el-Bosnavi.[53] The term *ferahiyye* refers to the final *maqam* (stage) of spiritual development, which gives contentment and joy caused by the anticipation of the long-sought union with God. This author saw alchemy as a branch of medical arts, and he pursued the *ilm-i nar* (science of fire) for a long time without any benefit. He was illuminated only after reading a medical text by the Mawlawi physician Ömer Şifai of Bursa. The terminology used in *Risale-i ferahiyye* indicates that Mehmed Fahreddin was a Sufi himself. He conceived of *taktir* (distilling), *tathir* (cleaning away impurity), and *tefsil* (dividing into portions or sections) as the main operations of the physician's art (*tabib zanaatı*).[54] He used the term *süluk eylemek* (to follow the path of godliness) to denote his pursuit of alchemy.

The book was written to enlighten the *salikîn*, the devotees who strive in holiness, and to assist in their difficulties. The chapter titles show this purpose: The first chapter is on *taharet*, the canonical purification of the body, which is an example of the *misal-ı şeriat* (sharia, true path). The second chapter is on the *tarikat* ("the way") and deals with the *teskiyye-i nefs*, the debasing of one's self. The third deals with the *hakikat* (reality) and the *taltif-i cesed* (gratifying of the body), the fourth is on *takavvi* (fortifying process), *riyazat-ı makam-i neşviyye* (growth), and *teskiyye* (irrigation).

The text is a medley of sentences describing in alchemical terms the transformation experienced by the spiritual guide and the aspirant striving for virtue and holiness within a cell of absolute seclusion. Their spiritual purification and their path to perfection are identified with the transformation of base metals to silver and gold. At the end of this spiritual-chemical process, the aspirant becomes the "Shah of silver" or the "Shah of gold" and declares he will "enslave all birds [sulfur] and revive dead bodies." The aspirant-turned-Shah issues an edict to turn all his domains into silver or gold.[55]

In sum, then, the flow of alchemical knowledge from the Middle East in the works of medieval Muslim alchemists to the West Anatolian town of Bursa reached its peak in the late sixteenth century through Jalwati Sufi Ali el-Izniki's intensive work. The copying of el-Izniki's works in subsequent centuries suggests a continous interest in alchemical texts in the region as well as in Istanbul.[56] In the seventeenth century, this knowledge was both enriched and challenged by texts from Europe, texts that themselves had come into being through other itineraries of the same medieval arabic works. The entanglement of alchemical practice, the recipes from European iatrochemical formularies, and the involvement of Sufi mystics in medical treatment led to the production of Turkish texts on iatrochemistry by dervish physicians of Bursa.

Ibn Sallum, Iatrochemistry, and Dervish Physicians of Bursa

The fifteenth- and sixteenth-century Sufi mystics of Anatolia were probably involved in medical treatment, including certain psychological conditions, but they produced no therapeutic texts. The alchemical treatises studied so far are devoid of references to therapeutics. However, the probate inventory of Lami'i Çelebi (d. 1532), the Naqshbandi sheikh and poet from Bursa, is indicative of a personal involvement in medicine and apothecary.[57] The inventory list included pharmaceutical preparations such as syrups, theriacs, and oils; mortars and scales; copies of treatises of the eleventh–thirteenth-century Muslim physicians; and Ibn al-Baithar's (1197–1248) pharmaceutical compendium *al-Kitab al-jami al-mufradat al-adwiya wa al-aghdhiya* (The comprehensive book on medicinal simples and nutrition). Some of the dervish lodges in Bursa possessed libraries to which books were endowed. Most of the books were on Islamic theology, Sufi literature and poetry, indeed, among the 62 titles (320 volumes) kept in the dervish lodges of Bursa, only a single volume pertains to medicine: *Hulasat al-tıbb* (Synopsis of medicine).[58] As a full list of books from dervish lodges in Bursa is not yet available, the involvment in medicine in the *tekke*s is still an open question.

Sufi mystics of Bursa seem to have embraced medicine as a profession only in the second half of the seventeenth century, and their entrance into the field coincided with the introduction of iatrochemical therapies to the Ottoman Empire, both through translations and through European physicians practicing in major cities. Madrasa-trained dervish physicians of Bursa, heirs of the alchemical literature accumulated by their Sufi forebears, were attracted by these new therapies and endeavored to disseminate them by compiling formularies in Turkish. Arabic texts introducing iatrochemical knowledge had been compiled earlier in Istanbul under the initiative and supervision of Salih b. Nasrullah ibn Sallum (d. 1669), a court physician and native of Aleppo. The most popular iatrochemical texts were *Tarjamat al-tıbb al-kimyai al-jadid* (The translation of the new chemical medicine) and *Al-Kimiya al-malakiya* (The chemistry of the emperors), which constituted a single volume. They were based on the formularies of Daniel Sennert (1572–1637), Oswald Crollius (1563–1609), and other established European medical compendia from which the Ottoman translators—including one Greek translator—borrowed mostly the preparation of cures by using alchemical methods.[59] In a *Medhiye* (laudatory poem) to Ibn Sallum recited by the Damascene scholar Muhammad al-Muhibbi (1651–1699), a reference is made to dervishes arising in their rituals like the bubbles springing within wine.[60] This reference to dervishes in the eulogy to Ibn Sallum suggests that he may have been part of a community of sufis, indicating that the use of iatrochemical drugs as therapeutics was accepted by both court physicians and dervish healers.

Since non-Muslim families were in closer contact with Europe, it is possible their members may also have endeavored to produce iatrochemical texts. An Ottoman physician dragoman from a well-established Greek family, Alexandros Maurocordatos (d. 1709) studied medical sciences in Padua and Bologna.[61] He is known for having penned a treatise on physics and iatrochemistry in Greek, but his practical activities are unknown.[62]

Iatrochemistry was called *tıbb-ı kimyai* (chemical medicine) or *tıbb-ı cedid* (novel medicine) in Turkish because it introduced pharmaceutical preparations containing chemicals and minerals not widely used in traditional Islamic or Galenic medicine. *Aqrabadin*s (formularies) compiled in Anatolia prior to the seventeenth century mostly recommend the use of herbal drugs.[63] These preparations including minerals and metals were generally prescribed for external and topical use, although iatrochemical texts recommended their oral administration to increase the drugs' effects.[64] These iatrochemical drugs and therapies soon became an alternative to Ottoman physicians' traditional remedies.

The Dervish Physician Ömer b. Sinan el-Izniki's Medicinal Preparations

Ömer b. Sinan el-Izniki (fl. end seventeenth and early eighteenth centuries), qualified as a *derviş-i hazıkî* (dexterous Sufi), was most likely affiliated with the Mawlawi order, and trained at the madrasa and/or the *darüşşifa* (hospital) in Bursa before heading for Istanbul.[65] As no information is as yet available on the circumstances of his employment, we can surmise that he either lived in a Sufi lodge or maintained a "shop" (consulting room) where he may also have met European physicians and acquired books on chemical medicine. The formulary he penned in Turkish is entitled *Künuz-i hayat el-insan kanun-ı etibba-i feylesofan* (Treasures of human life: Canon of the philosophical physicians, 1695).[66] It starts by stating that Paracelsus "transformed chemistry which aimed to transmute common metals into gold and silver into the *spagyric art* which seeks to cure diseases."[67] The text includes numerous pharmaceutical preparations, such as spirits, powders, pills, pastilles, cataplasms, digestives, oils, ointments, and so on. A preparation called *mıknatıs-ı hükema* (magnet of the philosophers) is said to have the property of extracting the poison from the body of a plague-stricken person as a magnet would attract iron. For this preparation, *antimoni kibrit* (antimony sulphide, or stibnite [Sb_2S_3]) and *semm el-far* (arsenic trioxide [As_2O_3]) 100 dirhems (drachms) of each, are melted in a pot and then impregnated into cataplasms for employment in plague afflictions.[68] Ömer b. Sinan borrowed the recipes "he tested and deemed useful" from the "esteemed books" that Arab, Persian, Greek, and European physicians referred to.[69] The introductory chapter about natural philosopy and the foundations of chemical medicine were borrowed from Salih b. Nasrullah's *Tıbb al-kimyayi al-jadid.*[70] However, Ömer b. Sinan mentioned his efforts to provide the largest number of preparations made by *taife-i efrenc*—that is, Europeans, indicating that he acquired texts from European physicians practicing in Istanbul.[71]

The Dervish Physician Ömer Şifai Dede between Alchemy and Iatrochemistry

Son of a Mawlawi sheikh, the dervish Ömer Şifai (d. 1742, Bursa) appears to have acquired his medical knowledge in Bursa. He traveled to Egypt to learn *ulum-i garibe* (occult sciences) from a Khalwati sheikh Hasan Efendi.[72] His medical background and his exposure to alchemy in Egypt may have attracted him to iatrochemistry. Chief physician to the *darüşşifa* of Bursa, he was reputed to have mastered the literature in French and Italian medicine and chemistry.[73] Ömer Şifai compiled more than fifteen treatises, most of them formularies, including many pharmaceutical recipes. All in Turkish, these works were instrumental in popularizing iatrochemical recipes in Anatolia and served in the development of a vernacular iatrochemical corpus.

The dedication of *Mürşid el-muhtar fi ilm el-esrar* (An excellent guide for occult sciences) to sheikh Hasan, Ömer Şifai's Egyptian master, expresses in its title the correlation between the occult sciences and iatrochemistry.[74] Moreover, Ömer Şifai's reference to the authors he drew from as *hakim zu fünun-i mektum* (philosophers in occult sciences) implies an identification of alchemy with iatrochemistry. The text begins by describing methods for preparing strong mineral acids called *miyah-ı harre* (literally, corrosive waters) such as *aqua regia*, *ervah* (acids, such as spirit of sulfur), *duhan* (oils, such as antimony oil), mercury amalgams, sulfur compounds, salts, commonly used distillates, the calcination and purification of metals, the preparation of oxides, and so on. It ends with recipes including elixir of antimony and powder of magnesia. The *hacer-i filozof* (*lapide philosophorum*) described in the text is not the substance that is supposed to transform base metals into gold but, rather, a mixture of saltpeter, alum, vitriol, rock salt, and red vitriol, to be employed as a laxative. Ömer Şifai's sources possibly include the pharmacopoeia of Moise Charas (1619–1698), physician and apothecary to the French and British kings.

Ömer Şifai's most popular formulary was *Cevher el-ferid fi tıbb el-cedid* (The unique gem of the new medicine).[75] Composed in Turkish, it included

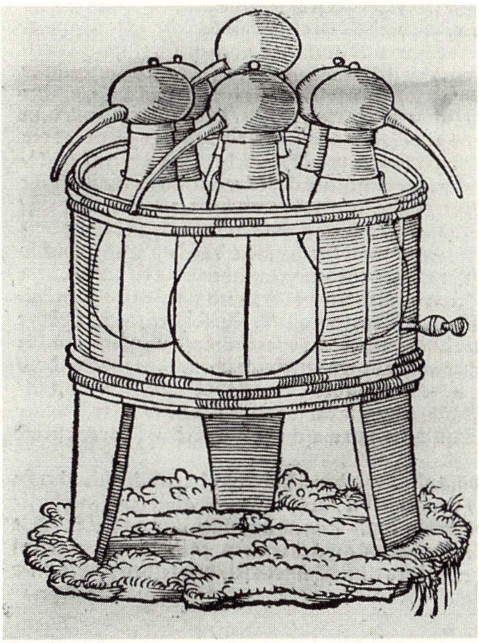

FIGURE 11.4. (*top*) Bain-Marie in Ömer Şifai, *Cevher el-ferid fi tıbb el-cedid* (The unique gem of the new medicine; 1698). Fol. 99a, MS Hamidiye 1020, Istanbul University Library, Istanbul. (*Bottom*) Walther Hermann Ryff, *Das new gross distillier Buch* (Franckfurt, 1545), p. X (image 36). http://reader.digitale-sammlungen.de/de/fs1/object/display/bsb11069395_00036.html?zoom=0.5.

about five hundred medicinal preparations, herbal in majority, although mineral remedies also appear in a good number of them. More than sixty recipes for powders, cataplasms, opiates/electuaries, cataclysms, pills, syrups, vines, herbal waters, ointments, and pastilles are appended to the main text. Besides borrowing recipes from Ibn Sallum's Arabic text *Tarjamat al-tıbb al-kimyai al-jadid*, Ömer Şifai also drew from other "recognized books and treatises of distinguished Latin and *efrenc* (Frankish) authors."[76]

The seventy figures of chemical utensils and apparatus in Ömer Şifai's *Cevher el-ferid fi tıbb el-cedid* appear to mostly comprise distillation apparatuses for the production of oils, spirits, and the herbal decoctions.[77] The *hamam-ı maria* (bain-marie) is similar to that depicted in Walther Hermann Ryff's *Das new gross distillier Buch* (1545), hinting at Ömer Şifai's borrowings from European chemistry books (see figure 11.4).

Figures depicted in both Ömer b. Sinan's and Ömer Şifai's treatises include water cooled *inbik el-hayye* (serpentine alembics), which do not figure in Ibn Sallum. Among Ömer b. Sinan's alembics (figure 11.5), the one demonstrating a barrel deserves special attention. It was used for fractional distillation to separate two components having different boiling points and for the purification of spirits (i.e., aqua vitae), which requires repeated distillations. These alembics appear to have first been used in the fifteenth or sixteenth century.[78] They apppear in Giambattista della Porta's *De Destillatione* (1608), which was published as an appendix to his *Magia Naturalis*.[79] They also appear in Libavius's *Alchymistische Practic* (1603) and *Syntagma selectorum* (1611). It is interesting to note that the barrel is placed on the right side in European texts and on the left side in Ottoman-Turkish texts. Arabic script is aligned from right to left, thus when rendering a European text to Turkish or Arabic, the translator must have copied the figures in the reverse.

Ömer Şifai's *Cevher el-ferid fi tıbb el-cedid* includes figures of three serpentine alembics that serve to separate the "bad humidity" emanating during the distillation of spirits. One of them, the *inbik el-hayye el-baride*, a serpentine alembic with cooler (see figure 11.6), has an open circular vessel filled with water in the middle of the pipe, which is changed as it warms. Whether the apparatuses depicted in both treatises were used for medicinal preparations is an open question. Moreover, we do not know whether the figures appended to Ömer Şifai's text were simple outlines of figures given him by Ömer bin Sinan, or perhaps both authors used a common source for the figures, employing them with more or less artistic skill.

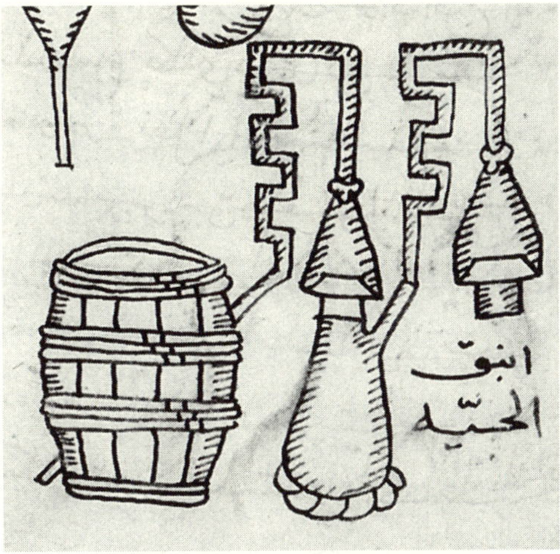

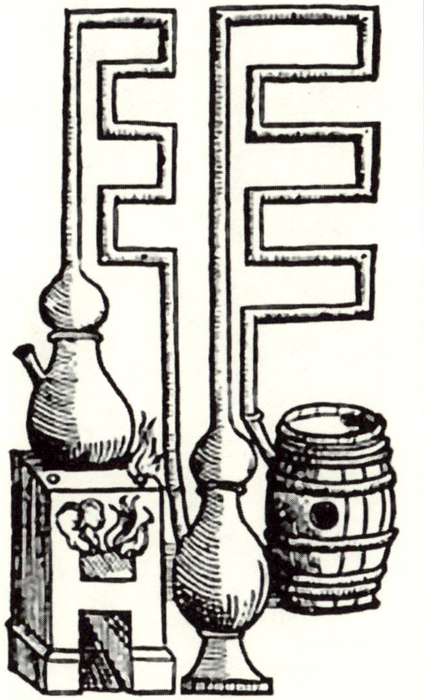

Fig. 41. Still with condenser and receiver, PORTA

FIGURE 11.5. Serpentine alembics for repeated distillation. (*Top*) Ömer b. Sinan, *Künuz-i hayat el-insan kanun-ı etibba-i feylesofan* (1695). Fol. 346a, MS T7083, Istanbul University Library, Istanbul. Note that the furnace under the vessel at lower right is not depicted. (*Bottom*) Della Porta's *De Destillatione* (1608), in Robert J. Forbes, *The Short History of the Art of Distillation* (Leiden: E. J. Brill, 1970), 119 (fig. 41).

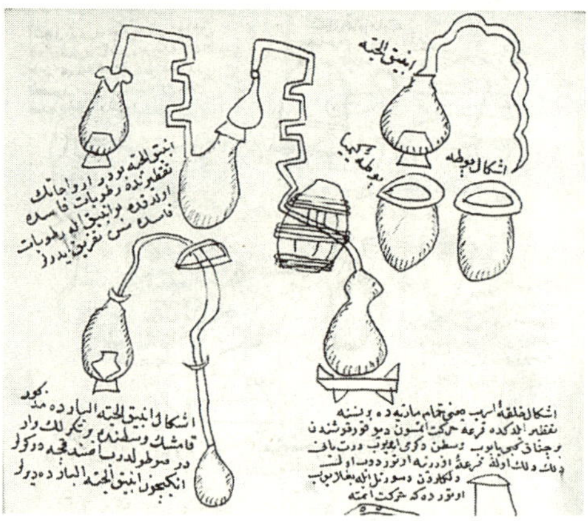

Figure 11.6. Serpentine alembics (*inbik el-hayye*). (*Top*) Ömer b. Sinan, *Künuz-i hayat el-insan kanun-ı etibba-i feylesofan* (1695). Fol. 346a, MS T7083, Istanbul University Library, Istanbul. (*Bottom*) Ömer Şifai, *Cevher el-ferid fi tıbb el-cedid* (1698). Fol. 99a, MS Hamidiye 1020, Süleymaniye Library, Istanbul.

Ali Münşi of Bursa, a Medrese-Trained Disciple of Ömer Şifai

Ali Münşi el-Burusavi (d. 1734), who studied medicine with dervish Ömer Şifai in Bursa, like his master was interested in European medical and iatrochemical texts. A member of the Bayramiyya Sufi order, Ali Münşi set off for Istanbul to become a court physician. He is known to have authored two *aqrabadin*s (formularies) in Turkish, both dated 1731, *Bizaat el-mübtedi* (Beginner's knowledge) and *Kurazat el-kimya* (Fragments of chemistry).[80] The German alchemist Adrien von Mynsicht (1603–1638) is frequently cited in the first formulary, and the second is said to be a translation of Mynsicht's *Thesaurus et armamentarium medico-chymicum*.[81] The content of Ali Münşi's two formularies may have been borrowed from the *Thesaurus*, which describes in detail the preparation of medicines containing mineral substances, discovered by Mynsicht himself or used by local apothecaries. Besides formularies, Ali Münşi compiled a number of treatises introducing exotic medicinal plants and their drugs (ipecacuanha, cinchona, zedoary root, sea coconut, and tea) that were hitherto unknown to the Ottomans.[82]

Ottoman Alchemical and Iatrochemical Texts on the Eve of Chemistry

Alchemical texts produced in the sixteenth and seventeenth centuries in and around Bursa were copied in subsequent centuries. Ali el-Izniki's works in Arabic seem to have been copied more frequently than his Turkish treatises, which points to a protracted interest in alchemy among enthusiasts proficient in Arabic who may have been attached to madrasas where alchemy was never officially sanctioned nor included in the curriculum. Nevertheless, some madrasa scholars did write about alchemy, and students grew interested in this art, against the admonitions of their masters.[83] The copying of alchemical texts among madrasa scholars or students may be related to the Sufis' eventual involvement in madrasa training over time. The penning of the abovementioned *Risale-i ferahiyye* by the physician Mehmed Fahreddin testifies to an ongoing adherence to alchemy in the eighteenth century as a guide in attainment of divine truth. The seventeenth-century iatrochemical formularies of Ömer Şifai also enjoyed popularity after the turn of the century. However, the Turkish formularies produced by Sufi physicians resident in Bursa could not compete with the production of Arabic works and the recipes of Ibn Sallum, the chief court physician. His connection to the imperial palace and the prestige of Istanbul as an intellectual and political center unavoidably eclipsed the works of Bursa iatrochemists although these formularies had many similarities.

From the mid-eighteenth century on, followers of Ömer Şifai and Ali Münşi moved to Istanbul where medical knowledge, both European or traditional, was more accessible and medical practice was more rewarding. Among their students was Derviş Abbas Tabib (d. 1760), the author of *Düsturü'l-Vesim fi tıbb el-cedid ve'l-kadim* (Vesim's code of new and ancient medicine) dated 1748, an encyclopedic work in Turkish dealing with some hundred diverse diseases.[84] The text refers to the cures and recipes of a number of seventeenth-century Italian, German, and Dutch physicians and naturalists. Abbas Vesim attributed the advanced state of European medicine to the lack of religious stipulations and consequently maintained that European remedies and physicians would not harm the Ottoman society.[85] The chief court physicians of the eighteenth century, the successors of Ibn Sallum, did not write on "new medicine," possibly regarding Ibn Sallum's work as definitive. An exception was Gevrekzade Hafız Hasan (d. 1801) who, by translating Ibn Sallum's *al-Tibb al-kimyai al-jadid* into Turkish, further contributed to the survival of Sennert's and Crollius's recipes among Turkish-speaking physicians.[86]

The dissemination of iatrochemical recipes in the eighteenth century did not hinder the translation of a Persian pharmaceutical-alchemical text into Turkish in Istanbul in 1733. This was the *Tuhfah al-muminin* (Gift to the believers) dated 1669 and written by the Safavid court physician Muhammad Mu'min at-Tankabuni (d. 1698–99).[87] A compilation from sixty-one medieval Islamic authors, the *Tuhfah* deals with drugs, remedies, and chemical processes, including the preparation of the elixir. Its wide use reveals that Ottoman physicians relied on both Eastern and Western pharmacopoeias throughout the eighteenth century.

The manuscripts of Sufi iatrochemists from the seventeenth–eighteenth centuries were more widely copied and circulated than the alchemical manuscripts of the fourteenth through sixteenth centuries, not only because the community grew larger and communication became easier but also because these manuscripts included a large number of pharmaceutical recipes. Copies, although very few, were still available in Anatolian towns such as Çankırı, Çorum, Izmir, Manisa, and Sivas as library catalogues attest. The circulation of manuscripts between Bursa and Istanbul could be expected to grow more intense as Sufi physicians of Bursa moved to the capital city to hold higher posts. The physical features of the manuscripts remained almost the same, but the page count increased. Illuminated manuscript copies were made to attract imperial patronage, as, in contrast to alchemy, iatrochemistry was not considered a harmful art.

Conclusion

Alchemical texts compiled in the fifteenth and sixteenth centuries demonstrate that Sufis, particularly in and around Bursa, endeavored to introduce medieval Islamic alchemy to Anatolia, especially the theories and recipes of Jabir ibn Hayyan, al-Razi, and al-Jildaki. These texts were mostly produced within Sufi orders that combined heterodox and orthodox elements in their teachings. The main paths for conveying this knowledge ran between Anatolia and Egypt, and also between Transoxania and the Iberian peninsula, up to the mid-seventeenth century when new itineraries emerged between Istanbul, Bursa, and Italy, Germany, and France following the ever burgeoning commercial and military contacts between Ottoman and European states. These new itineraries connecting Istanbul, Bursa, and the European cities paved the way for the Sufi physicians to adopt iatrochemical recipes from European formularies.

Ottoman Sufis turned to alchemy mainly to purify their soul to attain perfection and divine truth. They employed multistage transformation processes of base metals into gold or silver as a metaphor for their spiritual journey. They also hoped to unveil the secrets of the Creation of the universe and humankind. The lack of material and literary evidence makes it difficult to prove that Ottoman dervishes undertook experiments to search for God's reflection in nature. It seems there were no "laboratories" similar to those established in Europe in the early seventeenth century, resembling artisanal workshops, equipped with distillation and assay furnaces, laboratory utensils, and chemical substances and employing a number of workers.[88] The craftspeople who used distillation vessels—such as the *gülâb-ger* (rosewater producers) who produced rosewater and rose oil in Anatolia under Seljuk rule (1081–1307), and the many other artisans enumerated by Evliya Çelebi in the seventeenth century making therapeutic oils and arrack (spirits)—as well as ironsmiths, cauldron makers, glassmakers, and potters show that the material conditions for the investigation of nature and natural materials were present. But it seems Ottomans did not use these materials and equipment to investigate nature more broadly.[89] Alchemical activities seem to have been confined to textual and spiritual practices.

Although Ottoman Sufis regarded alchemy as part of a spiritual quest and an intellectual practice, their endeavors in creating an alchemical literature in both Arabic and Turkish in the sixteenth century along with the copies made from their works throughout the seventeenth century converged to create a textual and intellectual field of alchemy. Within this field were gathered together a multitude of recipes for distillation, sublimation, calcination, and the softening and solidification of various metals and chemical substances. The copying and distribution of these recipes brought Sufis familiarity with mineral drugs. Consequently, when iatrochemical recipes were introduced in the second half of the seventeenth century, the Sufi physicians, heirs of this literary alchemical tradition, made these recipes available by compiling formularies in Turkish for the community unfamiliar with Arabic, thereby introducing a practical dimension to what had previously been an apparently purely textual tradition.

The introduction of new chemical recipes did not create a conflict among Ottoman physicians as it sometimes did between Galenists and Paracelsian physicians in Europe, although English physicians merged chemical therapy with the Galenic system.[90] This lack of conflict among Ottoman medical practitioners may be because iatrochemical recipes largely stemmed from the works of Daniel Sennert (1572–1637) who was simultaneously a strong believer in humoral medicine and a proponent of Paracelsian principles. For Sennert, the new chemical medicines were an important addition to the traditional materia medica.[91] The coming of new chemical recipes to Ottoman Turkey did not disrupt the use of Galenic remedies, nor did it hinder the transmission of pharmaceutical remedies from the East. A few years after Ali Münşi, the physician of Bursa, rendered into Turkish the popular formulary of Mynsicht, the renowned German Paracelsist, a Persian alchemical-pharmaceutical text (*Tuhfat al-muminin*) was translated into Turkish in 1733. A group of Ottoman physicians—including the Sufis—incorporated the new recipes into their materia medica and prescribed them, together with the herbal and animal drugs they derived from medieval Islamic and Ottoman *aqrabadin*s. Some Ottoman physicians were resistant to iatrochemical remedies, however, perhaps because of their lack of knowledge or experience about the efficacy and toxicity of the mineral remedies or because of difficulties in deciding the dosage. Indeed, the shops of some Frankish physicians who harmed the patients by administering "new medicine" were ordered to close, and this, too, may have contributed to their reluctance.[92] Interest in iatrochemical cures seems to have gradually waned from the mid-eighteenth century onward, but recipes still permeated the traditional formularies composed in subsequent years.

Reliable evidence is lacking concerning Ottoman patronage of alchemy. In the eyes of his subjects, Bayezid II was a *veli* (*wali*), a saintly figure associated with miracles, and his sobriquet *kimya'i* is plausibly a derivative along the same lines. The availability of a dozen alchemical manuscripts in the palace may attest to Bayezid II's interest in alchemy.[93] However, there is no indication of the sultan's personal practical involvement in alchemy. Supposedly, an alchemist promising to turn lead into gold performed some operations before the sultan and subsequently, on discovery of his trickery, took his own life.[94] Accounts of regular alchemical experimentation within the palace in the chronicles of his reign or illustrations depicting Sufis and guilds engaging in alchemy have not been found. The only other historical reference to "alchemical" performance is a century later, when Murad IV (r. 1623–1640) summoned two "alchemists" to transmute base metals to gold and silver and allegedly had both of these persons executed after their failure.[95] While court physicians, pharmacists, and astrologers were regularly appointed, no position of court alchemist can be found, nor any allusion to a workshop for alchemical experimentation. Therefore, patronage similar to that by the Caliphs and the wealthy families in the eighth–tenth centuries of the alchemists Jabir ibn Hayyan and al-Razi did not exist in the Ottoman court. Nor was support of alchemical activities by royalty and well-off persons forthcoming, as was the case in Europe in the sixteenth and seventeenth centuries.[96] And, in periods of monetary crisis, the Ottoman rulers simply opted for the debasement of the coinage. Thus, despite the attitude of madrasa scholars who regarded alchemy as a harmful practice, and the ulema's commitment to theology, as well as the lack of any noble patronage, the esoteric tradition of alchemy still managed to flower among fifteenth- and sixteenth-century Ottoman sufis in Bursa and produced a textual tradition, which, in turn, with the growing exchange between the Ottoman Empire and Europe, fostered a practical dimension of iatrochemical practice in Bursa and Istanbul by the late seventeenth century.

Table 11.1. Alchemical Texts Referenced (14th–18th Century).

"Alchemists"	Ancestors, Sufi order	Title, date	Language	Sources	Notes
Aşık Ali Pasha (Sheikh Aşık) (1271-1332)	Khorasan, Wafaiyya	*Risale-i Aşık Paşa*	Turkish		An early Anatolian alchemical poem
Sheikh Safiyüd-din (d.1513)	Zayniyya	*Kaside-i Şeyh Safi*	Turkish		
Mohammad b. Omar al-Antaki (Mullah Arab) (d. Bursa, 1531)	Transoxania, Hanafi scholar	*Talhis al-burhan, 1512*	Arabic	al-Jildaki's (d.1342) *Kitab al-Burhan fi asrar ilm al-mizan*	Epitomised
		Talhis ar-rutbat al-hakim	Arabic	Maslama b. Qasim al-Qurṭubī's (d.964) *Rutbat al-Ḥakīm*	Epitomised
Ali Çelebi el-Izniki (d.1609)	Ashrafiyya (a branch of the Qadiriyya)	*Mecmuatü'l-mücerrebat*	Turkish		Modeled after al-Razi's *Kitab al-asrar* (10th c.) ?
		Sırr al-Rabbani fi ilm al-mizan	Arabic	Jabir ibn Hayyan's (9th c.) *Kitab al-Khawass al-Kabir, Kitab al-burhan fi asrar ilm al-mizan, and al-Taghrib fi asrar al-tarkib*	
		Durar al-anwar fi asrar al-ahcar	Arabic	al-Jildaki (d.1342)	Similarity of its title with al-Jildaki's *Kitab anwar al-durar fi idah al-hajar*
el-Seyyid Mehmed Fahreddin b. Tabib Ahmed el-Bosnavi (eighteenth c.)	Sufi	*Risale-i ferahiyye*	Turkish		Compiled after the lecture of Ömer Şifai's works

Table 11.2. Ottoman Iatrochemical Texts Referenced (17th–18th Century)

Iatrochemists/ Physicians	Ancestors, Sufi order	Title, Date	Language	Source	Notes
Salih b. Nasrullah ibn Sallum (d. 1669)	A dervish ?	*Tarjamat al-tıbb al-kimyai al-jadid; Al-Kimiya al-malakiya; Tarjamat al-aqrabadin al-jadid.*	Arabic	Sennert (d.1637); Crollius (d.1609) and other European Paracelsians	Completed and edited by other court physicians after ibn Sallum's death
Ömer bin Sinan el-Izniki (fl. end seventeenth early eighteenth c.)	Mawlawiyya	*Künuz-i hayat el-insan kanun-ı etibba-i feylesofan* Istanbul, 1695	Turkish	Arab, Persian, Greek and European physicians; ibn Sallum's *Tarjamat al-tıbb al-kimyayi al-jadid*	Author's copy; mentioned as Şifa el-Müminin in the catalogues
Ömer Şifai el-Burusavi (d. 1742)	Mawlawiyya	*Mürşid el-muhtar fi ilm el-esrar*	Turkish	Moise Charas (d.1698) ?	Studied in Bursa, fl. in Bursa and Istanbul
		Cevher el-ferid fi tıbb el-cedid		Ibn Sallum's *Tıbb al-kimyai al-jadid;* Latin and Frankish authors	
Ali Münşi el-Burusavi (d.1733/34)	Bayramiyya	*Bizaat el-mübtedi,* 1731	Turkish	Mynsicht's (d.1638) name is frequently cited	Student of Ömer Şifai; studied in Bursa, fl. Istanbul
		Kurasat el-kimya, 1731		Mynsicht's (d.1638) *Thesaurus et armamen-tarium medico-chymicum*	
Dervish Abbas Tabib (d.1760)	Khalwatiyya, Qadiriyya, Naqsbandiyya	*Düsturü'l-Vesim fi tıbb el-cedid ve'l-kadim,* 1748	Turkish	seventeenth c. Italian, German, Dutch physicians and naturalists.	Student of Ömer Şifai and Ali Münşi
Gevrekzade Hafız Hasan (d.1801)	Jalwatiyya	*Mürşid el-etibba fi tercemet-i ispagirya*	Turkish	Translation of Ibn Sallum's *Tarjamat al-tıbb al-kimyai al-jadid*	Istanbul, chief physician

Chapter 12

A Wooden Skeleton Emerges in the Knowledge Hub of Edo Japan

Chang Che-chia

In 1793, six years after the Irish Giant's skeleton was put on display in the Royal College of Surgeons in London, a Japanese bonesetter named Hoshino Ryōetsu 星野良悦 (1754–1802) welcomed visitors to his private house in Hiroshima to see a wooden skeleton from which they could learn about the structure of human bones.[1] The exhibitions' dates are close, but they appear in quite different social contexts. In Europe visitors were already used to exhibitions of human skeletons run by individuals, but in Japan hosting such an exposition by a private individual was unprecedented. Although the object on display appeared to be a genuine skeleton, Hoshino actually exhibited a wooden model. This model was intended to represent a "standard" human being rather than an exceptional case for public curiosity. By mounting such an unusual exhibition, Hoshino aimed to challenge the normal circulation of knowledge about bone setting, usually called *hiden* 秘伝. His act can be seen as intending to bring a new openness to the knowledge of bone setting. The term *hiden*, or *secret transmission*, connotes a certain body of knowledge or the circulation of skills that is confined within a closed group that is usually hereditary. This manner of passing down knowledge still survives in some fields in modern Japan, such as Buddhist practices and martial arts. In the Edo period (1603–1848), hiden practices were widespread.[2] In some specific fields such as acupuncture, the hiden system even regulated lessons by organizing them into several levels of access to knowledge. Only when a master approved a pupil's performance could the student then advance to a higher level of secrets. As for the most important secrets, the master passed these on only to his successor, usually his eldest son. This mode of knowledge transmission was the target of Hoshino's display.

Hoshino's skeleton, in fact, was not just the result of his opposition to the system of closed associations in bone setting. It also represented the convergence of three different systems of medical knowledge. The first was local bone setting, in which lineages of expert practitioners guarded their knowledge from outsiders. The second was Chinese medicine, in which large corpuses of texts prevailed. The third was Dutch Studies, recently arrived in Edo Japan, and which at first also largely followed the local hiden customs but later was characterized by its widely circulating texts crowded with anatomical images.

The Cultural Field of the Skeleton

Hoshino's skeleton was in part a response to the particular problems of exposing the human body in Edo Japan. Authorities permitted the exposure of criminals' carcasses in public for several days in order both to extend their punishment into the afterlife and to deter potential offenders from committing such crimes. Dead bodies were seen as impure, and after the tortures, the criminals' bodies were burned down to their bones. The dead bodies, however, were of interest. After one execution in Kyoto, for example, an eye specialist, Negoro Tōshuku 根来東叔 (1721–1787), was curious about the skeleton, and considering images made from it as possibly useful for medical studies, he depicted it in *Jinshin renkotsu shinkeizu* 人身連骨真形図 (Illustrations of the true shape of human bones, 1741). These are Japan's earliest depictions with medical intent of the internal landscape of a human body based on observations from life. The government did not prohibit Negoro from drawing his sketches, despite criticism that the act was cruel and ungentlemanly.[3]

The government reserved the right to decide when and to whom special permission would be granted for opening the human body. The Yamada family, who were the Shogun's special agents designated as executioners, held a monopoly to eviscerate those who had been executed in order to obtain their livers for making pills.[4] The first known individual outside this monopoly who applied for special permission was the Chinese-style physician Yamawaki Tōyō 山脇東洋 (1706–1762). Through his earlier experience in dissecting otters he had come to doubt the correctness of the anatomical knowledge in the Chinese medical canons. In order to investigate the truth of the canons, he presented a request to dissect a criminal's body, and this petition was granted by an administrator of Kyoto. Yamawaki authenticated his observations with colorful illustrations in the book *Zōshi* 蔵志 (Records of the organs, 1759),

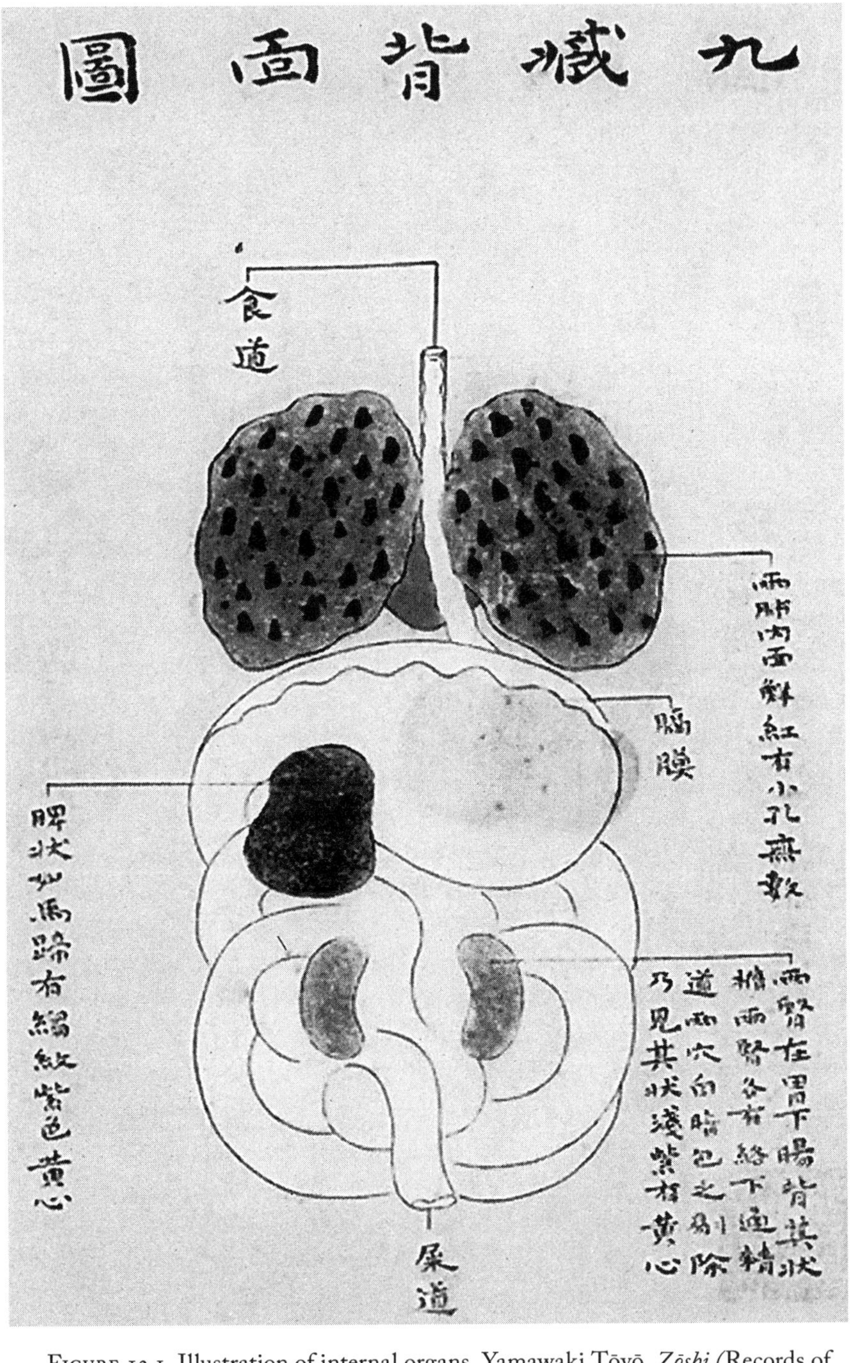

FIGURE 12.1. Illustration of internal organs. Yamawaki Tōyō, *Zōshi (*Records of the organs; 1759). Public domain.

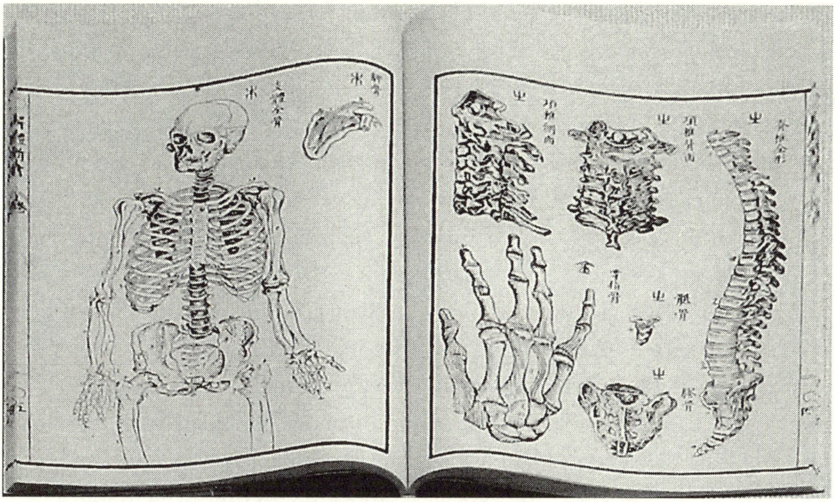

FIGURE 12.2. Illustration of the skeleton. Sugita Genpaku, *Kaitai shinsho* (New book of anatomy; 1774). Public domain.

pointing out that the Chinese orthodox understanding was problematic. This achievement earned him a place in medical history as the first dissector but also condemnations for cruelty from his contemporaries for the rest of his life (figure 12.1).[5]

Yamawaki's work stimulated Sugita Genpaku 杉田玄白 (1733–1817), who served the fief lord Obama 小浜 as attendant physician and who had devoted his early life to studying Chinese medicine. Seeing from Yamawaki's work that Chinese doctrine could be unreliable and having come into contact with Dutch anatomical illustrations, Sugita applied for permission from the shogunate in Edo to carry out a dissection of a criminal's body. When he went to the execution grounds, he carried a Dutch anatomical manual with him. Astonished by what he saw as the accuracy of the anatomical illustrations, he concluded that the Chinese canons had misled the Japanese for too long, and he resolved to make Western anatomy known to the public. Soon afterward he launched a project to translate the Dutch manual, which resulted in *Kaitai shinsho* 解体新書 (New book of anatomy) in 1774. Although at this moment Sugita could verify the Dutch superiority only in anatomy, he threw down the gauntlet to the Chinese medical orthodoxy in general, incurring strong resistance. One year after the publication of the Japanese translation, he wrote a pamphlet to give an account of the mutual bitter criticisms between himself and the Chinese-style physicians.[6] He insisted that his campaign against

Chinese medicine was correct but had to admit that some of his critics' comments on the incorrectness of his translation were fair. Later he assigned his student Ōtsuki Gentaku 大槻玄沢 (1757–1827) to refine his somewhat rushed translation of 1774 (Figure 12.2).

Sugita's efforts reverberated throughout Japanese society, and they aroused Japanese interest in Western learning. Young people came to Sugita and other scholars with access to Dutch knowledge, such as the official interpreters, to study Western knowledge. In the Edo period the knowledge was commonly called *rangaku* 蘭学 (Dutch learning).[7] In the beginning, the contents of Dutch learning were limited to products of trade, medicine, and technology, and they did not challenge the canonical mainstream mainly formed by the Chinese texts. The pupils of Dutch learning did not necessarily bear animosity toward Chinese learning as Sugita did. For example, in the field of medicine, some physicians adopted both Chinese and Dutch clinical techniques, and they were called eclectic thinkers sitting between Chinese and Dutch medicine (*Kan-ran secchū ha* 漢蘭折衷派).

The Field of Bone Setting

So, as this background shows, the treatment and representation of the human body and skeleton was a contentious arena when Hoshino exhibited the wooden replica in 1793. From Hoshino's perspective, however, he saw only an opportunity within the field of bone setting. Hoshino succeeded his father as a bonesetter, practicing in the castle town of Hiroshima. When he was a novice, someone in town dislocated his jaw and no bonesetter in the vicinity knew how to deal with the problem. Hoshino discovered that the townspeople knew whom to turn to when there was a difficult case—an illiterate healer, Tanaka Michinaga, who easily cured the dislocation. As Tanaka was about to manage the mandible disorder, he had his assistants hide the procedure with a large piece of cloth so that nobody could observe his healing technique. Admiring Tanaka's proficiency, Hoshino paid homage to him by inquiring about learning his procedures. Tanaka refused because he had acquired these techniques through *hiden* (or secret transmission), thus, any outsider was to be denied the opportunity to study his methods. As Hoshino later recalled, Tanaka's arrogance agitated him, which later triggered his desire to create the wooden skeleton.[8]

Within all the subfields of medicine, information on bone setting was the least available in the book market. Thus, unlike many other subfields in medicine, secret transmission was exceptionally common among bonesetters. At Hoshino's time, only one book had been published on this topic. There were two important causes for this: one was that the transmission of bone setting skills relied upon bodily knowledge learned through apprenticeship; the other was that physicians of high social status would avoid treatments of bodily injuries that required manual skills. Indeed, most elite physicians considered bonesetters as belonging to a lower social level.

While the lack of texts makes it difficult to trace the early history of bone setting, archaeological evidence makes clear that the practice goes far back into Japanese history.[9] In the Heian period (794–1185), a system of medical bureaus were formed in imitation of the Tang, and since that time Chinese medicine became the mainstream of medicine in Japan. The Heian period reforms categorized bone setting as a part of massage instead of medicine, and we know little about its practice.[10] The almost 150 years of civil war in Japan from the latter half of the fifteenth century dramatically raised bonesetters' visibility in society. The demand for capable surgeons on the battlefield saw a sharp increase, and monks and lower class samurai filled the ranks of surgeons, and they began to do the tasks of bone setting. These new specialists were especially concerned with keeping their expertise secret. Military surgeons developed into about thirteen schools during this period, and all of them without exception transmitted skills through hiden.[11]

At times on the battlefield, this preservation of secrecy was impossible to maintain. After a rout a surgeon might lose his belongings or even his life, and sometimes this opened up a window on surgeons' practices. For example, a prescription dated 1622 noted: "This formula was leaked from the school of Harii. Given that it should have been *transmitted in secret*, devoted healers passed the prescription around among themselves and made it quite well-known."[12] These "secret formulas" appear to have partial roots in some quite ancient written sources, among them Chinese sources from the earliest fifth-century *Xiaopin fang* 小品方 (Short collection of prescriptions) to the *Shiyi dexiao fang* 世醫得效方 (Hereditary physician's prescriptions proved to be effective) of 1338.[13] This indicates both that Chinese medical texts continued to be used during this time and that the bonesetters' secret knowledge when written down seemed to rely on well-known texts or at least borrow from them to express their knowledge in writing.

These secret recipes also show traces of popular practices: A remedy for a broken hand included mixing "horse dung and lotus seed, then grind them into powder, [and] take them with tea or hot water"; "before taking the medicine, [the healer] may chant the spell on the road eastward, going back and forth three times." A second recipe recorded the *Kinkotsu tsugi ju* 筋骨續咒 (Spell for linking tendons and bones), which translates as "Hurry! Hurry! Follow the Edict Orders" from the Chinese *Qianjin yifang* 千金翼方 (Supplementary prescriptions worth one thousand pieces of gold).[14] According to Andrew Goble, Japanese Buddhist priests maintained a network with the Chinese sangha during the medieval period. Because they were literate and possessed access to new texts, monasteries became an important source of textual accounts of the hiden of bone setting.[15] As treated in other essays in this volume, we see in this the interweaving of religious and medical practices.

After the end of civil war in Japan, continuing upheaval on the Eurasian continent caused further movements of population, which created new relational fields. Many refugees came to Japan in the mid-seventeenth century because of the wars between the Manchus and the Chinese, and they redefined the hierarchy of the bonesetters' trade. One eminent immigrant by the name of Chen Yuanyun 陳元贇 (1587–1671) escaped to Japan and taught various kinds of arts. Because he learned Shaolin kung fu when he was in China, it is believed that he taught the Japanese the skills that later developed into the *judō* school of martial arts in Japan, and as part of his courses he taught relevant practices of bone setting. Some bonesetters' lineages today regard this as their founding history, recounting that the Yoshihara 吉原 in Nagasaki, the Nembai 年梅 in Osaka, and the Nakura 名倉 in Edo all learned from Chen's school and became leading bonesetters in their respective cities. Chen himself may not have intended to keep the techniques of bone setting secret, but his disciples founded their own lineages of hiden. First competing with and then surpassing the local bonesetters, members of these new lineages became the dominant bonesetters in the Edo period. The lineage of Nakura still flourishes today.[16]

Nevertheless, because of the lively printing industry in the Edo period and the fact that Chinese medicine was transmitted by a textual tradition of *longue durée*, orthodox Chinese medicine was disseminated even more widely than previously.[17] Fields of practice such as bone setting that did not possess a canon of texts and illustrations continued to be marginalized in this medical marketplace. And when bonesetters' knowledge was written down, Chinese texts continued to underpin these books into the eighteenth and nineteenth centuries: for example, the work on bone setting *Seikotsu han* 正骨範 (Guidelines for bone setting) published in 1807, drew from the Chinese imperial compilation of the medical encyclopedia *Yuzuan yizong jinjian* 御纂醫宗金鑑 (Golden mirror of the orthodox lineage of medicine, 1742).

Increasingly in the early Edo period, Chinese texts were only one source of knowledge in the Japanese medical world. Japanese readers also began to encounter Dutch medicine and began to show an especial interest in Dutch anatomy and surgery. In the mid-seventeenth century, a Dutch physician serving in Nagasaki taught two official interpreters—Inomato Dembei 猪股伝兵衛 and Nishi Genbo 西玄甫—some bone-setting and surgeons' skills. In 1706, another interpreter, Narabayashi Chinzan 楢林鎮山 (1648–1711), translated the Dutch version of Ambroise Paré's (1510–1590) surgery book *De chirurgie, ende opera van allen de Wercken*, entitling it *Kōi geka sōden* 紅夷外科宗伝 (The master's transmission of Dutch external medicine).[18] Also, Hara Sanshin 原三信 (d. 1711) translated the first Western anatomy book in Japan. He noted on the manuscript that "[The secret] will only be passed on to one single son in the lineage, and always kept inside the door."[19] Although Narabayashi and Hara wrote down the knowledge, they did not intend to make it public. At this early stage only the interpreters had the ability to learn from the Dutch, which made it possible for them to form a monopoly in the medical marketplace. No matter whether drawing upon Chinese or Dutch texts and practitioners, the general practice of the field of bone setting arts was to follow the tradition of hiden.

Transmitting the Secrets

There was only one publication on the arts of bone setting before Hoshino's time. This first Japanese bone-setting book was entitled *Honedugi ryōji jūhōki* 骨継療治重宝記 (Treasured notes for healing bones, 1746). The author Kōshi Hōyoku 高志鳳翼 was a monk versed in Chinese learning, rather than a bonesetter. Nevertheless, he interested himself in the techniques and learned some bone setting. From the Namba village in Osaka where prestigious bonesetters gathered, he had easier access to the art. According to his prefaces, his ancestors owned a few secret volumes on bone setting. Furthermore, Kōshi himself had learned bone-setting skills through the traditional manner of hiden. As he boasted in his prefaces, he later encountered a rare Chinese manual, and he subsequently achieved skills so great that they went well beyond his peers' imaginings. On completion of his book, he was only about thirty, and he announced that he planned to publish another fifty-one titles—covering, among other subjects, the Confucian Classics, Buddhism, medicine, and astronomy. If a thirty-year-old had the knowledge to publish such a quantity of books on various topics, he clearly spent more time in reading than in practicing bone setting. Indeed, the power of Kōshi's book was in its literary scholarship: he cited seventeen titles of Chinese medical books up to the Late Ming, and some materials translated from Dutch.[20] The 350 pages of Kōshi's book included sections on treatments, manual skills, and prescriptions. Among the treatments, about forty pages were devoted to the human body, including a list of the names of the 365 bones.[21] This number of bones was deeply rooted in textual transmission since the time of *Huangdi neijing* 黃帝內經 (Yellow Emperor's Inner canon, first century BCE?) and was still faithfully followed by the Chinese physicians of Kōshi's time. He quoted all of the names of bones from the Chinese scholarly physician Wang Kentang's 王肯堂 (1549–1613) *Yangyi zhengzhi zhunsheng* 瘍醫證治準繩 (Guidelines for healing sores and wounds, 1608), which derived from a chapter of the Song dynasty's imperial compilation, the canon *Shengji zonglu* 聖濟總錄 (Medical encyclopedia, a sagely benefaction, 1117). Because the Song government conducted several dissections for medical investigation, the information had an empirical foundation, but nails and cartilage were also viewed as bones, which greatly increased the number. Most interestingly, Kōshi supplemented the textual descriptions with illustrations, which appear in surviving Chinese texts.

Kōshi also stated that a bonesetter should know some Dutch skills, and in his book's illustrations he depicted the bones in more than one dimension to show his object in the round, a practice unprecedented in any similar book in either Chinese or Japanese. In a short description of the skull, he shows the skull from the bottom and the side. In the bottom-up view, the various cavities of the skull as well as brain and spinal cord are explained as interlinked. In the view from the side, he explains that the nature of the eye is "cold and moist," which derives from Western medical theory of the time. In contrast, in the descriptions of the teeth, he states "The tooth is the bone's extension," (a quotation from Chinese medical canon). His text thus mixes Chinese and European medicine.[22] Kōshi apparently did not differentiate between these knowledge traditions, as he also used the term *kin* 筋 (by which he had previously named the nerves) to refer to the Twelve Meridians in Chinese medicine (figure 12.3). His book, which he paid to have published, was not reprinted until 1810. Despite this, Kōshi became known as the first *author* on this topic in Japan, known for translations from Chinese into Japanese, and for the

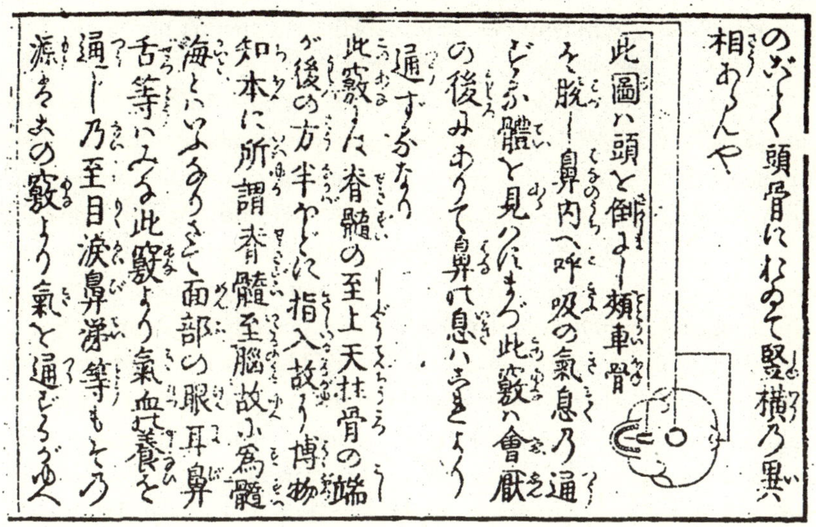

のごとく頭骨におゐて竪横乃異
相あらんや
此圖ハ頭と顋よし頬車骨
を脱し鼻内へ呼吸の氣息乃通
じる髏を見るにまづ此竅ハ會厭
の後みありて鼻の息ハこゝより
通ずるなり
此竅りは脊髓の至上天柱骨の端
が後の方半分に指入故より博物
知本に所謂脊髓至脳故ニ為髓
海といふよりさて面部の眼耳鼻
舌等ハみな此竅より氣血共養を
通し乃至目涙鼻涕等もその
源ハこの竅より氣を通じるがゆへ

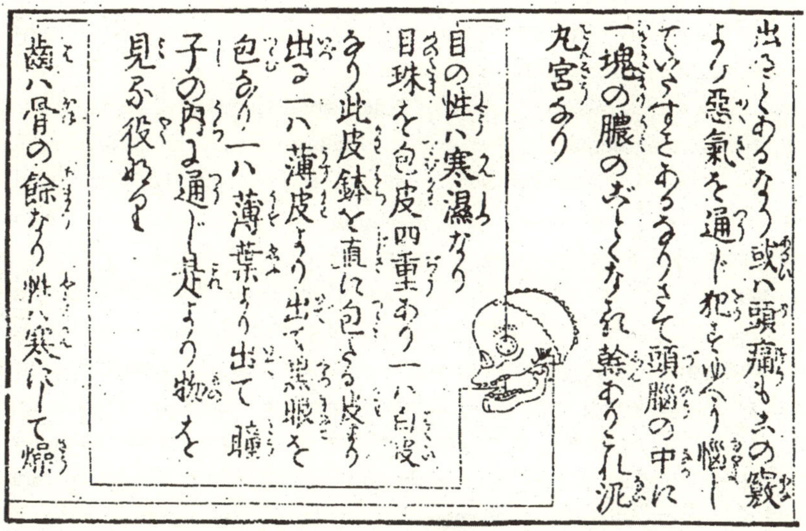

出ることあるなり或ハ頭痛もこの竅
より悪氣を通じ犯されゆへ悩し
ていたすことあるなりさて頭脳の中に
一塊の膿のごとくなれ餘ありこれ泥
丸宮なり
目の性ハ寒濕なり
目珠を包皮四重あり一ハ白皮
なり此皮鉢を直に包める皮なり
出る一ハ薄皮より出て黒眼を
包むなり一ハ薄葉より出て瞳
子の内に通じ是より物を
見る役なり
齒ハ骨の餘なり性ハ寒にして燥

FIGURE 12.3. Multidimensional presentation of the skull. Kōshi Hōyoku, *Honedugi ryōji jūhōki* (Treasured Notes for Healing Bones), 1746.

illustrations in his texts. Kōshi never mentioned which parts of his book were learned by means of hiden, nor did he explain why the masters allowed him to publish their knowledge.

Similarly, the next book to appear on bone setting—*Kahō Namba honedugi hiden* 家法難波骨継秘伝 (Family disciplines of the Namba secret transmission of bone setting, 1770), containing ten manuscripts from the Edo period—also had hiden origins but remained unpublished until modern times. The contents of this book were dictated by an anonymous author, and the recorder was named as Tanabe Hideo 田辺秀雄. The narrator noted that at a young age he petitioned a famous bonesetter in Namba, possibly the Nembai lineage, to learn from him, but he was refused. So he sold himself as a servant to that family. A few years later, when he had won the family's trust, he was allowed to assist in their practices. By taking advantage of such opportunities, he observed their manual skills. After several more years he learned their remedies. He then seduced a woman of the family, encouraging her to steal the family's secret medical book, and escaped with the knowledge. During his time in this family, the narrator also "secretly made friends with executioners and asked their help in finding some corpses," believing that a knowledge of actual bones could help him. The executioners showed him where they discarded the corpses, and he secretly collected bones.[23]

While his treatment on remedies is less than five pages, his manuscript contains twenty-one pages on anatomy.[24] He argues that the correct number of bones is 332 as opposed to 365, as was secretly transmitted. Detailed discussions of the different bones follow, along with illustrations showing the bones from various perspectives and explaining why 332 is the correct number. Next, he cites Kōshi's book and discusses its flaws. The author contrasts his book with the hiden.[25] In a description of the skull, he not only shows it from the front, the top, the side, and the bottom but also details the facial bones and cheekbones and, in the cavity, shows the brain linked to the spine (figure 12.4). Despite his stated contrast with hiden, the narrator decided to keep his discoveries within his family and opened a new lineage of hiden, announcing that "these are the secrets of our family and never pass them on to outsiders."[26] This written account of secret knowledge that was then kept secret serves to highlight the changing relations emerging between written and tacit knowledge within the field of bone setting.

Kōshi's and the anonymous manuscript thus contain the written corpus of bone-setting knowledge in the Edo period before Hoshino. Both texts sta-

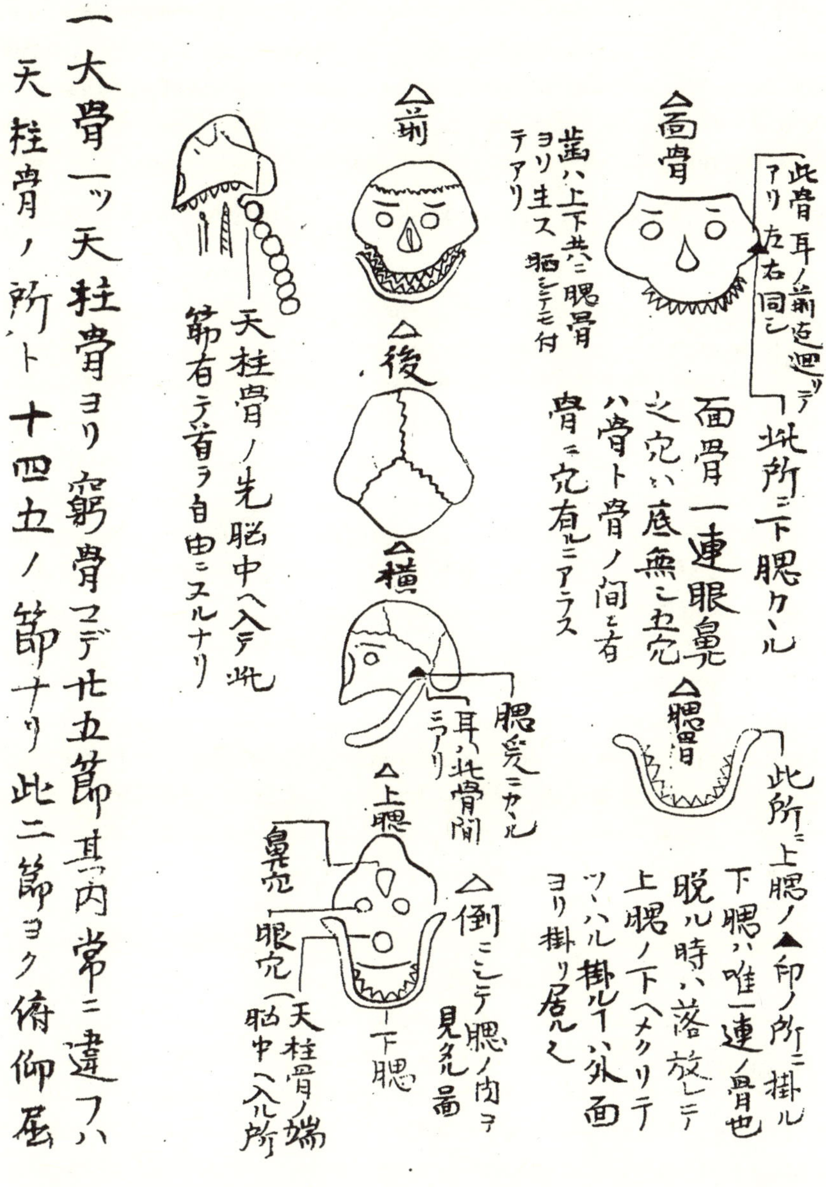

FIGURE 12.4. Three-dimensional presentation of the skull. *Kahō Namba honedugi* (Family disciplines of the Namba secret transmission of bone setting; 1770).

ted the importance of depicting the bones, providing illustrations from more than one perspective, thereby attempting to overcome the limitations of two-dimensional representation. The anonymous narrator even claimed that, without understanding the bones' shapes, he could not make full use of his manual skills. Hoshino's wooden model echoed these ideals of depiction.

The Itinerary of the Hoshino Wooden Skeleton

Frustrated at Tanaka's rejection of his plea for instruction, Hoshino decided to obtain a skull, believing that he might be able to solve the mystery of treating jaws by manipulating a real one. Fortunately, as he told it, not long afterward he came across a fresh skull on the grass on his way back from gathering herbs. Hoshino scrutinized the joints of the jaw and devised a technique that might be effective in managing the manipulation of the jaw. A few months later, he verified this by working on a female patient. To an amateur there were no clues for figuring out how to heal a person's jaw, no matter how long he may have observed a skull; however, a bonesetter with some practical grounding could simulate the process based on earlier experiences. This experiment built Hoshino's confidence in the usefulness of anatomical knowledge for developing bone-setting skills and awoke his ambition to study the entire skeleton. With the precedents of Yamawaki and Sugita, it was not difficult for Hoshino to figure out where to acquire the skeleton he desired. He allied himself with an attendant physician of the Hiroshima overlord to co-apply for permission to gather corpses. After much hesitation, the local fief lord approved their repeated pleas and granted them two criminals' bodies after execution in the summer of 1791.

Hoshino and the physician dissected one corpse together right at the site of execution, but dealing with the other corpse required more complicated procedures. Hoshino took the corpse to the Island Niho 仁保, six kilometers from the Hiroshima castle. As an important settlement of leatherworkers this island was perfect for disposing of the remains, because the inhabitants were already used to the smells of dead bodies. Before landing on shore, he took the body apart in preparation, and on the island he obtained a vacant workshop equipped with a kettle for boiling the remains. After one night's steaming Hoshino separated the bones from the cooked tissue. He washed and dried them in the sun and then carried out the challenging task of piecing them back together.[27]

When Hoshino first assembled a complete skeleton, he admired its sophistication and claimed that it finally demonstrated to him how the joints

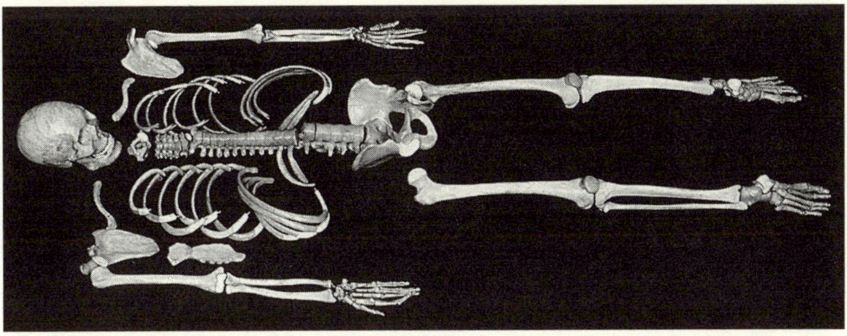

FIGURE 12.5. The current state of the Hiroshima skeleton. With permission of Hiroshima University.

functioned. Nevertheless, there remained several challenges. According to conventional deference to dead bodies (even a criminal's), Hoshino had to accord suitable treatment to the bones. Displaying actual remains of a body, as in the case of the Irish Giant, would not be tolerated. Furthermore, if he wanted to keep the bones for a long period, the skeleton would decay and deform, eventually becoming useless for the study of bone setting. Hoshino's solution was to persuade a carpenter, Harada Kōji 原田孝次, to make a model. It took Harada about three hundred days to carve all of the bones. Hoshino evaluated Harada's work to be so perfect that he did not find even the slightest difference from the original (figure 12.5). After the task was completed, Hoshino buried the corpse's remains with an elaborate funeral and erected a memorial tablet to honor the criminal's contribution to the art of bone setting.[28]

The fineness of Harada's craftwork continues to amaze connoisseurs even today. Having escaped the atomic bombing in 1945, the wooden skeleton is now preserved at Hiroshima University. It is a life-sized model of an adult male, around 161 centimeters in height, made mainly from paulownia wood with some parts, such as the teeth, of boxwood. Except for the hyoid bone and auditory ossicles, the carpenter replicated all the bones from the original. Each unit was carved separately, then assembled into a full skeleton. Most bones were connected by a pair of tenon and mortise joints, which are quite short, with the result that it is difficult to keep them connected to support the full skeleton.[29]

While Harada was carving, Hoshino also studied a copy of the *Kaitai shinsho* (New book of anatomy). As the carving progressed, Hoshino compared the illustrations with his skeleton and the wooden model and, early in 1792, wrote a ten-page commentary on the *Kaitai shinsho*.[30] In addition to offering praise, he also pointed out six faults, and this careful examination corrected genuine mistakes in the Dutch manual. For example, although the illustration correctly depicted the image, the original Dutch text misinterpreted the number of phalanges to be fifteen, as did the translation.[31] Hoshino clarified that the thumb has two instead of three phalanges; therefore, the correct number should be 14.[32]

Now that he had a complete skeleton, Hoshino might well have assumed that he could develop more effective skills regarding different parts of the body as he had done with the skull before, and thus he could found a new lineage of hiden. But, because his frustration at being prevented from acquiring the jaw-setter's knowledge had spurred him to create the model in the first place, Hoshino decided to do the opposite. Instead, he welcomed people, especially his peer bonesetters, to examine his model. An attendant physician of the Hiroshima fief, however, recorded that Hoshino had become notorious in the neighborhood for cooking a human body.[33] This raised the question as to how many local people would be willing to call on him to examine the model. Nevertheless, his reputation must have spread quite far because a master of Dutch-style surgery, Yoshio Kōgyū 吉雄耕牛 (1724–1800), showed up at Hoshino's door in 1797.

Yoshio was born into a family of official Dutch interpreters in Nagasaki. His duties provided many opportunities to learn in person about Dutch surgery and to meet knowledgeable travelers, such as Carl Peter Thunberg (1743–1828).[34] In addition to his formal occupation of translator, Yoshio practiced as a Dutch-style surgeon and opened a private school to teach Dutch knowledge. Admiring his authority in Dutch scholarship, Sugita had asked him to write a preface for the *Kaitai shinsho*.[35] Yoshio stayed with Hoshino for several days to examine the model closely. He believed the wooden skeleton to be highly valuable and persuaded Hoshino to go to Edo to show Sugita and the other Dutch-learning scholars what he had achieved. The compliments from a prestigious Dutch scholar such as Yoshio stimulated Hoshino's desire to impress distinguished scholars in the capital, and in the following year he brought three of his disciples with him to Edo.

On recognizing the imperfection of the first translation of the *Kaitai shinsho*in 1774, Sugita authorized his disciple Ōtsuki to improve the book. Unfortunately, Ōtsuki was an extremely painstaking worker, and before Hos-

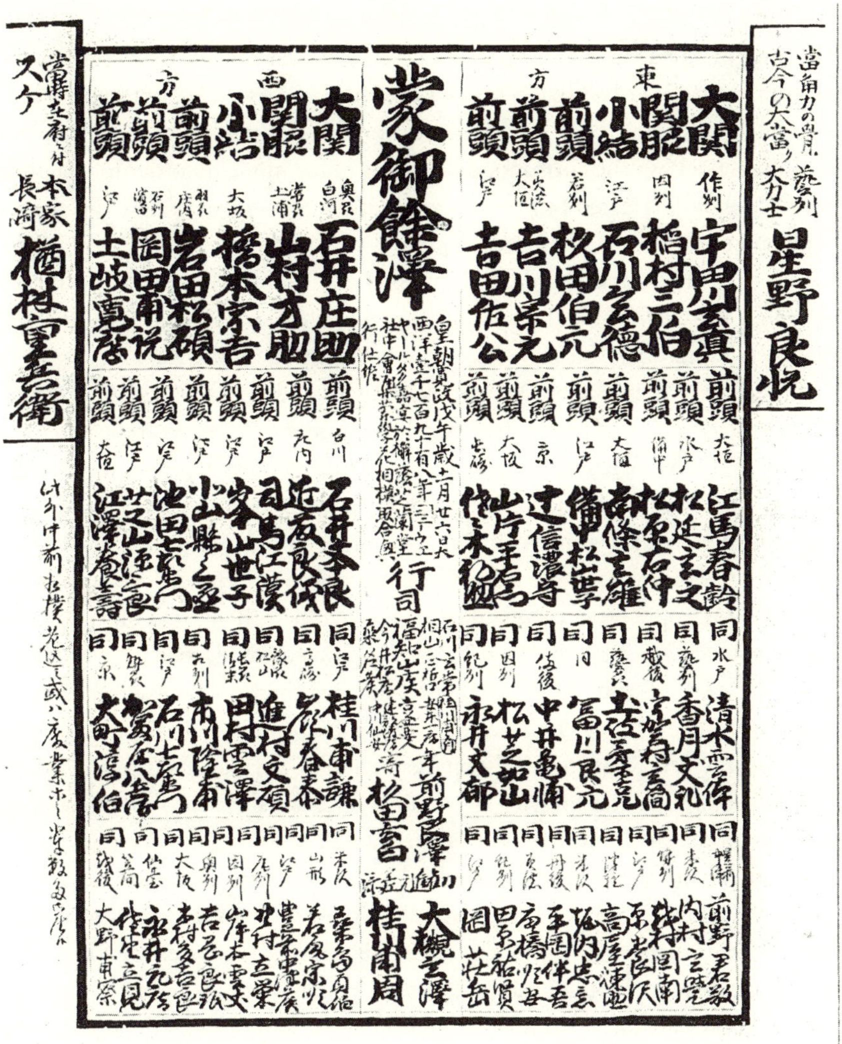

FIGURE 12.6. Dutch learners ranking in the form of a list of sumo wrestlers. With permission of Waseda University.

hino's arrival in 1798 he had been refining the translation for more than a dozen years without publishing any new results in support of his mentor.[36] Hoshino's arrival raised the morale of the Dutch-learning scholars, as they believed his model could strengthen their credibility in the court of public opinion. Among all the Dutch-style scholars, Ōtsuki professed to be the happiest about the skeleton because, as he confessed, an essential reason for the delay in retranslating Sugita's work was that he had not been quite sure of the meaning of the Dutch descriptions, and the model helped to crystalize the meaning of the words. Ōtsuki invited Hoshino to participate in the translation, which was carried out by Ōtsuki reading his draft aloud to Hoshino, and Hoshino checking the corresponding bones of his model in order to visualize what was being referred to in the book. Although Hoshino knew the shapes of the bones quite well, before this process he did not recognize many of the functions of their various indentations and holes. When completed, both men stated an admiration for the excellence of Dutch anatomy, and Hoshino left his three pupils with Ōtsuki where they could study Dutch learning.

In this way, the wooden skeleton not only provided visual evidence to support the legitimacy of Dutch learning but also contributed to improving the quality of this knowledge. To maximize the effects of this dissemination, the Dutch-style scholars made a ranking of their community in a poster, which, imitating a poster for sumo wrestling, placed Hoshino's name on a significant corner of the poster with the title: "Wrestler of Special Honor." Furthermore, although his pupils were merely novices in Ōtsuki's school, two of them were listed as *Makuuchi* 前頭, the fifth-highest within the ten levels in professional sumo wrestling (figure 12.6).[37] And Ōtsuki gave the wooden skeleton an elegant title, Apparatus of the Body's Mainstay (*Shinkangi* 身幹儀), which derived from the Dutch anatomical texts' assertion of the skeleton's fundamental role of supporting the body.[38] The Dutch-style scholars thus identified the wooden model as an integral part of their work.

At the same time, even Chinese-style physicians were impressed by the wooden skeleton. Taki Genkan 多紀元簡 (1755–1810), an attendant physician of the Shogun Ienari 徳川家斉 (in office 1787–1837), requested that Hoshino duplicate the model for educational display in the official medical school *Seijukan* 躋寿館 in Edo. Hoshino accepted the order and presented a replica two years later. As a reward, the shogun granted Hoshino thirty taels of gold and the privilege to attend the New Year's Greeting ceremony. Sugita also desired one, but Hoshino declined, which seems to have upset Sugita, with the result that in all of Sugita's surviving writings, he only left one line concerning Hoshino, in his diary for December 31, 1798: "Stayed at home. Hoshino Ryōetsu visited for discussions." Although Hoshino considerably enhanced the status of Dutch learning, Sugita did not want to talk about him.[39]

On his way back home to Hiroshima, Hoshino mounted an exhibition of his wooden skeleton in Osaka, a leading medical center. Its village Namba was famous for excellent bonesetters. An eminent Dutch-style physician came all the way from Kyoto for this exhibition; this was Koishi Genshun 小石元俊 (1743–1809), who studied with Yamawaki's student and who himself carried out a dissection in 1783. Koishi praised Hoshino highly and wrote a Chinese poem for him as a gift. In this exhibition Hoshino allowed everyone, including the masters of hiden in Namba, to see his model.[40]

After returning to Hiroshima, Hoshino continued to communicate with Ōtsuki. Two books from their collaboration survive: one a working draft from their collaboration in Edo, entitled *Shin kan seiteki* 身幹正的 (The righteous reality of the body's mainstay), dated 1798, and the other *Shinkangi setsu* 身幹儀説 (On the apparatus of the body's mainstay), dated 1802, which listed both as authors. Hoshino wrote the preface one month before his death. As indicated by the title, the wooden skeleton seems to have taken on the main role, while Ōtsuki's translation seems to have become a footnote. As it turned out, both texts remained unpublished for a long time. The slow worker Ōtsuki did not release his new translation until 1826, twenty-four years after Hoshino's death. In its appendix, Ōtsuki remembered to add a courteous note to memorialize Hoshino's contributions.[41]

The Aftermath of Hoshino's Exhibitions

Hoshino passed away only three years after returning to Hiroshima. His stepson, who had been his pupil and had also studied under Ōtsuki for some time, succeeded to his trade and name. Although Hoshino Ryōetsu II had received academic training superior to his stepfather and continued exhibiting the model, he never won fame for the excellence of his knowledge and practice. In spite of the fact that the model was displayed in the capital rather than in the countryside of Hiroshima, it did not immediately represent a challenge to the bonesetters' craft. The influence of the wooden skeleton only emerged over time. Rather than overturning the hiden system directly, Hoshino's exhibitions exerted influence via indirect routes. The most straightforward and important was the support it gave Ōtsuki's translation, which reinforced Dutch learning. In addition, Hoshino's open exhibition stimulated more bonesetters to follow in his footsteps of studying depictions of the human skeleton.

A possible but unverifiable consequence was the publication of *Seikotsu han*, mentioned previously. The author Futamiya Isao 二宮彦可 (1754–1827) first learned bone setting from Yoshio. Then, with his introduction to study under the best Chinese-style bonesetter Yoshihara Gentō 吉原元棟 (?–1800)

in Nagasaki, he learned Chen Yuanyun school's martial arts and bone setting.[42] As a result, his book covered both Chinese manual skills and Dutch therapeutic methods such as bandages. This book was considered the iconic work of bone setting in the Edo period.[43] Its most significant social meaning, however, was that for the first time a formal member of the Yoshihara hiden lineage systematically revealed internal knowledge—at least partially, by openly printing it.[44]

What caused Yoshihara to agree to have his secrets released? Behind Yoshihara's decision, we can see numerous signs of Yoshio's involvement, as Yoshio and Yoshihara had learned from each other, and they were friends.[45] In addition, Yoshio's pupils helped to compile Yoshihara's personal writings for publication after Yoshihara passed away. Having praised Hoshino's exhibition, Yoshio himself arranged to publish his own bone-setting book, although it did not come to fruition before his death.[46]

Hoshino's influence can also be seen in the production of a series of models imitating his wooden skeleton. During the Edo period, there were eleven attempts at reproducing wooden skeletons, including one by Hoshino's stepson. Having failed to acquire one from Hoshino, Sugita sought a carpenter to make a copy after the medical school model.[47] The most significant producer of models was Kagami Bunken 各務文献 (1755–1829). On April 22, 1806, a conflagration hit downtown Edo leading to over 1,200 deaths and the destruction of around 126,000 houses. Hoshino's model at the medical school was destroyed in this disaster. Kagami was able to present another wooden skeleton to the shogunate to replace the lost one.

Kagami had set his mind on becoming a physician in his youth. He first learned Chinese internal medicine. Because he enjoyed practical operations more than abstract theories, he switched to studying obstetrics, and he invented eight instruments for difficult births. Although he considered himself a good obstetrician, he had no hope of going beyond the greatest of obstetricians, Kagawa Genetsu 賀川玄悦 (1700–1777), so to satisfy his ambition, he then entered bone setting, which he believed had never been well developed. Kagami also claimed that he had requested to learn from a Nembai master but was refused because of the system of hiden. Reflecting on it later, he denounced those who were familiar with these skills: "all followed stale Chinese talk and spoke hollow words to cheat people," so he switched to the "right way" of finding evidence from real human bones so as to study their functioning in order to understand how to heal them.[48]

At this point, he became a radical Dutch-style physician and openly opposed Chinese medicine, as did Sugita Gempaku. Although Kagami never mentioned Hoshino in any of his surviving writings, it is hard to imagine that, as a mature physician who had strong interests in bone setting, he could possibly have refrained from attending Hoshino's 1799 exhibition in Osaka. This seems especially likely because, in the very next year, Kagami acquired permission to hold a dissection in the execution ground at Yoshijima 葭島. Two Dutch learning scholars also joined this dissection. Together they conducted several physiological experiments. The most famous of these was intended to compare and verify the theories of kidney functions according to Chinese and Western medicine. They injected ink into the renal artery and found that the ureter became dark. This experiment proved that the kidneys were urinary organs, as Western doctrine had described, and not for reproduction, as Chinese orthodoxy claimed. It inclined Kagami toward the study of Dutch medicine.[49]

Kagami and his wife carried on a prolonged project to collect human bones, continually watching for human bones along roads, and they sometimes snuck into Yoshijima in the middle of the night to convey discarded bodies and hide them at home.[50] At first, they hid the corpses under the bed to avoid detection, and in their free time removed the bones from the remains.[51] In addition to his wife, Kagami mobilized his sons and students to help his collecting. In 1848 his son wrote an article discussing the abnormal shaped bones he had collected in his youth.[52] There was a rumor stating that Kagami had generously taken care of an old beggar, and after his death Kagami had taken care of his burial. A few months later, however, someone looted the grave and removed the remains. It was rumored that Kagami himself had stolen them in order to examine the joints in detail.[53] When he had finally collected enough bones, he assembled complete skeletons. Unlike Hoshino, who had used only one individual body to make his model, Kagami collected bones from people of different genders, ages, and health conditions. He was especially interested in abnormal samples. When exactly his wooden skeleton was produced is unclear. However, in the preface to his unpublished manuscript, he mentioned that he had hired a carpenter named Tanaka to carve a skeleton for convenience in studying bones. In his writing Kagami also used the term *Body's Mainstay*, which would seem to refer to Hoshino's model. This shows the way in which both Hoshino's wooden skeleton and his role in the *New Book of Anatomy* contributed to changing the field of bone setting.[54]

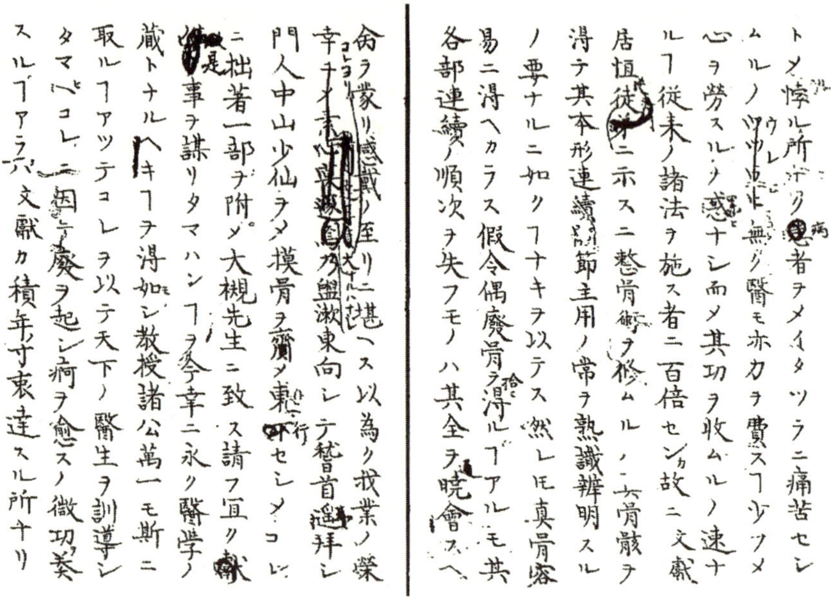

FIGURE 12.7. Draft of Kagami Bunken's petition for reforming the training of bonesetters. With permission of the Takeda Science Foundation.

Indeed, Kagami went on to push the field of bone setting into a textual discipline by producing publications to publicize his work. He hired skillful painters to depict in color the bones he had collected, and he published their paintings. He used the language of Western osteology in these publications, alongside the practical observations he had made based on his own experience.[55] In 1819 Kagami decided to dedicate his wooden skeleton to the shogunate. He wrote to Ōtsuki, the leader of his own school of Dutch learning, asking him to make an introduction to the official medical school to enable him to donate his model for aiding the education of bone setting (figure 12.7).[56] The government accepted Kagami's offer and granted him twenty taels of gold, but they ignored his pleas to make Dutch learning the basis of medical education. Although Kagami pushed this project tenaciously, his efforts were not persuasive to the government.

Instead of successfully promoting Dutch learning, Kagami's works received immediate attention from an official physician, a master of Chinese medicine, Ishisaka Sōtetsu 石坂宗哲 (1770–1842), who was inspired by Kagami to write a book entitled *Kotsu kyō* 骨経 (Canon on bones).[57] Ishisaka was interested in Kagami's works not because of his interest in bone setting. He was a master of acupuncture, and in the doctrines of Chinese medicine bones serve as the standard coordinates to position the locations of acupuncture points. Ishisaka considered Kagami's work excellent for illuminating that art.

Hoshino's posthumous influence continued. Kagami's student Okuda Manri 奥田万里 (flourished 1820) produced a third type of wooden skeleton. In order to carve out his own niche, Okuda made miniature female figures in sitting postures. Because his mentor's model already occupied the top position in the country, he decided to dedicate one of his to the highest noble feudal lord, the Lord of Nagoya. It traveled easily because of its light weight, which made it convenient for display in private local exhibitions. One of Okuda's models remains in the collection of the National Tokyo Museum of Science, where it is still displayed to demonstrate the disposition of the bones.[58]

Conclusion

Only the Meiji government's educational reforms between 1868 and 1885 caused the bonesetters' hiden lineages to collapse, and Hoshino's wooden skeleton, in the guise of Kagami's imitation, became the first teaching aid for instructing a new generation of orthopedists.[59] With this end of the skeleton's itinerary, it is easy to misunderstand Hoshino as being a member of the Western camp and, especially because Dutch learning scholars used Hoshino's skeleton as propaganda, as having cooperated with the Dutch scholars to induce the fall of old forces including both Chinese medicine and the bonesetters' hiden. However, the itinerary described in this chapter shows that this would be a misrepresentation of the process. The Dutch learning scholars used Hoshino's skeleton to support their arguments against Chinese medicine, but this was not, however, Hoshino's target. In fact, Hoshino saw no immediate effects of his initial goal of breaking down the tradition of hiden.

Hoshino was from the countryside, far from the highly skilled bonesetters' hub of Namba. Although his father was a bonesetter, he was not part of a bonesetters' secret transmission lineage. Compared to Kōshi or the anonymous author of *Kahō Namba honedugi hiden* (Family disciplines of the Namba secret transmission of bone setting), the resources he could have accessed were meager. The low level of his skills made him uncompetitive even with illiterate practitioners in his local region. Based on his experience of trying to figure out skills from a skull, he gained the confidence to use his skeleton to compete directly with established bonesetters' secret transmissions. By openly exhibiting his wooden skeleton in Namba, Hoshino staged a demonstration in which he claimed that effective skills could be discovered by studying the skeleton without any explicit teaching. Because experiential knowledge is most effectively transmitted through an apprenticeship system such as that of the secret transmission lineages, however, his claims proved hollow among the immediate field of bonesetting hiden lineages. He could not undermine established hierarchies by means of his model alone, no matter how much of a spectacle it proved to be.

This case turns out to suggest that the insiders of hiden lineages did not worry that publication would harm their interests. Or they may even have learned from Hoshino's example that some openness could enhance their reputation in the field. In any case, Kagami was the first bonesetter to take practical steps to follow Hoshino's example of making a skeleton to challenge the tradition. No matter how much he published, he did not shake the status of the hiden clans, his projects came to nothing, and the government ignored him.

Although Hoshino failed in achieving his personal goal, if we consider the whole case from a broader perspective, his skeleton influenced the medical field in an even more profound way. He overestimated the skeleton's effects on the arts of bone setting but underrated its potential meaning in the arena of medical competition. The Dutch studies scholars grasped the opportunity offered by the wooden skeleton to help them in their competition with Chinese medicine. Recognizing Hoshino's value, Dutch style scholars intentionally integrated him into their projects as their partner. This precedent also impelled Kagami to join the Dutch scholars, so he could utilize Ōtsuki's connections in the attempt to contribute his model to the government. This perspective makes clear why it seemed self-evident to previous historians of medicine that Hoshino was deeply linked to Dutch medicine.

Ironically, although Hoshino made friends with Ōtsuki and his peers, he himself probably inclined more to orthodox ideas of Chinese medicine. In the beginning he even evaluated this knowledge as superior to Dutch books. His rejection of Sugita's request for a skeleton also suggested that the consideration of Dutch learning was not foremost in his mind. The Chinese style physicians also valued his skeleton, and they too wanted a replica for the medical school. However, their major concern was neither bone setting nor anatomy itself but the anatomical information to help identify the location of acupuncture points. That Ishisaka praised Kagami's works again shows this point. Because the status of Chinese medicine in Japan was so much higher than Dutch learning, it did not need the skeleton to support its position, but the skeleton could still be useful. The Chinese-style physicians' attitudes show that they did not oppose anatomy; indeed, under certain conditions they could find a common ground with Western medicine. In fact, Ishisaka was seen as an eclectic thinker sitting between Chinese and Dutch medicine. Similarly, the interaction between Yoshio and Yoshihara in the field of bone setting also suggests that, in the Edo period, the border between different groups of medicine was not always so clear-cut and was sometimes negotiable. The case of Hoshino's skeleton treated in this chapter helps underscore this point. The narrative, with Hoshino's skeleton at its center, describes the cultural field of bone setting in premodern Japan, tracing the routes by which knowledge flowed into and around Japan—including from local regions in Japan, from China in different periods, and from Europe from the seventeenth century—and showing how monks, merchants, and refugees were the carriers of information and skills into Japan. This narrative of people, practices, and texts moving, circulating, and recombining shows how new knowledge and new objects emerge out of cultural crossroads, which themselves in turn then create new relational fields and new kinds of knowledge.

Notes

Chapter 1: Nodes of Convergence, Material Complexes, and Entangled Itineraries

Although written primarily by Pamela Smith, this introduction emerged out of two workshops hosted by the MPIWG, which involved many more participants than appear in this volume. Most of the volume's authors met for a final two-week intensive conversation. I thank Tara Alberts, Dagmar Schäffer, Tansen Sen, and Ronit Yoeli-Tlalim for their contributions to this introduction. All contributors wish to express deep gratitude to the MPIWG for generously hosting these stimulating workshops.

Epigraphs: Ingold, *Perception of the Environment*, 345; Bray, "Science, Technique, Technology," 323.

1. Chandra Mukerji, in "Tacit Knowledge," provides a case study of such movement. Arjun Appadurai, Homi Bhabha, and Prasenjit Duara offer theoretical statements, as treated in the next chapter.

2. I treat this in Pamela H. Smith, "Science," and Smith, "Knowledge in Motion."

3. An exemplary study that starts from the materials available in the environment is Susan Naquin, "The Material Manifestations of Regional Culture" (unpublished paper, 2016). My thanks to Sue Naquin.

4. See, for example, Horden and Purcell, *The Corrupting Sea*; Abulafia, *The Great Sea*. For an overview, see Horden and Purcell, "The Mediterranean and the New Thalassology"; Wigen, "Oceans of History."

5. Beckwith, *Warriors of the Cloisters*; Beckwith, *Empires of the Silk Road*; Beckwith, *The Tibetan Empire in Central Asia*; Forêt and Kaplony, *Journey of Maps and Images*; Forêt, *Mapping Chengde*; Hansen, *The Silk Road*; Liu, *The Silk Road in World History*; Liu, *The Silk Road: Overland Trade*; Liu, "Silks and Religions in Eurasia"; Millward, *The Silk Road*; Millward, *Eurasian Crossroads*; Millward, Dunnell, Elliott, and Forêt, *New Qing Imperial History*; Millward, Yasushi, and Jun, *Studies on Xinjiang Historical Sources*; Starr, *Lost Enlightenment*; Starr, *Ferghana Valley*; Starr, *The New Silk Roads*.

6. Frank, "The Centrality of Central Asia," 44. More recent work by Thomas Allsen and Nicola di Cosmo has also been influential for historians of science and medicine.

7. Starr, *Lost Enlightenment*, 39–40, in his study of what he calls such "cross-road civilizations" (69).

8. The publications in the series Transcultural Research—Heidelberg Studies on Asia and Europe in a Global Context, with editors Madeleine Herren, Axel Michaels, and Rudolf G. Wagner, aim for similar goals, albeit in a later period. Especially useful in rethinking historical narratives is Herren, Rüesch, and Sibille, *Transcultural History*. For the early modern period, Bleichmar and Martin, *Objects in Motion* provides useful case studies of the variety of meanings and dimensions "motion" can take. Grasskamp and Juneja, *EurAsian Matters* appeared after the present volume was complete.

9. In his examination of the city of Khwarizm, Starr in effect argues that intensive systems of water control gave rise to the skills of numeracy and techniques of algebra that were codified in the texts of Al-Khwarizmi (Starr, *Lost Enlightenment*, 66–67; for further examples of the effects of such hubs, see also 68–71, 94).

10. Norton, "Tasting Empire," 691. See also Norton, *Sacred Gifts, Profane Pleasures*.

11. Schiebinger and Swan, *Colonial Botany*.

12. Cook, *Matters of Exchange*; Allsen, *Culture and Conquest in Mongol Eurasia*.

13. Herren, Rüesch, and Sibille also treat this, in a chapter of *Transcultural History* aptly titled "Fighting Zombies: Methodological Challenges of Transcultural History."

Chapter 2: Trans-Eurasian Routes of Exchange

1. Some material in this chapter appeared in Smith, "Itineraries of Materials and Knowledge," 31–61.

2. Appadurai, "How Histories Make Geographies," 11.

3. Appadurai, "How Histories Make Geographies," 9–10.

4. Appadurai, "How Histories Make Geographies," 11.

5. Bhabha, "Introduction," 8, 9.

6. Duara, *The Crisis of Global Modernity*, 73.

7. On the Austronesians, see Bellwood, "Austronesian Prehistory in Southeast Asia." For early connections through the overland routes of Central Asia, see Kuzmina, *Prehistory of the Silk Road*.

8. V. Selvakumar, "Contacts between India and Southeast Asia." On the contribution of the nomads in the transmission of ideas and technologies across Eurasia, see Kuzmina, *Prehistory of the Silk Road*; Golden, *Central Asia in World History*.

9. Hans Ulrich Vogel and Sabine Hieronymus have detailed the import of cowries into China in "Cowry Trade (Part I)" and "Cowry Trade (Part II)." See also Yang, "Horses, Silver, and Cowries." For global trade in cowries, see Yang, *Cowrie Shells*; and Boomgaard, "Early Globalization."

10. Vogel and Hieronymus, "Cowry Trade." See also Yang, "Horses, Silver, and Cowries," and Deyell, "Cowries and Coins."

11. A detailed record of these findings is Xiong's "The Hepu Han Tombs and the Maritime Silk Road."

12. Stern, "Early Roman Export Glass."

13. See specifically, Glover and Bellina. "Ban Don Ta Phet and Khao Sam Kaeo."

14. Parker, *The Making of Roman India*, 170–1.

15. On the trade in horses between the Persian Gulf and China, see Yokkaichi Yasuhiro, "Chinese and Muslim Diasporas"; Kauz "The Maritime Trade of Kish."

16. Digby, "The Maritime Trade of India."

17. See Mills, *Hsing-ch'a sheng-lan*, 69.

18. Chakravarti, "Early Medieval Bengal."

19. See Deyell, "Cowries and Coins."

20. Sima Qian, *Records of the Grand Historian of China*, 2:235–6.

21. Franck and Brownstone, *The Silk Road*; Liu, *The Silk Road in World History*.

22. Frank, "The Centrality of Central Asia," 74.

23. On the possible movement of South Asian artisans to Southeast Asia for the purposes of trade, see Glover and Bellina, "Ban Don Ta Phet and Khao Sam Kaeo."

24. On the Sogdian commercial networks, see Vaissière's *Sogdian Traders*. The role of Sogdians in the transmission of Buddhism is examined in Walter's "Sogdians and Buddhism."

25. For the connections between Buddhism and commercial activity, including the role of Buddhism in facilitating transregional trade, see Liu Xinru, *Ancient India and Ancient China*; Sen, *Buddhism, Diplomacy, and Trade*; Jason Neelis, *Early Buddhist Transmission and Trade Networks*.

26. One of the best studies on the Muslim pilgrimage activity in the transregional context is Tagliacozzo, *The Longest Journey*.

27. Flecker, "Shipwreck in Indonesia."

28. Qin, "Archaeological Investigations," 87–88. Herman argues that by at least the second century CE East Africa was already incorporated into a system of trade that depended on monsoon routes. "Mutual Exchange," 148.

29. Rong Xinjiang, "History of Sino-Arabic Relations." A more recent study is Schottenhammer, "Yang Liangyao's Mission of 785."

30. Sen, "Military Campaigns of Rajendra Chola."

31. The availability and transregional trade in bulk goods are evident from the imports of Song China. See Hartwell, "Foreign Trade."

32. On the trade in porcelain, see Finlay, *The Pilgrim Art*.

33. Prazniak, "Siena on the Silk Roads," 178.

34. Sen, "Formation of Chinese Maritime Networks."

35. Wade, "The Zheng He Voyages"; Sen, "The Impact of Zheng He's Expeditions"; Finlay, "Portuguese and Chinese Maritime Imperialism."

36. For examples of such connections over the *longue durée*, the writings of Sanjay Subrahmanyam, especially the collected essays *Explorations in Connected History*.

37. Frank, "The Centrality of Central Asia," enumerates these peaks of trade back into the early millennia BCE. See Frank, "The Centrality of Central Asia," 72.

38. For a critical examination of the term "Silk Road," see Chin, "The Invention of the Silk Road." For the maritime silk route, see Miksic, "Introduction," for a useful introduction and summary of recent research. For the fur route, see Potts, "Technological Transfer and Innovation in Ancient Eurasia." For the musk route, see Akasoy and Yoeli-Tlalim, "Along the Musk Routes."

39. Crosby, *The Columbian Exchange*, 199 (peanut), 200 (sweet potato), 189 (maize).

40. Günergun, "Travellers in the Field," citing Blench, *Archeology, Language and the African Past*, 222.

41. See Smith, "Chinese Sugar."

42. Allsen, *The Royal Hunt in Eurasian History*.

43. Hartwell, "Foreign Trade, Monetary Policy," 466–68.

44. Bréard, "Problems of Pursuit," 57, 65–67.

45. Bayly, "'Archaic' and 'Modern' Globalization," 50–53.

46. Lu Gwei-Djen and Sivin, ed., *Medicine*, 38–45. Lloyd and Sivin see great differences in basic concepts used to articulate ideas about nature in ancient Greece and China, stating that the Greeks focused on "nature and the elements, the Chinese on qi, yin-yang, the five phases, and the Way." Lloyd and Sivin, *The Way and the Word*, 6.

47. Schafer, *Golden Peaches*, 219.

48. Anderson, *Gems*.

49. For discussion of Greek sources, see Xydopolous, "Alexander's Historians"; Laufer, *The Diamond*, 11.

50. Daston and Park, *Wonders and the Order of Nature*, 21, 75.

51. Alam and Subrahmanyam, *Indo-Persian Travels in the Age of Discoveries*, 4; ibn Shahriyār, *Kitab ʿajāyib al-Hind*, 47ff and 120ff (snakes), 157 and 173 (lizards), 7–8 (crayfish), 171 (ointment). Muḥammad ibn Aḥmad Bīrūnī's (973?–1048) *Kitab al-Hind* (Book of India) has a very different tone than these books of wonders, although he mentions oddities such as "in shaking hands they grasp the hand of a man from the convex side," and "they spit out and blow their noses without any respect for the elder ones present, and they crack their lice before them," and "in playing chess they move the elephant straight on." Sachau, *Alberuni's India*, 1:182. He also relates several examples of charms but concludes that "most of their charms are intended for those who have been bitten by serpents," about which he says, "I, for my part, do not know what I am to say about these things, since I do not believe in them" (1:194). About the people in India in general, he notes, "in all manners and usages they differ from us to such a degree as to frighten their children with us, with our dress, and our ways, and customs, and as to declare us to be devil's breed, and our doings as the very opposite of all that is good and proper. By the bye, we must confess, in order to be just, that a similar depreciation of foreigners not only prevails among us and the Hindus, but is common to all nations towards each other" (1:20).

52. Laufer's pioneering philological research on folk beliefs shows how ancient some of these tales are, as well as tracing the remarkable accretions they acquired as they moved over time and space. Laufer, *The Diamond*; Laufer, "The Story of the Pinna"; Laufer, "Asbestos and Salamander." These three works trace "exotic" concepts and objects back to the Hellenistic-Roman East, whence, Laufer argues, they spread throughout Eurasia. Aesop's fables were also told along the Silk Roads.

53. Prazniak, "Siena on the Silk Roads," 198–99.

Chapter 3: The Silk-Roads as a Model for Exploring Eurasian Transmissions of Medical Knowledge

This chapter is based on material from my forthcoming book, *ReOrienting Histories of Medicine: Encounters along the Silk Roads* (Bloomsbury, forthcoming). Research for this paper has been conducted as part of my Wellcome Trust project "Re-Orienting Early Medicine: Bridges of Knowledge between East and West" (grant no. 088251).

1. For an overview on these views, see the posts on "Secrets of the Cave" on Sam van Schaik's early Tibet blog: www.earlytibet.com (last accessed October 25, 2018).

2. For more details on the Dunhuang collections as well as digital images of some of the manuscripts, see the website of the International Dunhuang Project, based at the British Library: idp.bl.uk (last accessed October 25, 2018).

3. For studies on the Chinese medical manuscripts from Dunhuang, see Lo and Cullen, *Medieval Chinese Medicine*; Despeux, *Médecine, religion et société*; and numerous works by Ming Chen, such as, "The Transmission of Indian Ayurvedic Doctrines" and "La médecine dans la région de Turfan." For an overview of the relevance of the Dunhuang manuscripts to the history of science, see Rong Xinjiang, "Dunhuang Studies and the History of Science and Technology." The only Tibetan medical texts from Dunhuang that have been studied up to now in any Western language are those dealing with horse veterinary medicine, which have been studied by Blondeau, in *Matériaux pour l'étude de l'hippologie*.

4. See Lo and Jianmin, "Manuscripts, Received Texts and the Healing Arts."

5. See Yoeli-Tlalim, "Between Medicine and Ritual".

6. Richthofen, "Über die zentralasiatischen Seidenstrassen"; Waugh, "Richthofen's 'Silk Roads'."

7. For a concise and illuminating overview of many of these areas, see Millward, *The Silk Road.*

8. "The spaces in between" was a phrase used by the British explorer and politician Rory Stewart. See Millward, *The Silk Road*, 7.

9. Christian, "Silk Roads or Steppe Roads?"

10. Rezakhani, "The Road That Never Was."

11. On the concept and history of lingua franca in the singular and the plural, see European Commission, Directorate-General for Translation, *Studies on Translation and Multilingualism.*

12. Dakhlia, *Lingua Franca.*

13. For a published version of this paper see, Yoeli-Tlalim, "On Urine Analysis."

14. This collaborative work has been published in a special Silk Roads issue of *Asian Medicine* 3, no. 2 (2007), coedited with Vivienne Lo, and in two coedited volumes with Anna Akasoy and Charles Burnett entitled *Islam and Tibet* and *Rashīd al-Dīn.*

15. For "enlargement of scale," see William McNeill, "Foreword," in Frank and Gills, *The World System.*

16. There is no conclusive dating of when Ravigupta lived. Emmerick, in "Ravigupta's Place in Indian Medical Tradition," places him ca. 650 CE.

17. Emmerick has suggested that Ravigupta, author of the *Siddhasāra*, is the same as the Ravigupta who, according to the *Blue Annals*, founded a cult of Tārā in Kashmir. Emmerick, *The Siddhasāra of Ravigupta*, 1:2. For Emmerick's other studies of the *Siddhasāra*, see "The Sanskrit Text of the Siddhasāra," "On Ravigupta's 'Gaṇas'"; "New Light on the 'Siddhasāra.'"

18. The *Siddhasāra* is one of two major Sanskrit medical compendia (the other is Vāgbhaṭa's *Aṣṭāṅgahṛdayasaṃhitā*) that were translated into Tibetan and incorporated into the *Tanjur*. On the Arabic version of the *Siddhasāra*, see Emmerick, "Ravigupta's *Siddhasāra* in Arabic." Emmerick mentions that al-Rāzī frequently quotes the *Siddhasāra* in its Arabic translation.

19. Zieme, "Notes on Uighur Medicine."

20. The Khotanese manuscripts and fragments date from the fifth–tenth centuries. Most of the Khotanese texts are Buddhist. The Khotanese Buddhist texts include both local compositions and translations from known texts (mostly Sanskrit). Maggi, "Khotanese Literature." On Hoernle as the primary decipherer of Khotanese, see Sims-Williams, "Hoernle, Augustus Frederic Rudolf."

21. Kumamoto, "Khotanese Official Documents," 10–11.

22. Rong Xinjiang, "Official Life at Dunhuang," 60–61.

23. Takata, "Multilingualism in Tun-huang."

24. This introduction was translated and discussed first by Bailey and then by Emmerick. See Bailey, "The Preface to the Siddhasāra-śāstra." Emmerick published a new edition, translation, and commentary in "Some Remarks on Translation Techniques."

25. Van Schaik and Galambos, *Manuscripts and Travellers*, 119n25. The suggestion of linking Daoyuan in Pelliot chinois 2893 with the monk who features in an ordination certificate from Khotan from the year 961 (S.6264/Or.8210) was raised by Zhang Guangda and Rong Xinjiang in their "Les noms du royaume de Khotan."

26. See Chen Ming, "The Transmission of Indian Ayurvedic Doctrines," 203n7. This Sogdian signature is mentioned by Bailey, "The Preface to the Siddhasāra-śāstra."

27. Hansen, "Introduction."

28. On the Sanskrit sources of al-Rāzī's *Kitāb al-Ḥāwī fī ṭibb*, see Kahl, *Sanskrit, Syriac and Persian Sources.*

29. See Zieme, "Notes on Uighur Medicine."

30. Van Bladel, "Bactrian Background of the Barmakids," 77; Beckwith, *Warriors of the Cloisters*, 85. Alī b. Sahl Raban al-Ṭabarī is sometimes referred to as Sahl Raban al-Ṭabarī, in short—and he is to be distinguished from Muhammad b. Jarīr al-Ṭabarī, the Arabo-Persian historiographer, who is said to have read the other Ṭabarī, 's medical compendium when he was sick.

31. Bosworth, "Merv," 403; see also Yazberdiev, "The Ancient Merv and Its Libraries."

32. Meyerhof, "'Ali aṭ-Ṭabarī's 'Paradise of Wisdom'."

33. Siddiqi, "An Early Arabian Author."

34. See Van Bladel, "The Bactrian Background of the Barmakids."

35. Shefer-Mossensohn and Abou-Hershkovitz, "Early Muslim Medicine and the Indian Context."

36. The Third Société Européenne pour l'Étude des Civilisations de l'Himalaya et de l'Asie Centrale (SEECHAC) International Conference held in the Vienna Academy of Sciences in 2013 was dedicated to "Interactions in the Himalayas and Central Asia," dealing with processes of transfer, translation, and transformation in art, archaeology, religion, polity, and medicine. Within Silk Road visual cultures, it is worth mentioning here the Iranian and Sogdian influences on early Tibetan art, demonstrated by Heller and Melikian-Chirvani. See Heller, "Preliminary Remarks on Painted Coffin Panels"; Melikian-Chirvani, "Iran to Tibet."

37. Nasr and Mohaghegh, *Al-As'ilah wa'l-Ajwibah*; Bīrūnī, *Al-Bīrūnī*. For Ibn-Sīnā's life, see William E. Gohlman, *The Life of Ibn Sina*.

38. Bosworth, "Bukhara ii."

39. Starr, *Lost Enlightenment*, 4.

40. Hansen, *Silk Road*, 26; Stein, *On Central-Asian Tracks*, 1–2; Escauriaza-Lopez, "Aurel Stein's Methods and Aims."

41. See Takata, "Multilingualism in Tun-huang."

42. Shaked, "Judaeo-Persian Letter from Dandan-Uiliq." For numismatics, see Skaff, "Sasanian and Arab-Sasanian Silver Coins."

43. Ming Chen, "La médecine dans la région de Turfan."

44. Sen, *Buddhism, Diplomacy and Trade*; Akasoy and Yoeli-Tlalim, "Along the Musk Routes."

45. See Takeuchi, "Sociolinguistic Implications of the Use of Tibetan"; Uray, "L'emploi du tibétain."

46. The following section is based on my "Central Asian Mélange."

47. For a study of this work, see Martin, "An Early Tibetan History of Indian Medicine."

48. For a detailed analysis of the two Western components of this and similar lists (i.e., *stag gzig* and *phrom*), see Martin, "Greek and Islamic Medicines' Historical Contact with Tibet"; Yoeli-Tlalim, "On Urine Analysis"; Yoeli-Tlalim, "Re-visiting 'Galen in Tibet.'"

49. Pelliot tibétain 127, ll. 174–75. Luo Bingfen et al., *Tun hong nas*, 222.

50. IOL Tib J 756, l. 17. Luo Bingfen et al., *Tun hong nas*, 32.

51. IOL Tib J 756, l. 33. Luo Bingfen et al., *Tun hong nas*, 133. A similar expression is found in the *Li yul lung bstan pa*, where it appears in the account dealing with a Chinese princess who married the Khotanese king and brought silkworms with her. As the Khotanese ministers get suspicious, the queen produces some Kashmiri silk in order to demonstrate to the king the secret of silk production. This type of silk appears as *kha cher dar*. See Emmerick, *Tibetan Texts concerning Khotan*, 33.

52. Pelliot tibétain 127, l. 182; Luo Bingfen et al., *Tun hong nas*, 223, 233n158. The word *phug* can mean "pierce." Previous translations of this sentence have appeared in Blondeau, *Matériaux pour l'étude de l'hippologie*, 7; Uray, "The Old Tibetan Sources

of the History of Central Asia," 304. Both were unsure about the meaning of the term but proposed iron cautery.

53. Ligeti, "Rapport sur les rois."

54. Pelliot tibétain 1044, l. 52; Luo Bingfen et al., *Tun hong nas*, 238. The following paragraph is based on Lo and Yoeli-Tlalim, "Travelling Light," 282–84.

55. Indian cauterization practices are fundamentally different from the moxibustion practices found in the Tibetan and Chinese sources. These are delineated, for example, in the chapter devoted to cauterization in the *Aṣṭāṅgahṛdaya saṃhitā*.

56. One of the most interesting discoveries regarding Zhang zhung texts from Dunhuang was recently made by Takeuchi and Nishida, who have pointed out that the Zhang zhung texts from Dunhuang are medical texts. See Takeuchi and Nishida, "Present Stage of Deciphering Old Zhangzhung."

57. Pelliot tibétain 127, ll. 183–84.

58. Lo and Yoeli-Tlalim, "Travelling Light," 277, 280; Pelliot tibétain 1044, ll. 26–27.

59. For an overview of references to Confucius in the Tibetan Dunhuang manuscripts, see Lin, "The Tibetan Image of Confucius." It should be noted, however, that the Tibetan designation Kongtse does not always refer to Confucius.

60. Rolf Stein, "Tibetica Antiqua VI: Confucian Maxims in Two Dunhuang Manuscripts," in *Rolf Stein's Tibetica Antiqua*, 273–84.

61. Kalinowski, "Mantic Texts in Their Cultural Context."

62. See Uray, "Earliest Evidence," 354.

63. This quote has been commented on by Macdonald and by Stein. See Macdonald, "Une lecture"; Rolf Stein, "Tibetica Antiqua VI: Confucian Maxims in Two Dunhuang Manuscripts," 276n3; Uray, "Earliest Evidence."

64. Uray, "Earliest Evidence." 358–59.

65. Rolf Stein, "Tibetica Antiqua III: Apropos of the Word *Gtsug lag* and the Indigenous Religion," in *Rolf Stein's Tibetica Antiqua*, 117–90.

66. Rolf Stein, "Saint et divin, un titre tibétain et chinois des rois tibétains," *Journal asiatique* 269 (1981): 269; Rolf Stein, "Tibetica Antiqua I: The Two Vocabularies of Indo-Tibetan and Sino-Tibetan Translations in the Dunhuang Manuscripts," in *Rolf Stein's Tibetica Antiqua*, 41–42. IOL Tib J 748 ends with the words "*cu yag gyi yi ge rdzogs sho [s+sho]*." I thank Brandon Dotson for sharing his transliteration of this text with me. The more well-known later term in Tibetan for the Yijing is *spor thang* (or *spor thang nag rtsis*). For a list of Tibetan sources dealing with the Yijing, see Dorje, *Tibetan Elemental Divination Paintings*, 46.

67. Macdonald, "Une lecture," 284.

68. Laufer, "Loan-Words in Tibetan," 509–11.

69. Lo, "Manuscrits de Dunhuang et de Khotan."

70. Takata, "Multilingualism in Tun-huang."

71. Mair, "Kinesis versus Stasis," 3.

72. Bala, *Dialogue of Civilizations*, 45. See also Yoeli-Tlalim, "Review of Arun Bala, *The Dialogue*."

73. Lo, "Manuscrits de Dunhuang et de Khotan."

74. See Yoeli-Tlalim, "Re-visiting 'Galen in Tibet.'"

75. For Gutas's analysis of similar Sasanian texts on the origins and transmission of knowledge, see Gutas, *Greek Thought, Arabic Culture*, esp. 34–40. See also Van Bladel, "Arabic History of Science." For a discussion of the Jewish case, see Reed, "Ancient Jewish Sciences."

76. On origin narratives in Islamic sources, see Abbou-Hershkovitz, "Historiography of Science"; Brentjes, "Narratives of Knowledge."

Chapter 4: Things (*Wu*) and Their Transformations (*Zaowu*) in the Late Ming Dynasty

1. Brook, *The Confusions of Pleasure*; Clunas, *Superfluous Things*, 103.

2. Handler, *Austere Luminosity*, 208.

3. The *wuchan* chapters of local gazetteers (*Difang zhi* 地方志) reflect such notions, as do tax registers. On the diversity of local gazetteers, see Dennis, *Writing, Publishing, and Reading*.

4. *Zhouyi zhengyi* 周易正義, 86.

5. Zhu Xi 朱熹, *Huian xiansheng zhuwengong wenji* 晦庵先生朱文公文集, juan 15, 15b, 220.

6. Elman, *On Their Own Terms*, 89.

7. The *Yijing* is a core philosophical and cosmological text, and part of the classics. During the Late Ming multiple versions with comments by core thinkers such as Zhu Xi or successful exam candidates for teaching were circulating. The renowned bibliophile Zhu Mujie 朱睦 (1518–1587), for instance, specialized on Yijing studies and lists 115 titles in his catalogue *Wanjuang tang shumu* 萬卷堂書目 庚午 (1570, preface).

8. Kim Young-Oak, "The Philosophy of Wang Fu-chih"; Fu Liming 付黎明 and Xie Yunxia 謝云霞, "Lun Li Yu."

9. Ye Kangning 叶康寧, "Cong 'qi yi zai dao' dao 'qi ti dao yong'." Wang Fuzhi discussed the topic in his *Zhouyi waizhuan* 周易外傳, ch. 5.

10. The imperial compilation *Gujin tushu jicheng* 古今圖書集成 (completed ca. 1725) and the *Shoushi tongkao* 授時通考 (ca. 1742) copied parts of the Song Yingxing's *Tiangong kaiwu*, which referred to acknowledged fields such as metallurgy, salt, and agricultural production.

11. Schäfer, *The Crafting of the 10,000 Things*, ch. 6.

12. Huang Cheng, *Xiushi lu* 髹飾錄. See also Huang Cheng 黃成, *Xiushi lu tushuo* 髹飾錄圖說.

13. Clunas, *Superfluous Things*, 34–36.

14. Song Yingxing, *Tiangong kaiwu* (1997), 49b.

15. Huang Cheng's name indicates his affiliation to production (*cheng* 成). He shares this idiosyncrasy with another contemporary author of the practical arts, the garden architect ("mountain master," *shanshi* 山師) Ji Cheng 計成 (b. 1582), who

compiled the *Yuan Ye* 園冶 in 1631. For an English translation of *Yuan Ye*, see Hardie, *The Craft of Gardens*; also Hardie, "Ji Cheng's Yuan Ye in Its Social Setting."

16. Wang's updated edition follows the Dingmao version. For Yang Ming, see Suo Yuming 索予明, "Tihong kao 剔紅考." His courtesy name was Yang Qingzhong 楊清仲. The Mietang 蔑堂 edition spells his name as 揚, whereas the Dingmao Zhu Shi 丁卯朱氏 edition describes 揚 as a typo, replacing it with 楊.

17. The epigraph is from Huang Cheng, *Xiushi lu*, preface. I used the Wang Jie 王解 edition and then relied on Chang Bei's 長北 comparison of the various editions. See Chang Bei 長北, "*Xiushi lu* banben jiaokan ji" "髹飾錄"版本校勘記.

18. See, for instance, Esherick and Rankin, *Chinese Local Elites*.

19. McDermott, "'Noncommercial' Private Publishing," 128.

20. Zhang Binlun, "Scientific Development and Its Causes," 339–40. See also Yongtao Du, *The Order of Places*, 54–56, for a discussion of translocal practices.

21. Song Yingxing, *Tiangong kaiwu* (2002), 46, 207.

22. For all quotes in this paragraph, see *Zhouyi zhengyi* (2003), 78.

23. All of these quotes are from Huang, *Xiushi lu tushuo*, preface. For an overview of Hetu and Luoshu, see Dorofeeva-Lichtmann, "Spatial Organization," 31. For detailed discussions of the role of Luoshu, see Granet, *La pensée chinoise*, 177–208; Henderson, *Development and Decline of Chinese Cosmology*, 82–87. For Hetu, see Seidel, "Imperial Treasures and Taoist Sacraments."

24. According to Wang Chong 王充, *Lunheng*, 244, "skillful" here refers to a method without a script (*wuwen* 無文).

25. Huang, *Xiushi lu tushuo*.

26. Clunas, *Superfluous Things*.

27. Sun Yirang 孫詒讓, *Kaogong ji* 考工記.

28. Bray, *Technology and Gender*.

29. Bray, *Technology and Gender*, 185.

30. Authors often explained deviations from the existing literature with regard to each author's focus. Some emphasized that their writings only filled an evident gap, omitting issues of changed emphasis by focusing on the importance and details of regional diversity or by expanding the overall scope of a topic. See Chuan-Hui Mau, "Les progrès de la sériciculture."

31. See Rowe, *Saving the World*, 236–37.

32. *Zhouyi zhengyi*, ch. 6, 12.

33. Waley, *The Analects of Confucius*, 147–48.

34. For nobles being given garments, see *Tiangong kaiwu* (2002), 43. According to the *Sancai tuhui* 三才圖會, compiled by Wang Qi 王圻 in 1607, such a decorated garment was the highest ritual gown of the Song and Ming. See also Huang Nengfu 黃能馥, *Zhongguo fuzhuang shi* 中國服裝史.

35. Song, *Tiangong kaiwu* (1997), Naifu entry.

36. Song, *Tiangong kaiwu* (1997), 1a.

37. Song's omission of cotton in this context would be a valuable subject for further research.

38. Song, *Tiangong kaiwu* (1997), 305.

39. Song, *Tiangong kaiwu* (1997), 74. For a translation, see Sung Ying-hsing. *T'ien kung k'ai-wu.*

40. Golinski, *Making Natural Knowledge*, 172. See also Chemla, *History of Science*; Günergun and Raina, *Science between Europe and Asia.*

41. See Bourguet, Licoppe, and Sibum, *Instruments, Travel and Science*, 1–19.

42. Janelle, "Global Interdependence," 49–81.

43. For an overview of Ming travel times, see Brook, *Confusions of Pleasure*, 173–85; Twitchett, and Mote, *The Cambridge History of China*, 8:632–34.

44. The first mention of this concept is in Giddens, *Critique of Historical Materialism*, 1:90.

45. See, for instance, Finlay, *The Pilgrim Art.* See also Ma, "The Great Silk Exchange," 26.

46. Golas, "Like Obtaining a Great Treasure," 574.

47. The original was published in Kangxi 25 (1685). Sun Pei was part of a compilation team for the local gazetteer. For a discussion of this work and translation into German, see Piontek-Ma, *Der Bericht von Sun Pei.*

Chapter 5: Curative Commodities between Europe and Southeast Asia, 1500–1700

1. On Malacca's role in the region and colonization, see especially Sandhu and Wheatley, *Melaka.*

2. Soledade, *Historia Serafica*, 5:497.

3. Soledade, *Historia Serafica*, 5:497.

4. Archivio Segreto Vaticano, "Processus in causa Aloysii a Cruce," f. 146r. It is unclear why Juliana herself did not testify, as other women attested to their own miracles.

5. See Soledade's *Historia Serafica*, 480–523; Cardoso, *Agiologio Lusitano*, 416–17, 420; Trinidade, *Conquista espiritual do Oriente.*

6. In Italian: *se li Religiosi di detto conuento [de Madre de Dios] l'hanno pagato per deporre in questo tribunale cosa alcuna circa di Frà Luigi della Croce Frate Laico del detto ordine*. This is the formula used in each deposition throughout the manuscript.

7. Chen, "Traditional and Modern Medicine," 128, 131. On the *bidan* as repositories of spiritual power in nineteenth-century western Malaya, see Peletz, "Poisoning, Sorcery and Healing Rituals," 134–35.

8. Boxer, *Women in Iberian Expansion Overseas*; see also Stoler, *Carnal Knowledge*, 47–70.

9. Soledade, *Historía Serafica*, 496–97, 503.

10. Trinidade, *Conquista espiritual do Oriente*, 3:401.

11. Trinidade, *Conquista espiritual do Oriente*, 3:401.

12. Andaya, "From Temporary Wife to Prostitute," 12–16.

13. See Taylor, *The Social World of Batavia*, esp. 34–45, also appendix 3; Noncelli,

"Erotics of Mercantile Imperialism," 137–38; Stoler, *Carnal Knowledge*, 41–78; Few, "Medical *Mestizaje*."

14. See Souza, Joaquim, Canço, Pires, and de Castro, *O rosto feminino*; Sanceau and Sampaio, *Mulheres portuguesas no ultramar*.

15. Sarabia, "The Culture of Peyote," 23.

16. See Gianno, "Women Are Not Brave Enough," 55; Andaya, *The Flaming Womb*, 209–10.

17. Jakob de Bondt, *An Account*. On Bondt's sources, see especially Sargent, "Global Trade and Local Knowledge," and Cook, "Global Economies."

18. On "specific" remedies in early modern medical theory, see Cook, "Trading in Medical Simples."

19. Bondt, *An Account*, 210. Motherby, in *New Medical Dictionary*, defines "Borri Borri" as an "East Indian" name for turmeric and for an ointment made from the same.

20. See especially Andaya, *Flaming Womb*, 208–15.

21. Manderson, "Shaping Reproduction," 27. See also Ortiz Gómez and Cabré I Pairet, "Mujeres y salud." On the "imaginary" of childbirth and pregnancy in early modern Europe, see especially Santo-Tomás, "Offspring of the Mind," 155–56.

22. Boer and Cotingting, "Medicinal Plants for Women's Healthcare," 748, 759–60.

23. Manderson, "Roasting, Smoking and Dieting," 509–10); Manderson "Traditional Food Classifications."

24. Niehof, "Traditional Medication at Pregnancy," 251.

25. See Churchill, *Female Patients*; McClive, *Menstruation and Procreation*.

26. Erédia, "Description of Malaca," 47.

27. *Daya* or *dayah* also means "wet nurse" or "foster mother." See Baxter and de Silva, *Dictionary of Kristang*, 24. Note: among the Portuguese creole community in Macao at the beginning of the twentieth century, Portuguese-language education was credited with causing considerable shifts in vocabulary. This included the replacement of the local term for midwife (*daya*) with *partera* (from the Portuguese *parteira*). Caudmont, "Présence de la langue portugaise, " 24n39).

28. Andaya, *Flaming Womb*, 210.

29. Andaya, *Flaming Womb*, 210.

30. Chen, "Traditional and Modern Medicine," 128.

31. See Peletz, "Poisoning, Sorcery and Healing Rituals," 134–35; also King, "Cursing, Special Death, and Spirits," 138–39.

32. Desouza, "Tradition, Colonialism and Modernity," 197.

33. Gracias, *Kaleidoscope of Women in Goa*, 120; Desouza, "Health Practices and People's Identity," 457.

34. See Carrillo, "Nacimiento y muerta."

35. Boomgaard, "Development of Colonial Healthcare," 88–89; Manderson, "Shaping Reproduction," 28–30.

36. Desouza, "Tradition, Colonialism and Modernity," 197.

37. Barbosa, *The Book of Duarte Barbosa*, 173.

38. Erédia, "Description of Malaca," 47. The English common names given in square brackets are approximate and mainly based on Mill's annotations to Erédia's text (173–75). See Seidemann, *World Spice Plants*, and discussions of Malay terms in Mahdi, *Malay Words*.

39. Boxer, "Note on Portuguese Reactions"; Casale, "Ottoman Administration"; Subrahmnayam, "A Note on the Rise of Surat," 25–28.

40. Jung, "Cultural Biography of Agarwood."

41. See also Ralph Bauer's discussion of the shifting meanings of gold and "dragon's blood," in "The Blood of the Dragon," esp. 74–75, 87.

42. See Rees, "Gall-nut."

43. Pereira, *Elements of Materia Medica*.

44. Rees, "Gall-nut."

45. On these routes, see Özbaran, "The Ottoman Turks"; Casale, *Ottoman Age of Exploration*.

46. "Do Principio do Reyno de Ormus," in Rego, *Documentaçao ultramarina portuguesa*, 2:93.

47. Azevedo, "Apontamentos pera Vossa Magestade," in Rego, *Documentaçao ultramarina portuguesa*, 2:163.

48. Subrahmanyam, "A Note," 29.

49. Diogo de Couto, cited by Casale, "Ottoman Administration," 191.

50. Villiers, "Aceh, Melaka."

51. Bishop of Malacca, cited by Casale, "Ottoman Administration," 193.

52. See Widrlechner, "History and Utilization of *Rosa damascene*."

53. Andaya, *Flaming Womb*, 209.

54. Ghazanfar and Fisher, *Vegetation of the Arabian Peninsula*, 259.

55. Mohd Shafri, "Makkah and Medicine," 149.

56. Erédia, "Description of Malaca," 232.

57. Stein explores the varied meanings of rosewater according to nineteenth-century periodicals that shaped Ottoman Jewish tastes and habits, in "Creating a Taste for News," 15–16.

58. Andaya and Andaya, *History of Malaysia*, 16; Boomgaard "Sacred Trees."

59. Boomgaard, *Frontiers of Fear*; Sevea, "Pawangs on the Malay Frontier."

60. Andaya, *Flaming Womb*, 209.

61. Erédia, "Description of Malaca," 47.

62. Erédia, "Description of Malaca," 48–49.

63. Erédia, "Description of Malaca," 41, 48.

64. Erédia, "Description of Malaca," 41.

65. Howes and Classen, *Ways of Sensing*, 42–44.

66. Schafer, *Golden Peaches*, 164. Cammann, "Magical and Medicinal Woods," 120–21.

67. Nguyễn and Hưng, *Monts et merveilles*, 68–69, 101.

68. Pliny the Elder, *Natural History*, 102.

69. Orta, *Coloquio*, 2:53.

70. See, for example, Reid, "Elephants and Water."

71. See also the "Xavier-water" discussed in Johnson, "Blood, Tears and Xavier-Water."

72. Archivio Segreto Vaticano (ASV), "Processus" vol. 1, ff. 7v–8r.

73. Suh Soyoung, "Herbs of Our Own," 397–99, 412–20.

74. Suh Soyoung, "Herbs of Our Own," 398–99.

75. See also Reyes, "Glimpsing Southeast Asian Naturalia," 113–14.

Chapter 6: Translating the Art of Tea

My warmest thanks to Georg Freise, who provided invaluable help with the initial stages of research for this chapter.

1. "Naturalization" was the term used by Bentinck, Chairman of the Tea Commission of 1834, arguing for the introduction of tea to British India; see Liu, "Birth," 80. On "skill and science," see Walker, *Memorandum*, 11.

2. Antrobus, *Assam Company*, v. The East India Company (EIC) was granted its monopoly in 1668. The Charter Act of 1833 revoked the monopoly, opening the lucrative China trade to competition from other British companies. The EIC was shorn of its commercial functions and charged instead with the administration of India. The 1833 charter legally formalized India's status as a colony, thus British ownership of land became possible.

3. Antrobus, *Assam Company*, 30, 163. The first phase is covered in chapters 7–9, the second in chapters 10–12, and the third in chapters 13–21

4. Hauser, *Tea*, 11. For an example of today's global tea company websites, see http://www.finlays.net/about-finlays/history (accessed April 15, 2016).

5. Antrobus, *Assam Company*, 30, 163.

6. For exports from 1856 to 1885, see Gardella, *Harvesting Mountains*, 59, 61 (table 7), 62 (table 8). The streamlined clippers—which transported only light goods such as tea, opium, and spices and whose rigging allowed them to sail fast without relying on monsoon winds—were able to deliver new teas fresh from the premonsoon harvest. After 1895, see Gardella, *Harvesting Mountains*, 164 (table 22), 11 (table 3). It is important to remember that even in the heyday of Chinese tea exports to Britain, these Fujian black teas represented only a tiny fraction of the tea grown for the internal Chinese market. Chinese smallholder tea producers and small-scale processers could easily switch between types of tea, or they could withdraw temporarily from tea production to cultivate other crops until markets improved. Gardella provides an excellent account of the long-term resilience and flexibility of Fujian's tea districts.

7. Lemonnier, *Technological Choices*; Schlanger, "Mindful Technology"; Hodder, *Studies in Human-Thing Entanglement*, 115–20. In archaeological terms an "assemblage" is a group of artifacts associated with one another in a particular context, such as the goods found in a grave. Here I use "technical assemblage" to denote the cluster of components, whether material, social, or ideological, assembled to perform a technical task such as manufacturing tea, within a specific context (1830s Fujian, 1880s

Assam), in an approach to the materialities and politics of technological choice similar to Ong and Collier, *Global Assemblages*.

8. Finlay, "Pilgrim Art"; Riello and Roy, *How India Clothed the World*; Bray, "Technological Transitions"; Smith, chapter 1 (this volume).

9. The STS concept of *technological closure* was first proposed by Pinch and Bijker in "Social Construction," as a key element for investigating the Social Construction of Technology (SCOT). For an excellent account of the complexities of convergence between material practice and philosophical consensus, see Hilaire-Pérez, *La pièce et le geste*, on the transition to industrial culture in late eighteenth- and early nineteenth-century London.

10. For more on how Indian ownership was impeded and limited, see Sharma, "British Science," 452–53; Sharma, "'Lazy' Natives," 1316–17. The Indian plantations were producing tea for a market of consumers half a world away. Indians themselves did not develop a taste for tea until well into the twentieth century (one possible factor being that nationalists condemned tea as an iconic product of colonial exploitation). As late as 1930, "only a tiny fraction of the Indian population had tasted tea, and more than 90 percent of the huge crop continued to be exported" (Lutgendorf, "Making Tea," 14).

11. The full set of chaînes opératoires analyzed in this chapter are available online, in a set of seven tables and eight figures or diagrams, at https://www.francescabray.co.uk/tea/ and referenced as they occur in the notes.

12. The exception is a project in progress by John Bosco Lourdusamy of the Indian Institute of Technology, Madras, on the mechanization of the Indian tea industry. See Lourdusamy, "Tea and Technology."

13. See Huang, *Science and Civilisation*, 515–18, on the *Chajing*; also *Sishi zuanyao*, 69–70 (possibly quoting an earlier work), and *Nongshu*, 162–65.

14. See Huang, *Science and Civilisation*, 535, for possible explanations. Local gazetteers, which were revised every few decades, are one potential source of information on production practices in the tea districts, though the information they provide usually lacks detail. They may, however, be the technical and "statistical" works to which Ball refers. In terms of modern histories of tea production based on Chinese sources, key works in Chinese include Chen Zhugui and Zhu Zizhen, *Zhongguo chaye*, and Zeng Xiongsheng, *Zhongguo nongxue*, 491–98, 645–47. In English, H. T. Huang's study systematically treats the evolution of tea manufacturing (though not of cultivation skills) in China up to the nineteenth century; see Huang, *Science and Civilisation*. Gardella, *Harvesting Mountains*, traces the economic and many elements of the technical development of tea production in the main export region, Fujian, through into the mid-twentieth century. Both Huang and Gardella refer to the exchanges and competition between the Chinese and British tea industry, but from the Chinese perspective.

15. Ball, *Account*; Fortune, *Three Years*; Fortune, *Journey*.

16. Ball, *Account*, vii.

17. Fortune's first foray into China was in 1842, when the Chinese government still strictly prohibited Westerners from leaving Canton. Even during his longer, later

visit between 1848 and 1852, Westerners were not supposed to go beyond the walls of the newly opened Treaty Ports.

18. For histories, see Gardella, *Harvesting Mountains*; Huang, *Science and Civilisation*. Tables 4 and 5 at https://francescabray.co.uk/tea/ analyze Chinese chaînes opératoires of tea cultivation and tea processing, respectively, as detailed in Ball, *Account*; Fortune, *Three Years*; and Fortune, *Journey*.

19. See, for instance, Macfarlane and Marfarlane, *Green Gold*; Ellis, Coulton, and Mauger, *Empire*. See also four chapters in Berg et al., *Goods from the East*: Nierstraz, "Popularization"; Hodacs and Müller, "Chests, Tubs"; Mackillop, "A North Europe World"; and Blondé and Ryckbosch, "Arriving to a Set Table."

20. For example, Money, *Tea*.

21. Sharma, "British Science." Sharma notes that taking on this scientific mission helped to redeem the EIC from the disgrace of successive scandals over its corrupt and exploitative rule of the Indian territories it controlled.

22. McEwan, "Tea," 477; Bruce, *Report*.

23. See, for example, Fortune's evaluation of the Indian tea gardens that he visited around 1850, on returning from his second visit to China (Fortune, *Journey*). A glance at the lists of sources cited by recent studies of the early Indian industry confirms how sparse technical accounts of the tea industry were for the first two or three decades; for example, Nitin Varma, "Producing Tea Coolies."

24. See, for example, Barker, *A Tea Planter's Life*; McGowan, *Tea Planting*.

25. Dey, "Of Planters," 113. Labor history is the key motif in much Indian-based scholarship on Indian tea. See Breman, *Labour Migration*; Hayami and Damodaran, "Towards an Alternative"; Behal, "Coolie Drivers"; Sen, "Commercial Recruiting"; Joshi, "Review"; Besky, *Darjeeling Distinction*. Consumption history frames much of the literature on tea in Britain. See Macfarlane and Macfarlane, *Green Gold*; Hanser, "Teatime"; Berg et al., *Goods from the East*; Ellis, *Empire*. Company or regional histories include "The Old Campaigner," *History*; Antrobus, *Jorehaut Tea Company*; Antrobus, *Assam Company*; Forrest, *A Hundred Years*; Baig and Henderson, *A Centenary*.

26. Tables 2 and 3 at https://francescabray.co.uk/tea/ analyze the chaînes opératoires of the various methods of tea cultivation and production that Bruce and his Chinese "tea men" experimented with in Assam, based on Bruce, *Report*. Tables 4 and 5 draw on the treatises of Ball and Fortune to analyze the original Chinese methods from which Bruce's experiments derived.

27. Figure 4 at https://francescabray.co.uk/tea/ is a diagram showing the chaîne opératoire of tea production according to Money, *Tea*.

28. Figure 5 at https://francescabray.co.uk/tea/ is a diagram showing the chaîne opératoire of tea production according to Crole, *Tea*.

29. Baig and Henderson, *A Centenary*, 15.

30. Figure 6 and Table 7 at https://francescabray.co.uk/tea/ show the chaînes opératoires of tea cultivation and tea processing, respectively, according to Robertson, *A History*. The typescript of this two-volume memoir, along with the original illustrations, is in our family collection.

31. Figure 7 at https://francescabray.co.uk/tea/ compares the chaînes opératoires of Money and Robertson to assess how far the ambitions of the 1880s had been fulfilled in actual practice half a century on, in the 1930s.

32. Bray, *Technology, Gender and History*, 7–9.

33. Coupaye, "Ways of Enchanting"; Akrich, "A Gazogene."

34. Bray, "Thinking with Diagrams in the History of Technology."

35. Tea has also generated a number of complex arts for making, enjoying, or sharing the beverage, which alas I cannot include in this short study, but which likewise could usefully be organized through the device of the chaîne opératoire. The loopback between taste and production techniques was, however, an essential element in the development of the Assam tea industry and features prominently in the story as I tell it here.

36. For diagrams and tables go to https://francescabray.co.uk/tea/.

37. Sharma, "British Science," 433, 437–40. On early tea production in Java, see Jacobson, *Handboek*; Ball, *Account*; Antrobus, *Assam Company*, 13–15.

38. Nitin Varma, "Producing Tea Coolies," 17; McEwan, "Tea," 477. The first Anglo-Burmese war lasted from 1824 to 1826; on defeat, Burma also gave up Manipur and Arakan to the EIC. All three states then became part of the Bengal Presidency. Manipur later proved to be another source of wild teas and was, like Assam, developed into a lucrative tea region.

39. Smith, *Taxing Heaven's Storehouse*; Gardella, *Harvesting Mountains*; Huang, *Science and Civilisation*; Hinsch, *The Rise of Tea Culture*, and so on.

40. Matthee, "From Coffee to Tea"; Gardella, *Harvesting Mountains*. Chinese exports of tea to Central Asia date back to the eighth or ninth century; Smith, *Taxing Heaven's Storehouse*.

41. On European ignorance of Chinese tea, see Bentinck, *Measures Adopted*; Ball, *Account*; Fortune, *Three Years*, 197–225.

42. See Fortune, *Three Years*, 212; Gardella, *Harvesting Mountains*, on up-country quality control by merchants.

43. Notably by growing opium in India and exporting it to China (Liu, "Birth"). This eventually prompted the Chinese government to ban opium imports (and tea exports) by closing down the Chinese ports and burning seized opium shipments; the British government retaliated by declaring war (the first Opium War) in 1839.

44. Antrobus, *Assam Company*, 30. Recent scholars accurately describe these forays as "smuggling" and "spying" (Sharma, "British Science," 442). Jacobson's Chinese assistant was arrested on one of his clandestine expeditions to the tea regions (Antrobus, *Assam Company*, 14), while Fortune found it necessary to disguise himself as a Chinese in order to travel up-country (Fortune, *Journey*, 19).

45. Bentinck, *Measures Adopted*, 5–6. See also Nitin Varma, "Producing Tea Coolies," 17. Jacobson had set up his Java estate with tea workers "imported or impressed from China" (Antrobus, *Assam Company*, 255).

46. Antrobus, "Chinese Labour," in *Assam Company*, 374–83; Sharma, "British Science"; Sharma, "'Lazy' Natives"; Liu, "Birth."

47. Ball, *Account*, 366; McGowan, *Tea Planting*. On the *Asia Felix* recruitment fiasco, see Antrobus, *Assam Company*, 378–81; on successfully settled Chinese tea workers, see Fortune, *Journey*, 373.

48. Bruce, *Report*, 464, Fortune, *Journey*, 373. Tea chests are a Chinese invention that, like other packing techniques, has attracted less attention than it deserves from historians of technology. To survive long sea voyages without spoiling, tea had to be most carefully packed and sealed. Timber planks were later replaced with plywood, and lead lining with aluminum foil and parchment paper, which made the chests lighter and less expensive to produce and thus reduced their share in the costing of tea production. But procuring the materials and skilled workers to make good-quality boxes remained a constant preoccupation of estate managers. Bruce devotes a long section to the intricacies of making and lining chests (*Report*, 470–71).

49. For the relative wages of Chinese and Indian workers, see Antrobus, *Assam Company*, 383; Nitin Varma, "Producing Tea Coolies," 25.

50. See Tables 2 and 3 at https://francescabray.co.uk/tea/.

51. Antrobus, *Assam Company*, 249. See Tables 4 and 5 at https://francescabray.co.uk/tea/.

52. Fortune, *Journey*, 392.

53. The quotation is from Fortune, *Three Years*, 203. See Table 5 at https://francescabray.co.uk/tea/.

54. Ball, *Account*, 111.

55. Gardella, *Harvesting Mountains*; Lu, "Beyond the Paradigm."

56. Bruce, *Report*, 470. See also Antrobus, *Assam Company*, 277.

57. Antrobus, *Assam Company*, 260, 269.

58. Bruce, *Report*, 464. *Jhat* (*jat*, or *jāt*) was a term denoting North India's ethnic groups or kingdoms, applied here by extension to tea stocks or populations of different "ethnic" origin.

59. Ball, *Account*, 108; Huang, *Science and Civilisation*, 538–39.

60. Fortune, *Three Years*, 215; Ball, *Account*, 108; Huang, *Science and Civilisation*, 539. See Table 5 at https://francescabray.co.uk/tea/.

61. Compare Table 2, "Land" and "Tea Plants," with Table 4, "Sowing and Seedlings" and "Transplanting," at https://francescabray.co.uk/tea/.

62. Antrobus, *Assam Company*, 251–54; Bruce, *Report*, 468.

63. Bruce, *Report*, 468.

64. Fortune, *Journey*, 92, 383; Bruce, *Report*, 466.

65. Bruce, *Report*, 469.

66. Walker, *Memorandum*, 11; Hauser, *Tea*, 11.

67. See Table 6 at https://francescabray.co.uk/tea/.

68. Antrobus, *Assam Company*, 289–92.

69. In 1861 Assam teas amounted to a mere 1 percent of British imports from China. But Indian teas added attractive body: grocers began to offer blends where small amounts of Indian tea were added to the predominant China leaf; Antrobus, *Assam Company*, 271; Ellis, *Empire*, 238.

70. Antrobus, *Assam Company*, 293. Several of the main companies building tea machinery were based in Norfolk, others in the Lothians, both centers of "improved farming" since the eighteenth century; see, for example, Tea Gazette, *Vade Mecum*, 167–79.

71. See Figure 4 at https://francescabray.co.uk/tea/. The second edition of Money's book was published in 1874, the third in 1878.

72. After a disastrous coffee plague, Ceylonese planters began switching to tea in the mid-1870s. With a perfect climate for tea, Ceylon soon emerged as "a formidable competitor"; Money, *Tea*, 183.

73. Money, *Tea*, 7; Antrobus, *Assam Company*, 144; Money, *Tea*, 328.

74. Money, *Tea*, vii.

75. Money, *Tea*, 35. Watson (*Essay*) shows different kinds of pruning knives and methods of pruning, suggesting that the practice was more routine in the 1870s; see Antrobus, *Assam Company*, 254.

76. See Figure 6 and Table 7 at https://francescabray.co.uk/tea/. Money held a dim view of the prospects for less productive regions, including Darjeeling (*Tea*, 21).

77. Money, *Tea*, 125, 110–13; Table 5 at https://francescabray.co.uk/tea/ shows the Chinese methods as described by Fortune. The *Tea Planter's Vade Mecum*, however, cites experts who argue for reintroducing panning as a method to reduce spoilage of finished teas (Tea Gazette, *Vade Mecum*, 195). Money also describes green tea methods, again identical to the Chinese methods used in Bruce's factory (see Table 3 at https://francescabray.co.uk/tea/). Money believes that, where China plants are grown, the labor-intensive green teas are still worth making since they fetch high prices in North America; the sustained success of Darjeeling teas up to the present bears out this view.

78. In the first edition of 1871 Money states that: "(1) Indian Tea is not known to the public. (2) Except in one or two shops in London and Glasgow, unknown to the mass of the people, not an ounce of pure Indian Tea can be bought in all England. (3) That India is even a Tea producing country is scarcely known in England." In a note to the 1883 edition Money observes that the situation has somewhat improved: a fall in world tea prices in 1877 brought Indian teas to British attention and started a market for pure Indian teas (*Tea*, 174).

79. As noted earlier, Fujian's exports of semi-fermented teas to Britain almost tripled between 1856 and 1885. See Gardella, *Harvesting Mountains*, 59, 61 (table 7), 62 (table 8).

80. Money, *Tea*, 261, 250.

81. See Figure 4 at https://francescabray.co.uk/tea/.

82. In fact, Jackson's Rapid Rollers, developed in the 1880s, superseded Kinmond's and dominated the market for many years. Their great advantage was that they ran themselves: "It is a remarkable tribute to the quality of both workmanship and materials put into these machines that in the hands of unskilled coolie labour during long hours they have continued in service up to fifty years after being installed" (Antrobus, *Assam Company*, 294). The *Tea Planter's Vade Mecum*, like Money, rates

Kinmond's machine highly because it comes in small sizes and can therefore "meet the requirements of very small estates" (Tea Gazette, *Vade Mecum*, 175).

83. Nitin Varma, "Producing Tea Coolies," 85. Dating back many centuries in China, the cheap, versatile, and extremely effective winnowing-fan spread through Asia and Europe in the seventeenth and eighteenth centuries; Bray, *Science and Civilisation*, 373.

84. Money, *Tea*, 122.

85. Tea Gazette, *Vade Mecum*, 131–67.

86. Tea Gazette, *Vade Mecum*, 162.

87. Tea Gazette, *Vade Mecum*, 167.

88. Behal, "Coolie Drivers," 41. See also Knowles, *Economic Development*, 84; Sharma, "British Science," 430.

89. One reviewer criticizes Crole's bias toward Indian and Ceylonese teas and his rosy-tinted view of labor conditions; he also contests Crole's understanding of the chemistry of tea, while conceding the value for planters of the technical chapters on cultivation and manufacture; Anonymous, "*Text-Book*."

90. Crole, *Tea*, ix (the syllabus is given on p. 114). See Figure 5 at https://francescabray.co.uk/tea/.

91. Crole, *Tea*, 153. Turbines were especially attractive for remote gardens. Robertson notes (see Figure 6 at https://francescabray.co.uk/tea/) that local streams initially powered the factories in the High Range tea gardens, set up around the time Crole was writing; eventually they were connected up to Finlay's central power station (Robertson, *A History*, 1:17–18).

92. Letter quoted in Money, *Tea*, 223; Crole, *Tea*, 62.

93. McEwan, "Tea," 480. For tea machine companies, see Crole, *Tea*, 159, endpapers xvi; Tea Gazette, *Vade Mecum*, 171. For the 1930s, see Antrobus, *Assam Company*, 293; Lutgendorf, "Making Tea," 22.

94. Money, *Tea*, 224; Tea Gazette, *Vade Mecum*, ch. 11.

95. Crole, *Tea*, 60, endpapers vii.

96. Tea Gazette, *Vade Mecum*, 188–93; Anonymous, "*Text-Book*"; Crole, *Tea*, 190.

97. Nitin Varma, "Producing Tea Coolies," 196.

98. Jeffery, "Merchant Capital."

99. Glasgow University Archives, *Guide*. Finlay's also managed extensive tea holdings in Ceylon, with an office in Colombo.

100. Jeffery, "Merchant Capital," 242, Table 2.

101. See Figure 6 and Table 7 at https://francescabray.co.uk/tea/ for Robertson's chaînes opératoires of production and processing. Table 6 shows how the pace of innovation slowed down after 1890.

102. Robertson, *A History*, 2:17–18. "Technological closure" is defined as consensus about desired outcomes and the best technological procedures for reaching them

103. Robertson's notional tea estate is fifteen hundred acres, of which twelve hundred are brought under tea; see Figure 6 at https://francescabray.co.uk/tea/.

104. See Figure 7 at https://francescabray.co.uk/tea/: "50 years of mechanization: 1880–1930."

105. Robertson, *A History*, 2:16.

106. For transplanting seedlings, see Robertson, *A History*, 2:11; also the Tamil-language instructions in Ferguson, *Iṅgē Vā*, 23.

107. Robertson, *A History*, 1:21, 2:13, 16.

108. Bruce, *Report*, 473.

109. Robertson, *A History*, appendix. Iron drag-hoes like the *mamoti* were also the "Universal Tool" across East and Southeast Asia (Bray, *Science and Civilisation*, 207–12). Figure 6.4a–b in this chapter corresponds to Figure 8A and 8B at https://francescabray.co.uk/tea/, where the explanations from Robertson's glossary are also shown.

110. Nitin Varma, "Producing Tea Coolies," 178, citing *Standing Instructions for the Tea Estates Department of Messrs Finlay Muir & Co. Calcutta* (Glasgow 1900).

111. Ferguson, *Iṅgē Vā*, 34.

112. Robertson, *A History*, appendix; Ferguson, *Iṅgē Vā*, 24.

113. Money, *Tea*, 135.

114. Those anxious to develop a modern tea industry in China in the 1930s feared that mechanization was unsuited to Chinese leaf type and the short season. See Gardella, *Harvesting Mountains*, 151.

115. Crole had called for a shift to cuttings, together with research on how to hybridize and graft the tea bushes (*Tea*, 66), but his call went unheard in India, even though the technique of propagation by slips or cuttings was being used elsewhere. It is recorded in parts of China in the eighteenth century and was introduced from Amoy to Taiwan in the 1860s, when a Scottish merchant and an Amoy compradore set up a partnership and launched a local tea industry; see Gardella, *Harvesting Mountains*, 64.

116. Besky, *Darjeeling Distinction*. The Darjeeling tea factories continued to hire seasonal Chinese workers well into the 1950s.

Chapter 7: The Itinerary of Hing/*Awei*/Asafetida across Eurasia, 400–1800

The authors would like to thank Pamela Smith and the two anonymous reviewers of the chapter for their excellent comments and suggestions.

1. See Anderson, Dunlop, and Smith, "Introduction," in their edited volume, *The Matter of Art*, 2–12; Rodgers, "Cultures in Motion," esp. 8–12. See also one of the pioneer works on traveling commodities, Appadurai, *The Social Life of Things*.

2. Yule et al., *Hobson-Jobson*, 418.

3. Dalby, *Dangerous Tastes*, 110–12; Balfour, *Asafoetida Plants*, 367.

4. Dymock, Warden, and Hooper, *Pharmacographia Indica*, vol. 2, p. 148.

5. Li Shizhen, *Bencao gangmu*, Taibei ed., ch. 34, p. 35.

6. Laufer, "Three Tokharian Bagatelles." Ming Chen revises Laufer's proposition by suggesting *ankwas(t)* as the Tokharian word that *awei* transcribed, which was a word with some affinity to the Persian term *anguzad*.

7. On the complex history and ethnicity of the Tokharians, see Hansen, *The Silk Road*, 70–80 (esp. 79–80). Basing her work on earlier works by Eric Trombert, Hansen confirms the role of Kuchen traders in turning Kucha into a trading center of the time.

8. On the importance of Kumārajīva's translation in Chinese culture, see Hansen, *The Silk Road*, 56–70.

9. The story of this path-breaking translation enterprise is recorded in the Chinese Buddhist text *Chu Sanzang ji ji* 出三藏記集 (Compilation of notes on the translation of the Tripitaka, ca. 502–519), chs. 2 and 3 (Chinese Buddhist Electronic Text Association [hereafter CBETA], T55, no. 2145, p. 11, a25–27; p. 20, a21–b21).

10. *Shi song lü*, ch. 21 (CBETA, T23, no. 1435, p. 157, a1–3), ch. 26 (CBETA, T23, no. 1435, p. 194, a12–14).

11. *Sui shu* 隋書 (Official history of the Sui dynasty), 83:1857.

12. The Tang monk and translator Xuanzang (Hvenasāmga, 602–664) recorded his seventeen-year travel in India and Central Asia in his famous *Da Tang Xiyu ji* 大唐西域記 (Great Tang record on the Western Regions, 646) in which he mentioned the State of Cao as having a distinct language and being a producer of many herbs and flowers, including the famous *xingqu* herb (CBETA, T51, no. 2087, p. 939, b17–25).

13. The word is *chong* (literally "creeping bugs"), which could imply in this period both insects and bugs or disease causing bugs inside the human body.

14. Okanishi Tameto, *Chongji xinxiu bencao*, 231–32. Su Jing compiled this compendium based on an extant sixth-century text that did not include awei.

15. See, for example, chapter 3 of the eighth-century medical text by Wang Tao, *Waitai miyao*.

16. Sun Simiao, *Qianjin yifang*, 232. On the ancient notions of disease transmission in China, see Leung, "Evolution," 31–32; Li Jian-min, "They Shall Expel Demons."

17. Leung, *Leprosy in China*, 54–55.

18. Sun Simiao, *Qianjin yifang*, 250–51. The recipe was called "Qipo's recipe for the malicious disease." Qipo was the Chinese name for Jīvaka the Ayurveda doctor; Ming Chen claims that awei was used by both Chinese and Indian doctors to exterminate bugs, detoxify, warm the body, and cure coughs. See Ming Chen, *Yindu fanwen yidian*, 62–64.

19. Sun Simiao, *Beiji qianjin yaofang*, 310. The recipe is based principally on garlic with awei and other constituents newly introduced into China such as milk and pepper (*fructus piperis longi*).

20. Wang Tao, *Waitai miyao*, 131–32, 164, 366, 369.

21. Tanba Yasuyori, *Ishinpò*, 560.

22. For his biography, see Huijiao, *Gaoseng zhuan*, ch. 29, pp. 890–92. Other Buddhist writers who wrote on the resin include Atigupta, who translated the *Dhāraṇīsamuccaya* sutra in 654, and Yijing (625–713), in *Nanhai jigui neifa zhuan*.

23. Huilin, *Yiqie jing yin yi*, ch. 1 (CBETA, T54, no. 2128, p. 311 a3–b8, b11; p. 312, a4); Zanning, *Song gao seng zhuan* (CBETA, T50, no. 2061, p. 738, a22–b5).

24. Huilin, *Yiqie jing yin yi*, ch. 67 (CBETA, T54, no. 2128, p. 750, b04). Unless otherwise noted, all translations from the original are our own.

25. Ming Chen, *Yindu fanwen yidian*, 63.

26. "Tianbao er-nian Jiaohe jun shigu an." It was considerably cheaper than clove, another imported ingredient (thirty-five coins for one *liang* of superior quality).

27. One zhang was about 360 centimeters. Eight zhang was about 2.9 meters.

28. Duan Chengshi, *Youyang zazu*, 101. Diego Santos argues that Duan's informant, the "Fulin" monk Wan, was probably a Melkite Christian who introduced plant knowledge in China in Greek-Syriac-Arabic-Persian vocabularies. Santos, "A Note."

29. Robert Hartwell estimated that international commerce at Chinese sea ports amounted to around 1.7 percent of GNP, or 10–20 percent of national nonagricultural income. See Hartwell, "Foreign Trade," 453–54.

30. Lin Tianwei, *Songdai xiang yao maoyi shigao*, 199–228. By 1133 there were already more than two hundred types of spices that reached the port of Quanzhou in southern Fujian. See Li Yukun, "Song-Yuan shiqi," 58.

31. Hartwell, "Foreign Trade," 478.

32. Hinrichs and Barnes, *Chinese Medicine and Healing*, ch. 4.

33. Wang Huaiyin ed., *Taiping shenghui fang*, 3163. The other ingredients include *Angelica Sinensis*, Cassia bark, dried orange skin, several rhizomes (white *atractylodes*, *qiongxiong*, *turtschaninovii coydalis*), *Aucklandia root*, aconite, *curcuma zedonary*, *evodia rutaecarpa*.

34. Leclerc, *Traité des simples*, 143.

35. *T-S Ar.39.458* (recto) in Lev and Chipman, *Medical Prescriptions*, 52–54.

36. Robert Carrubba, "First Report," 456. The 1360 text was translated from Latin to English by John Trevisa in 1398.

37. Song Xian, introduction to *Huihui yaofang kaoshi*, vol. 1, ch. 1, pp. 1–31. See also Schottenhammer, "Transfer of *Xiangyao*," 128.

38. Other than *awei*, transcriptions of Arabic words such as *sikbínaj*, *sakhínaj* (*sāghā fyín*), *Sagapenum*, *hiltit*, *anjudhan*, *al-kasakibanaj*, *alpiltiti* are used in this Chinese text. Some of the terms seem to indicate the roots of the plant. Recipes are found mostly in chapters 12, 30, 34, three of the extant chapters of the book. See volume 2 of Song Xian's *Huihui yaofang kaoshi*.

39. Nie Hongyin, "Xixia 'Tiansheng lü ling'"; Li Yingcun, Li Jintian, and Shi Zhenggang, "E cang Dunhuang"; Wang Shizhen, "E cang wenxian."

40. *Sibu yidian*, 57. For aphrodisiac use in Tibetan, Arabic, and Indian cultures, see Gyatso and Hakim, *Essentials of Tibetan Traditional Medicine*, 137.

41. *Song shi*, ch. 490, p. 14107; Yan Guirong, "Qian yi Songdai Long."

42. *Song shi*, ch. 489, p. 14099.

43. Huang Shu (1019–1058), *Fa tan ji*, ch. 2, ff. 16b–17b. The moral lesson that Huang drew from this fact was that poisonous drugs were in demand as most people consulted a doctor only when the illness was already well developed, requiring more aggressive treatment. He compared this to policies that were efficient in the short run but ruinous to the body politic in the long run.

44. The biography of Monk Huiri (seventh–eighth centuries), in Huijiao, *Gaoseng zhuan*, ch. 29, pp. 890–92.

45. In a Dunhuang manuscript "Zaji shi yaoyong zi," awei was listed under the category "Vegetables," and its root was listed as a medicine, showing that awei was consumed both as a vegetable and as a medicine in western China in this period. The medical text *Sheng ji jing* (ca. 1111–1118), attributed to the emperor Zhao Ji, claimed that the stench of awei disqualified it as a food. The author referred to Confucius's *Analects* in which food with a strong bad smell was considered not fit for consumption (*Sheng ji jing*, ch. 10, p. 179).

46. Leclerc, who translated and annotated Ibn Al Baytar's pharmacopeia of the thirteenth century, noted the particular use of asafetida as a condiment by the "Orientals," which was never the case in Europe because of its fetid odor (*Traité des simples*, 144–45).

47. For an appraisal of the impact of Mongolian cuisine in China, see Allsen, *Culture and Conquest*, esp. 131 on the use of spices. See also Hu Sihui, *Yinshan zhengyao*, 20, 21, 29, 35, 251–52, and the annotated translation of this text by Buell, Anderson, and Perry, *A Soup for the Qan*, 106, 159, 272, 287. For the discussion on the short-lived Mongolian culinary culture in China, see Mote, "Yuan and Ming," 207–27.

48. Whyte, Geest, and Hardon, introduction to *Social Lives*, esp. 7.

49. See Raj, *Relocating Modern Science*, esp. ch. 1.

50. Garcia da Orta, Seventh Colloquy, in *Colóquios dos simples e drogas he cousas medicinais da India*, 47. Garcia da Orta's contemporary Cristóvão da Costa (1515–1594) added that the resin was used as an aphrodisiac in Arabic regions and India. The German adventurer Johann A. von Mandelslo (1616–1644) who traveled in Persia and India in the late 1630s noted that the "Hingh" mostly came from Persia. See Laufer, *Sino-Iranica*, vol. 15, no. 3, p. 356, quoting Cristóvão da Costa's *Tractado de las drogas, y medicinas de las Indias orientales* (Burgos, 1578), 357.

51. For the background of Kaempfer, see Haberland, *Engelbert Kaempfer*, and especially the chapter by Roelof van Gelder, "Nec semper feriet quodcumque minabitur arcus—Engelbert Kaempfer as a Scientist in the Service of the Dutch East India Company" (Haberland, *Engelbert Kaempfer*, 211–25).

52. Carrubba, "First Report," 456.

53. Brief biography of Kaempfer in Carrubba, "First Report." Kaempfer's description of asafetida can be found in observation 5 of Kaempfer, *Amoemitatum exoticarum*, 535–52.

54. Carrubba, "First Report," 451–61.

55. Royle, *Xiyao dacheng*, 462–66.

56. An account of such naturalists is in Balfour, *Asafoetida Plants*, 361–63.

57. Balfour, *Asafoetida Plants*, 366–68.

58. Otsuki Bansui and Otsuki Banri, *Ran'en tekihō.*

59. Royle, *Xiyao dacheng*, ch. 5, pt. 2.

60. Chen Cheng, *Xiyu fan guo zhi*, 69. See Rossabi, "A Translation." Samarqand is in present-day southern Uzbekistan.

61. Garcia da Orta, *Colloquies*, 45–46, 48; Carrubba, "First Report," 455.

62. Royle, *A Manual*, 467.

63. Leclerc, *Traité des simples*, 144.

64. Kou Zongshi, *xuli* (introduction) to *Bencao yanyi*.

65. This significant change is discussed in Unschuld's "Traditional Chinese Pharmacology."

66. Zhao Ji, *Shengji jing*, see ch. 10, p. 180n47, the interpretation was provided by Li Ti, contemporary annotator of the book. For the fourteenth-century saying, see Li Shizhen, *Bencao gangmu*, Taibei ed., ch. 34, p. 35.

67. Such as Chen Jiamo's *Bencao mengquan*, ch. 1, p. 8. About the changes in the discussions on the nature of drugs in late imperial Chinese recipe books, see Bian He, "Assembling the Cure," 94.

68. This quality was described in some detail in the Ming doctor Miao Xiyong's *Shennong bencao jing shu*, 190.

69. Tang Shenwei, *Zhenglei bencao*, 253. See also Anonymous, *Buyi Leigong paozhi bianlan*, ch. 4, p. 632.

70. On Dioscorides's claim, see Dymock, Warden, and Hooper, *Pharmacographia Indica*, vol. 2, p. 141; on Avicenna's view, see Balfour, *Asafoetida Plants*, 367; Royle, *A Manual*, 461.

71. Guṇabhadra *Yang que moluo jing*, ch. 2 (CBETA, T02, no. 120 p. 530, a16–19).

72. Early bencao books with "Guangzhou awei" illustrations include Su Song, *Bencao tujing* (1061; and its 1159 edition), and Liu Wentai, *Bencao pinhui jingyao* (1505). Li Shizhen's entry on awei indicated the various Chinese localities that some claimed to be the native place of awei. See Li Shizhen, *Bencao gangmu*, Taibei ed., ch. 34, p. 35.

73. Anonymous, *Buyi Leigong paozhi bianlan*, 4:630–31.

74. Zhao Rukua, *Zhu fan zhi jiaoshi*, 198; Li Shizhen, *Bencao gangmu*, Taibei ed., ch. 34, p. 35.

75. Dymock, Warden, and Hooper, *Pharmacographia Indica*, vol. 2, p. 151.

76. Chen Jiamo, *Bencao mengquan*, ch. 4, pp. 262–63; Miao Xiyong, *Xianxing zhai*, 767; Chen Shiduo, *Bencao milu*, 187.

77. One example is the famous early seventeenth-century jottings on miscellaneous matters, the *Wu za zu*, by scholar Xie Zhaozhi (born 1567), ch. 10. Xie mistook the plant *huangqin* for "gold" (*huanjin)*, and the saying "true gold" overshadowed "true scutellaria root" after the seventeenth century.

78. Tang Kaijian and Tian Yu, "Wanli sishiwu nian," 131.

79. Garcia da Orta, *Colloquies*, 46–47.

80. Otsuki Bansui and Otsuki Banri, *Ran'en tekihō*, ch. 1, 29a–b.

81. Osbeck, *A Voyage to China*, 1:257.

82. Dymock, Warden, and Hooper, *Pharmacographia Indica*, vol. 2, pp. 147, 152; Watt, *Dictionary*, 3:331.

83. According to Dymock, Warden, and Hooper, *Pharmacographia Indica*, vol. 2, pp. 147, 149, 151, hing usually contained undue portion of the root and was adulterated with gum Arabic or sliced potato. The normal hingra contained some portion of

the root, whereas the most expensive Kandahari hing was adulterated with red clay. Adulteration with wheat flour, small stones, and other cheap oleoresin is still practiced today; see Shah and Zare, "Asafoetida (heeng)," 33.

84. As Dymock, Warden, and Hooper described *Kandahari hing*: "very high-priced Asafoetida obtained by wounding the leaf-bud of the plant which produces ordinary Asafoetida; our article is generally mixed with numerous leaf-buds . . . its price is also much higher than that of any other kind." (*Pharmacographia Indica*, vol. 2, p. 151).

85. Dymock, Warden, and Hooper, *Pharmacographia Indica*, vol. 2, p. 147; Watt, *Dictionary*, 3:333. Eighty rupees correspond to around 1,760 euros of today's value, or at least 35 euros per kilogram. See 1890 Indian Rupee to Euro (INR to EUR) currency converter: http://www.likeforex.com/currency-converter/indian-rupee-inr_eur-euro.htm/1890 (accessed April 12, 2014).

86. Watt, *Dictionary*, 3:333.

87. Lu Yitian, *Lenglu yihua*, 138.

88. Watt, *Dictionary*, 3:332. Indian trade statistics show that Bombay was the port that "supplies Europe and China."

89. *Andjodan* (or *anjuden*, *anjoodan*, etc.) for example was said to designate the tree, the fruit, or the leaves; *altiht* (or *haltit*) could indicate the tree, the seed, or the resin. There was no agreement among the naturalists.

90. Fryer, *A New Account*, vol. 2, p. 195–96.

91. This last point was raised by Yule et al. in the 1903 glossary, under the term "hing" as "asafetida": "This product affords a curious example of the uncertainty which sometimes besets the origin of drugs which are the objects even of a larger traffic" (*Hobson-Jobson*, 418).

92. Balfour, *Asafoetida Plants*, 361.

93. See Raj, *Relocating Modern Science*, 38–59. The same process occurred in Canton, China, where "gardener-collectors and trader-naturalists collected new plants, curiosities, and other scientific data and transported them back to Britain" (Fan Fa-ti, *British Naturalists in Qing China*, 26).

94. Guibourt, *Histoire naturelle*, 3:223.

95. Dymock, Warden, and Hooper, *Pharmacographia Indica*, vol. 2, pp. 143–48; Watt, *Dictionary*, 3:329–30.

96. Watt, *Dictionary*, 3:330.

97. On the discussion of medicines as specific mobile materials, see Whyte, Geest, and Hardon, *Social Lives*, esp. 5–8.

Chapter 8: Smoke and Silkworms

Research for this essay was supported by NSF Research Grants #1430843 and #1734596. The authors also acknowledge the generosity of the Gladys Krieble Delmas Foundation and the Florence Gould Foundation.

1. Ms Fr. 640, available at http://gallica.bnf.fr/ark:/12148/btv1b10500001g/f1.image/, with the library title "Recueil de recettes et secrets concernant l'art du mouleur, de l'artificier et du peintre," held by the Bibliothèque nationale in Paris (BnF). For research on this manuscript, see "The Making and Knowing Project" at www.makingandknowing.org.

2. The binding stamped and gilded with the arms and device of Philippe de Béthune (1565–1649) indicates that the manuscript was part of the bequest of the Béthunes to the royal library (now the BnF).

3. Female authorship of BnF Ms. Fr. 640 is not impossible but appears unlikely in light of the manuscript's script and its accounts of travel.

4. The term *godet* has been translated as "bowl" (of a pipe); see Randle Cotgrave, "godet," in *A Dictionarie*. For a later use of *godet* as a pipe, see *The Athenæum Journal, 1844*, 873. Thanks to Joel Klein for this reference.

5. Fol. 77r, "Contre les rougeurs du visage" and "Terre graisse" respectively.

6. Fol. 77r, "Fais secher du rosmarin au moys de may." Rosemary is often picked in May, when the herb produces flowers, and is commonly distilled then for perfume. See Lockwood, "Perfumes," 859.

7. Fol. 77r, "puys emplis ce godet de pouldre diceluy Et mects un charbon allume dessus."

8. Fol. 77r, "Et recois la fumee par la bouche bien serree Et une partye te sortira par le nes Mays si tu veulx purger la teste sarre aussy le nes Contre morfondiments rheumes & aultres malladies."

9. For various definitions of smoke, see *Oxford English Dictionary*, s.v., "smoke"; and of "inhale" as "to breathe in; to draw in by breathing; to take into the lungs," s.v., "inhale."

10. Stolberg, "You Have No Good Blood," 69–70. Stolberg uses the term "catarrh." Cotgrave defines "rheume" as "catharre" (*A Dictionarie*, s.v., "rheume").

11. *An Herbal*, 69 (known as the "Banckes Herbal" for its original printer).

12. According to Stolberg, dyers used "infection" to mean the tinging of clear water with a small amount of dye. Michael Stolberg, October 31, 2016, Skype lecture, Columbia University, New York.

13. Dugan, *Ephemeral History*, 102–3.

14. Fol. 102v, "anthos ou rosmarin"; "c'est pour les vieulx." *Anthos* is Greek for "flower" and was also used to denote rosemary itself. Steam distillation is thought to have been the discovery of the Persian alchemist and physician Avicenna (980–1037). See Kate Fox, *The Smell Report*. In the entry for "Atr" ("perfume"), the *Encyclopædia Iranica* states: "In Europe, Arnold of Villeneuve (1235–1312) had isolated rosemary oil and used it in an alcoholic solution as a medicine" (Aubaile-Sallenave, "Atr").

15. For memory, see Meyer, Trueblood, and Heller, *The Great Herbal*, 447. Also see *An Herbal*, 70. For rosemary's global medicinal uses, see Lev and Amar, *Practical Materia Medica*, 266–67; Oliver Kahl, *Sābūr ibn Sahl's Dispensatory*.

16. Leach and Fried, *Standard Dictionary*, 957.

17. Larkin, "Virtues of Rosemary."

18. Leach and Fried, *Standard Dictionary*, 957.

19. Mattioli, *Commentaires*, 464–66. Also Meyer, Trueblood, and Heller, *The Great Herbal*, 447.

20. Mäkinen, "Henry Daniel's Rosemary," 305. Possibly also called "sea dew," "because of its resistance of the plant to mists and salt." See Lev and Amar, *Practical Materia Medica*, 266.

21. Meyer, Trueblood, and Heller, *The Great Herbal*, 447. Rosemary in Greek (*libanotis*) means frankincense, and rosemary's scent is often likened to incense. D'Andrea, *Ancient Herbs*, 73.

22. Mäkinen, "Henry Daniel's Rosemary."

23. *An Herbal*, 69.

24. For Shakespeare, see Stratmann, *Tragicall History*, 83. Rembert Dodoens's herbal was first published in Flemish in 1554, then translated from French to English in 1578, as *A nievve herball; or, Historie of plantes*, 264.

25. Mattioli, *Commentaires*, 464–66. In the margin of fol. 35v of Ms. Fr. 640, a single word "Mathiol" is recorded, apparently a reference to Mattioli, one of the very few references in the text to other authors.

26. See Meyer, Trueblood, and Heller, *The Great Herbal*, 447; *An Herbal*, 70.

27. Quoted in Mäkinen, "Henry Daniel's Rosemary," 318.

28. Meyer, Trueblood, and Heller, *The Great Herbal*, 447.

29. Dodoens, *A nievve herbal*, 263.

30. Gerard's *Herbal*, quoted in Mäkinen, "Henry Daniel's Rosemary," 321.

31. According to the botanical historian Agnes Arber, "the school of botany at Montpellier . . . was second to none," and "most of the principal herbals published in other languages appeared in French versions quite early in their history." Arber, *Herbals*, 118–19. For the French transmission of Fuchs's *De historia stirpium*, see Pettegree, Walsby, and Wilkinson, *French Vernacular Books*, 653. Michael Stolberg and Elaine Leong note that medical knowledge during this period was often transmitted through word of mouth, thus texts are not the only object to be considered when tracing knowledge transmission. See Stolberg, "You Have No Good Blood," 63–82; Leong, "Making Medicines."

32. Hippocrates, *Ancient Medicine*, 91, 93, 361. On the relationship between smell and medicine since antiquity, see Palmer, "In Bad Odor."

33. Palmer, "In Bad Odor," 62.

34. Palmer, "In Bad Odor," 63–64.

35. Palmer, "In Bad Odor," 66. See also Willis, *A Practical Flora*, 206.

36. Dugan, *Ephemeral History*, 102–3. Also see Palmer, "In Bad Odor," 67.

37. Göttler and Neuber, *Spirits Unseen*, xx–xxi. Piper discusses the very long history of employing fumigation for religious illumination in Piper, "The Milk of the Goat Heidrun," 220. Cummins discusses the role of fumigation in practices of natural magic in early modern Europe in Cummins, "Textual Evidence," 171.

38. Harvey, *Scenting Salvation*, 99–100.

39. "The smell of Bitumen, Rew, or the smoak of it burning, is of wonderfull force against the fits of the mother" (53), and "Take Amber and bruise into gross powder,

put it upon a chafing dish of coales, and receive the smoake up into the Nose with a funell" (105) (Culpeper, *Culpeper's last legacy*).

40. Culpeper, *The English physician*, 320.

41. This historiographical focus on tobacco's diffusion out from the Americas may have obscured accounts of the use of other types of herbs in pipe smoking in the sixteenth century. For further discussion of tobacco as a medicine in the sixteenth century, see Dickson, *Panacea or Precious Bane*.

42. Gilman and Zhou Xun, *Smoke*, 9.

43. Porter, *Greatest Benefit*, 164.

44. Harriot, *A brief and true report*, 43–44. Also Gilman and Zhou Xun, *Smoke*, 11.

45. Grehan, "Smoking," 1352. Also Gilman and Zhou Xun, *Smoke*.

46. Gilman and Zhou Xun, *Smoke*, 12. Also Evans, "Fumigating for Health."

47. Entry recorded for the year 1573, in William Harrison, "Great Chronologie," 1593, quoted in Oswald, *Clay Pipes*, 3. Thanks to Melissa Morris for this reference from her dissertation, "Cultivating Colonies."

48. Hume, "Hunting for a Little Ladle."

49. Evidence for the medical use of burning herbs in other types of pipes in Europe before the sixteenth century is contained in Almeida, *Middle English Remedy Book*: "For þe tooþ ache of wormes. Tak henne bane seed, leek seed & powder of encens, ana. Ley hem togedere vp on an hoot glowyng tylston & make a pipe of laten, þat þe nether ende be wide, þat it may ourclose þe poudres & holde þe mouth opon ouer þe pipes ende so þat the eyre may go into þe sore tooþ. þat will sle þe wormes & do awey þe ache" (85).

50. Cotgrave, *A Dictionarie*, s.v., "oriental." The only other use of the "oriental" in Ms. Fr. 640v is on fol. 8v as "jacinthe orientale," a precious stone.

51. The term "notopterygium root" is the Latin name of the botanical genus. In Chinese it called Qianghuo 羌活. No English name is available for this plant.

52. Rosemary is not indigenous to China. The author of an eighth-century work, Chen Zangqi, evaluated it as having "pungent taste and warm property, nontoxic." The early tenth-century China-based Persian drug merchant and author of the book dedicated to exotic drugs, *Haiyao bencao* (Foreign materia medica), stated that rosemary had a balanced character rather than warm, and it did not remedy illness but could exorcise evil spirits. It was also used as a mosquito repellent. Both these books were lost, but most of their contents were preserved in a state-sponsored book of materia medica in the twelfth century. See Tang Shenwei, *Chongxiu zhenghe*, 240. Li Shizhen remarks that its use only for sachets indicates the limited availability and usage of rosemary in early modern China. Li Shizhen, *Bencao gangmu*, Beijing ed., 899.

53. Gilman and Zhou Xun, *Smoke*, 14.

54. For a discussion of Eastern trade and travel connections of early modern physicians, apothecaries, and merchants, see Poliquin, "Vegetal Prejudice," 170–71. In an email to Professor Pamela Smith on December 5, 2016, historian Dorothy Ko stated that the earliest mention of tobacco in a Chinese medical treatise was most likely in the late sixteenth century. In this context, smoke was sucked into the lower abdomen

so that the *qi* (material vitality of the smoke) could move up into the heart and lungs to warm the body. For a discussion of the beginning of tobacco smoking and tobacco growing in China, see Benedict, *Golden-Silk Smoke*, 17–18.

55. As noted in the introduction to this volume, Fuchs's herbal called the New World foods names that might indicate an Eastern origin: "Turkish maize" and "Calcutta peppers." When describing maize in North America, Harriot says that "English men call it Guinney wheate or Turkie wheate, according to the names of the countreys from whence the like hath been brought" (Harriot, *A brief and true report*, 43–44).

56. As a Catholic, the author-practitioner's dates would likely have corresponded to the Gregorian Calendar instituted in 1582.

57. Vida, *The silkworm*, 65.

58. Sasha Grafit gratefully acknowledges Tianna Uchacz, postdoctoral fellow in the Making and Knowing Project (2016–2019), for her help in translating Isnard's poem.

59. Isnard, *Memoires et instructions*, 283.

60. See fols. 55r–v,

61. Arie Wallert in 1990 identifies basilisk ash as an "alchemical 'cover name' (a secret term known only to the adepti)" for mercury; see Smith, "Vermilion, Mercury, Blood, and Lizards," 46.

62. Smith, "Knowledge in Motion," 109–33.

63. Smith, "Vermilion, Mercury, Blood, and Lizards," 46.

64. Smith, *Body of the Artisan*, 117–21.

65. See Smith, "Vermilion, Mercury, Blood, and Lizards," 47.

66. Or *Ku* in the Wade-Giles system. The character *Gu* first appeared in an oracle script dating to 1400 BCE, at which time, according to Chinese historians, *Gu* likely denoted insects in a granary. The most famous *Gu* sorcery happened in the Han court in the first century and caused political turmoil, which ended with the death of the crown prince. The Gu practice in this incident, however, harmed the target by cursing a puppet. See Yu Gengzhe, "Xu gu zhi di," 191–93.

67. Feng and Shryock, "Black Magic," 8.

68. See Fèvre, "Drôles de bestioles"; Nappi, *The Monkey*, 96–108. *Chong* has been provisionally translated into English as "wug" by anthropologist Cecil H. Brown who coined the portmanteau "wug" from the words "worm" and "bug." According to Brown, this corresponds to the Chinese folk conception of *Chong*, which means "small residual creatures that are neither worms nor bugs. The wug life-form of Chinese encompasses small reptiles other than snakes" (Brown, "Folk Zoological Life-Forms," 808).

69. From de Groot, *Religious system of China*: "Chen Tsang-khi, the distinguished authority on Materia Medica in the eighth century . . . no stranger to the secrets of ku sorcery . . . says, 'a stupid characteristic of the ancients to make ku. People bent on the acquisition of wealth put all sorts of insects or reptiles into a jar, and when a year had passed away they opened it, to find only one insect or reptile which had devoured the others to the last. This one they called ku'" (5:847).

70. Fox, "Precious Bodies."

71. Su E, *Duyang za bian*.

72. Cai Tao, *Tie wei shan cong tan*, 4b–5b.

73. Lu Yinglong, *Xian chuan kuo yi zhi*, 16. The significance of silkworms in Chinese literature really begins with the most detailed passage in ancient literature dealing with the silkworm—*tshan*, a riddle by Hsiin Tzu, a philosopher of the third century, which shows insight into the relationship between humans and silkworms: "There is a certain thing; its body is naked but has the power to transform itself many times, like a god. Its merit can clothe and ornament everything under heaven, to the 10,000th generation. Through this, rites and music are completed and the noble and base are distinguished. Nourishing the aged and bringing up the young—if you do this, they will live. Its name is not beautiful, being a close neighbour to tyranny. When its merit is established, its body is destroyed; when its work is completed its house is overthrown; its elders are thrown out and its posterity is taken. This is that from which men profit and which birds hate. I am stupid and do not know what it is, and so I ask the oracle Wu Thai" (Kuhn, *Textile Technology*, 301).

74. De Groot, *Religious system of China*: "Our knowledge of the matter is also increased by the following notice concerning the Kwang-cheu region. . . . The people breed poison in the following way: on the fifth day of the fifth moon (the theoretical apogee of summer heat) they collect all sorts of reptiles and insects, none bigger than snakes or smaller than lice, and place them in a pot, to devour each other . . . the one called 'that which draws out life,' and the other 'the gold caterpillar'" (5:851).

75. Diamond, "The Miao and Poison." Duanwu Festival is conventionally translated as "Dragon Boat Festival" and, at present, is celebrated as a day in memory of the death of the poet Qu Yuan (ca. 340–278 BCE). It has also been associated with several cultural taboos in traditional China because it came at midsummer, when the evil *qi* reached its annual zenith, which led to attempts to prevent disease and expel evil spirits such as hanging wormwood in the house and consuming wine laced with realgar.

76. Smith, "Knowledge in Motion." European alchemical knowledge is derived in part from Arabic alchemical texts that describe works and procedures similar to the "Algiers" recipe. Robert Halleux affirms this point, citing Julius Ruska's study of Razi: "We can never stress enough that the alchemy of the Latin West owes nothing to the Greeks; to the Arabs it owes more or less everything." In the same article, he refers to the "famous manuscript Palermo 40 Qq A 10 (early fourteenth century). This treatise using animal substances, especially eggs and worms, for transmutation, seems to have been translated from an Arabic original, probably belonging to the Jabirian corpus" (Halleux, "Reception of Arabic Alchemy," 3:888).

Chapter 9: Itineraries of Images

1. On the circulation and materiality of Buddhist objects, see Liu Xinru, *Ancient India and Ancient China*; Kieschnick, *Impact of Buddhism*; Sen, *Buddhism, Diplomacy, and Trade*; Neelis, *Early Buddhist Transmission*; and Rambelli, *Buddhist Materiality*.

2. Pollock, *Language of the Gods*.

3. On the issue of Buddhist identities, common and diverse, that emerged as a result of the movement of people, objects, and ideas, see Heirman, Meinert, and Anderl, *Buddhist Encounters*. On the diversity of the Buddhist world encountered by the Chinese Buddhist traveler Yijing, see Sen, "Yijing and the Buddhist Cosmopolis."

4. Key arguments with regard to the circulation of images and the integration of the Buddhist space, especially in East Asia during the seventh and eighth centuries, have been recently made by Dorothy C. Wong in *Buddhist Pilgrim-Monks*.

5. The term "India" in this essay is used to indicate what in the present-day are the Republic of India, the Islamic Republic of Pakistan, and the People's Republic of Bangladesh. The term "South Asia" is used when implying the larger space of the Indian subcontinent, including Sri Lanka, Afghanistan, and Nepal.

6. Rhi, "Reading Coomaraswamy."

7. DeCaroli, *Image Problems*, 13.

8. DeCaroli, *Image Problems*, 27.

9. Schopen, "On Monks, Nuns and 'Vulgar' Practices."

10. DeCaroli, *Image Problems*, 93.

11. On the Sātavāhanas and Buddhism in the Deccan region, see Ray, *Monastery and Guild*. Also relevant is Ray's *Winds of Change*.

12. It is perhaps important to note the distinction between the function of the corporal remains of the Buddha (i.e., śarīras or relics, which also played a key role in the dissemination of Buddhism) and Buddhist images. As Kevin Trainor points out in the case of early Buddhism: "A relic is regarded as an appropriate means of representing the Buddha because of its historical connection with him through space and time. . . . Images, on the other hand, gain their authority by their capacity to represent the Buddha visually." Trainor further explains that, "images, unlike relics, can be reproduced endlessly, and they are accepted as worthy of veneration because they embody basic iconographic conventions." This "ease of reproducing images," Trainor argues, "allows for their proliferation outside monastic control to an extent that distinguishes them from relics, which are usually confined within the ritually defined boundaries of monastic complexes" (Trainor, *Relics, Ritual, and Representation*, 30–31).

13. Zürcher, *Buddhist Conquest*, 22.

14. Rhie, *Early Buddhist Art*, 27–47.

15. Edwards, "Cave Relief"; Rhie, *Early Buddhist Art*, 47–56.

16. Wu Hung, "Buddhist Elements," 273.

17. Wang, "Early Chinese Buddhist Sculptures."

18. See Pelliot, "Artistes des Six Dynasties," esp. 238–40; Bagchi, *India and China*, 195.

19. *Lidai minghua ji*, 9:298; Acker, *Some T'ang and Pre-T'ang Texts*, 255.

20. *Fayuan zhulin*, 503a.6–12; Sen, *Buddhism, Diplomacy, and Trade*, 38.

21. On the role of Buddhist monks such as Xuanzang in the circulation of images, see Wong, *Buddhist Pilgrim-Monks*.

22. Kieschnick, *Impact of Buddhism*, 53, 85.

23. This issue of duality was also expressed in poems written by Chinese Buddhist travelers to South Asia. For a detailed examination, see Deeg, "Wailing for Identity," 227–52.

24. Appadurai, "How Histories Make Geographies."

25. These three aspects—of circulation, convergence, and circulatory history—theorized by Arjun Appadurai, Homi Bhaba, and Prasenjit Duara, respectively, are outlined in chapter 2 (this volume).

26. Appadurai, "How Histories Make Geographies," 11–12.

27. See Tanabe and Tanabe, *The Lotus Sutra*.

28. This has been pointed out by several scholars of Chinese cave paintings. See, for example, Duan Wenjie, "Shiliu guo," 31–32, who believes that Confucian norms led to the rejection of such Indic images.

29. See Howard, "Buddhist Cave Sculpture."

30. For paintings on cave walls, see for example Ning Qiang, *Art, Religion and Politics*. On the use of Buddhist art by Empress Wu, see Karetzky, "Wu Zetian"; Rothschild, *Emperor Wu Zhao*.

31. Wong, *Buddhist Pilgrim-Monks*, 1.

32. Wong, *Buddhist Pilgrim-Monks*, 1.

33. See Woodward, "Tantric Buddhism"; Woodward, "Bianhong."

34. Woodward, "Bianhong."

35. Sen, "Astronomical Tomb Paintings." On the origin of the Star Maṇḍalas, see Takeda Kazuaki, *Hoshi mandara no kenkyū*.

36. For a detailed study of the transmissions of images related to Mañjuśrī between India and China, see Wang Jingfen, "Zuowei."

37. Sen, *Buddhism, Diplomacy, and Trade*.

38. Shigeo, "Transmission of Paekche Buddhism."

39. See Duan Yuming, *Zhikong*; McKeown, "From Bodhgayā to Lhasa."

40. Rambelli, "The Idea of India."

41. See Jahn, *Rashīd al-Dīn's History of India*, xxi–lxxvii; Akasoy, "The Buddha"; Yoeli-Tlalim, "Rashīd al-Dīn's *Life of the Buddha*."

42. The temple dedicated to these two Chinese Buddhas in Kolkata is discussed in Zhang Xing, "Buddhist Practices." For a wider circulation of these images and the temples dedicated to the Ruan and Liang buddhas, see Sen, "Temple Heritage."

43. Some of these issues have been examined by Tythacott, *Lives of Chinese Objects*; Guha-Thakurta, *Monuments, Objects, Histories*; Ray, *Return of the Buddha*.

44. See a report on the decision by an Australian art gallery to return a sandstone image of the Buddha to India: https://indianexpress.com/article/india/india-news-india/australian-art-gallery-to-return-2000-year-old-buddha-idol-to-india (accessed June 24, 2018).

45. See the essays in Appadurai, *The Social Life of Things*.

46. For example, Latour, *Reassembling the Social*.

Chapter 10: Itineraries of Inkstones in Early Modern China

1. Although archaeologists have discovered clay or stone tablets for grinding pigments from Neolithic sites, the inkstone assumed its present format during the Eastern Han to Wei-Jin period (c. first to fifth centuries CE) with the appearance of a rod- instead of ball-shaped ink, or the inkstick in today's parlance. The collecting of inkstones began in earnest during the Northern Song period (960–1127 CE) when the first dedicated connoisseurship literature appeared.

2. Schäfer's chapter (this volume) elucidates the discourses of *wu* (things; affairs) in Chinese history. Perhaps "material-emotional complex" can be a translation for *wu*?

3. There are in fact scant "pictograms" in the written Chinese language. For the fascination with the Chinese ideograph, see Porter, *Ideographia.*

4. According to Okajima, the Japanese linguist Kono Rokuro coined the term "Kanji bunka ken" (also rendered as the cultural zone of China) in 1963. The term is controversial because a similar view was part of the imperialistic ideology that justified the colonialization of Korea (1910–1945) and Taiwan (1895–1945). Okajima, "Kanji bunka ken to wa."

5. The itinerary of inkstone-making techniques among China, Korea, and Japan awaits further research. One entry from the *Veritable Records* of Chosŏn Korea states that two Ming envoys in 1460 were given a set of purple-stone inkstones in a cinnabar lacquered box, for summer use, as well as another set that consists of a bronze burner and a purple-stone inkstone for winter use, along with inksticks and writing brushes. *Sejo sillok* 19:33a (1460/03/07). The inverse is also true. Emperor Qianlong, for example, gave the king of Korea two She inkstones and two boxes of tribute inksticks, among other things. *Chŏngjo sillok* 18:52b (1784/11/20). I thank Sixiang Wang for these references.

6. McCormick, *Tosa Mitsunobu*, 172–205. The first extant version of the story appeared in *Konjaku monogatari* (Tales of times now past), a collection of tales from the late Heian period (794–1185).

7. Ghosh, *In an Antique Land*, 237, 157; also 339. Caches of these traders' letters and papers were preserved in the Cairo Geniza. Francesca Trivellato's study of Sephardic diasporic traders has similarly called for "global history on a small scale" (Trivellato, *Familiarity of Strangers*, 271).

8. Informed by quantum physicist Niels Bohr's Principle of Complementarity, Barad speaks of entanglement and intra-action as modes of posthuman knowledge making, but they seem apt descriptions of the premodern or nonmodern and non-Western modes expounded in the present chapter as well. Barad, *Meeting the Universe*, ix, 180.

9. The importance of merchants and diplomats as intermediaries in this process is highlighted in the chapters by Alberts, Leung and Chen, Sen, and Yoeli-Tlalim (this volume).

10. Ingold, *Lines.*

11. The distribution is highly specific. The iconic Song treatise *Inkstone Chronicles* (*Yanshi*) by Mi Fu listed twenty-one stones (including two jades) from quarries scat-

tered throughout the empire, in addition to sieved clay retrieved from two locations and one type of rock crystal. Mi discussed the suitability for ink grinding of each but did not rank them. His contemporaries debated the virtues of She and Duan, the two stones vying for first place. With the decline of She mining in the Yuan and especially after the (re)opening of the Old Pit in the Duan area in the Ming, Duan stones became the unsurpassed leader up until the present day, to the chagrin of She natives and collectors.

12. Administrative boundaries on the city and county levels shifted multiple times in the area. By the "Duan area," I mean the general area of Zhaoqing walled city, the city of Gaoyao across the Xijiang River, Huanggang (and Baishi) villages on the eastern outskirts of the walled city, as well as the quarries in nearby Mount Lanke and Mount Beiling.

13. Sixteenth- and seventeenth-century writings give the iconic number of five hundred households in Huanggang. Today, Huanggang is divided into an upper and a lower village; Baishi (4,100 residents) is part of the latter. In contrast to quarrying, knowledge about inkstone making has never been textualized; in inkstone-making families, knowledge passed from generation to generation without written manuals. The ethnographer Chen Yu has found rubbings of carved inkstones in loose sheets dating back to the Ming (1368–1644 CE). Passed down through the generations within the family, these "models" were also given to apprentices. Chen, *Duanyan minsu kao*, 31, 80.

14. An ancient road over thirty kilometers long that traversed the Mei Ridge separating Guangdong from the Central Plain was opened in the eighth century, around the time Duan stones became valued in the empire. Beyond the pass, the Duan stones were conveyed by land and the Grand Canal to the core of the empire and to the north. Chen, *Duanyan minsu kao*, 11, 148–49.

15. Translation adopted from Van Gulik, *Mi Fu on Ink-Stones*, 32. Van Gulik praised Mi Fu's insistence on personal observation, independent investigation, and eschewing supernatural phenomena associated with mining as "scientific" (58, 62).

16. Geologically, Duan stone is a shale. Its main contents include mica, clay, and silica with traces of mineral salts—mostly iron, potassium, and magnesium—that are the source of its characteristic deep purple colour. The foliated rock, formed by metamorphic compression over 600 million years ago, is fine-grained and of medium hardness (about 3.5 on the Mohs scale,) deemed the most suitable for grinding ink.

17. For itinerant assistants (*muyou* or *mufu*), see Folsom, *Friends, Guests, and Colleagues*. For "logistical power," see Mukerji, *Impossible Engineering*, 216–23. For the genesis of this technocratic culture from the beginning of the Qing empire, see Chen, "Rise of Technocratic Culture."

18. For the rise of philological scholarship and the decline of speculative moral philosophy, see Elman, *From Philosophy to Philology* (esp. ii, for translation of the slogan).

19. Agora, or urban hubs of multidirectional information exchange, are important for the material itineraries in several chapters in this volume: Chang'an and Quan-

zhou (Leung and Chen's awei), Malacca (Alberts's curative medicine), Dunhuang/Tibetan language (Yoeli-Tlalim), and Bursa (Günergun). The four interrelated spheres of practical knowledge circulation in the empire described by Schäfer are Internode, Imperial Court, Agora, and Scholarly Arts. Schäfer, *Cultures of Knowledge*. The circulation of inkstone knowledge in the Qing Empire involves all four modes of communication, appropriation, aggregation, and documentation.

20. The shrine on the lower edge of the image is dedicated to the Heavenly Queen (Tianhou 天后), a patroness of seafaring in southern China who had been incorporated into the official pantheon in the Song. The shrine smack dab in the center of the image is labeled "Shrine of the Inkstone Pit" (*Yankeng miao* 硯坑廟) and guards the entrance to the Underwater Lode. It is dedicated to the tutelary spirits residing in the area who would cause the tunnels to collapse unless placated. Although officials left inscriptions, the rituals were initiated by and centered on the stoneworkers. A stone tablet remains standing. Chen, *Duanyan minsu kao*, 87–92.

21. The Shuiyan is so named because the usable lode lay at a level below the bed of the Xijiang River. The shaft had to be pumped dry before mining could ensue. Before the modern convenience of electric pumps, teams of workers were assembled with the onset of the dry winter months in November to December. Descending into the diagonal shaft, they formed an assembly line conveying jugs of water in clay pots. Manual removal of water required a team of over two hundred workers alternating in day and night shifts for over a month, leaving a mining season of three to five months. By May, rain would swell the river and the pit would again be submerged.

22. The treatise by Qian Chaoding (1647 *jinshi*), which discusses stones from the Underwater Lode exclusively, was instrumental in the canonization of this valuation. Qian, a native of Changshu, Jiangsu province, served as an education official in Guangdong. In his treatise Qian claimed to have learned of the supreme value of Shuiyan from the "natives," whose views he used to correct the old hierarchy of the "scholars from the Central Plain." Qian, *Shuikeng shiji*.

23. Bray, "Introduction," 1. To make a *tu* means "to order" and "to position in space" (Bray, "Introduction," 13). The *tu* on the first inkstone is entitled "Map of [the caves] inside the Old Pit" (*Laokeng neitu*); the second reads: "Map of Duanxi" (*Duanxi tu*). Duanxi is the name of the tributary of the Xijiang River that is next to the entrance to the Underwater Lode.

24. Included in this genre are three of the most influential inkstone treatises: Cao Rong 曹溶(1613–1685), *A Record of Inkstones* (Yanlu 硯錄), who served as Guangdong Provincial Commissioner in 1673; Jing Rizhen景東暘 (*jinshi*-metropolitan graduate 1691), *A Description of Inkstone Quarries* [in the Duan area] (Yankeng shu 硯坑述), who served as magistrate of Gaoyao in 1697–1703; Wu Yingnian 吳繩年, *A Gazetteer of the Inkstones of Duanxi* (Duanxi yanzhi端溪硯志, 1757), who served as prefect of Zhaoqing in 1752–1755 and wrote his treatise during his tenure in Guangdong. The trend of an out-of-province official dispensing knowledge about Duan inkstones predates the late eighteenth century but it becomes more intensified with the rise of philology.

25. He Chuanyao, *Baoyantang yanbian*. The only outsider who viewed his collection is Huang Peifang 黃培芳, a scholar from Xiangshan, Guangdong, who came to Gaoyao to collect information for a local gazetteer project. Huang, no expert on inkstones, urged He to commit his knowledge to writing. Huang himself drew the maps, presumably using information supplied from He and perhaps also from interviewing the stoneworkers. He Chuanyao completed the manuscript in 1827 but could not finance the publication until he found a patron, a visiting *muyu* by the name of Gao Hong. Gao arrived in 1833 and made a handsome profit buying quality stones from a major quarrying operation in 1834 and hiring teams of local craftsmen to carve them into inkstones (Gao preface, pdf counter 9).

26. I say the printed map is likely to be the model for the one carved on the back of the inkstone and not the other way around because it is a scholar's god's eye view of the quarry, and that the drawer of He's map is the same visiting gazetteer fieldworker who executed the map based on the information given to him by He. He's map of the caves in the Underwater Lode became iconic and widely circulated exactly because it depicts a view that no one had imagined before.

27. This discourse of manpower was first introduced in writing by the geographer Li Chaoluo 李兆洛 (1769–1841), who wrote *Notes on the Quarries of Duanxi* (*Duanxi yankeng zhi* 端溪硯坑記) during his sojourn in the Duan area 1820–1822.

28. I added the numbering and the parameters of the four caves from He's text (*Baoyantang yanbian*, 15) for clarity of presentation. Further references to this work will be set parenthetically in the text.

29. I have not been able to establish a direct connection between He Chuanyao and the burgeoning philological movement, but the link is tantalizing. The term I render "discerning inkstone" in the title of his treatise (*yanbian* 硯辨) can also read as "debates on inkstones" or "refuting other views on inkstones," which is common in the titles of philological treatises. Governor-General Ruan Yuan (1764–1849), a major figure in the movement, established the Ocean of Learning Academy in Guangzhou in 1824 by merging four regional academies—including the Duanxi in Zhaoqing—into one.

30. Wu Chongyao, "Afterword," 1.

31. These three subject positions may have resonances outside China. Pamela Smith, for example, has analyzed the traveling cleric Albertus Magnus (1193?–1280) who formulated a theory of metals by interviewing stonemasons and miners, a move similar to that of the "out-of-province scholar." Pamela H. Smith, "The Matter of Ideas in the Workings of Metal in Early Modern Europe," in Anderson, Dunlop, and Smith, *The Matter of Art*.

32. Barad, *Meeting the Universe*.

33. How texts can facilitate, hinder, or complicate itineraries of materials is a theme in virtually all the chapters in this volume. Bray suggests that textual representation is limited because it is linear. What counts as a "text" and its relationship with visual representation is worthy of further consideration.

34. Crossley, *Translucent Mirror*. For the protracted campaigns among the Manchu Qing, Zunghar Mongols, and Moscovite Russians, see Perdue, *China Marches West*.

Chapter 11: Convergences in and around Bursa

The author wishes to thank Dr. Şeref Etker who has given up time for the critical reading of the draft manuscript and for providing his valuable comments and suggestions.

1. Ömer Şifai el-Burusavi, *Cevher el-ferid fi tıbb el-cedid*, fol.99a.

2. Peacock and Yıldız, *Islamic Literature*; Ágoston, "Most Powerful Empire"; Inalcık, *Ottoman Empire*.

3. Yıldırım, "Sunni Orthodox vs Shi'ite Heterodox?"; Dressler, "How to Conceptualize." For the Sufi *tekkes*, see Lifchez, *Dervish Lodge*.

4. Shaw, *History of the Ottoman Empire*, 1:158; Çizakça, "Short History."

5. Faroqhi, "Bursa at the Crossroads"; Faroqhi, "Ottoman Empire's Industrious Core," 365–67.

6. Denny, *Iznik*; Atasoy and Raby, *Iznik*; Hitzel, "Iznik"; Raby, "Seventeenth-Century Description."

7. Atçıl, "Mobility of Scholars."

8. Günergun, "Science in the Ottoman World," 75; Günergun, "Ottoman Encounters."

9. Karamustafa, *God's Unruly Friends*; Köhbah, "Vom Asketen zum Glaubenskämpfer"; Cahen, "Futuwwa," 961–65.

10. Le Gall, *Culture of Sufism*, 80–85, 124–125. For the Mawlawiyya, see Tek, "Mevleviliğin Bursa'daki Izleri." For the Zayniyya, see Öngören, *Tarihte Bir Aydın Tarikatı*. For the Khalwati, see Karataş, "Ottomanization," and Curry, *Transformation*. For the Ashrafiyya order, see Zarcone, "Qadiriyya."

11. Evliya Çelebi, *Günümüz Türkçesiyle*, vol. 2, bk. 1, 18–19; Clayer, "Confréries."

12. These volumes are now kept at the Bursa Inebey Library for Manuscripts and Old Prints. See Türk, "Tasavvuf Kültüründe," 40–41.

13. *Risale-i Aşık Paşa* (83 couplets), MS Karaçelebizade 359/11, fols. 65b–68a, Süleymaniye Library, Istanbul; (81 couplets), MS Muallim Cevdet K.180/16, fols. 84a–86a, Atatürk Library, Istanbul.

14. Binbaş, *Intellectual Networks*, 114–22, 154; Artun, "Bâyezîd-i Kimya'î," 184.

15. Tek, "Osmanlı Tasavvuf," 220. The number of couplets slightly differs according to copies. For a mid-seventeenth century copy of forty-five couplets, see MS Karaçelebizade 359/10, fols. 64b–65b, Süleymaniye Library, Istanbul.

16. Özcan, "Molla Arap," 240–41.

17. Mullah Arab, *Talhis al-burhan*, fols. 66b–117a (for nonexperts, fol. 66b).

18. Callataÿ, "Magia en al-Andalus." *Rutbat al-hakim* was formerly attributed to the mathematician Maslama al-Majriti (950–1007) by Ibn Khaldun and al-Jildaki. Currently, its author has been identified with Maslama b. Qasim al-Qurtubi (d. 353/964). See Fierro, "Bāṭinism in Al-Andalus."

19. Callataÿ and Moureau, "Towards the Critical Edition," 391.

20. Holmyard, "Maslama al-Majriti."

21. For a seventeenth-century copy of *Talhis ar-rutbat al-hakim li Majriti* by Mullah Arab, see MS Ali Emiri 2835/2, fols. 8–23, Millet Library, Istanbul.

22. Diaz-Mas, *Sephardism*, 39.

23. For the various names given to Ali Çelebi el-Izniki, see Artun, "Hearts of Gold," 102–41; for a discussion of Ali el-Izniki's lineage, see Artun, "Hearts of Gold," 142–85. For works attributed to Ali el-Izniki, see Carl Brockelmann, *Geschichte der arabischen Literatur*, 667–68; Ihsanoğlu et al., *Osmanlı Tabii ve Tatbiki Bilimler*, 1:53–81.

24. Artun, "Hearts of Gold," 175.

25. Ali Çelebi el-Izniki, *Mecmuatü'l-mücerrebat*, MS Hekimoğlu Ali Paşa 541, fol. 2a.

26. Adıvar, *Osmanlı Türklerinde Ilim*, 113–14; Artun, "Hearts of Gold," 33, 62.

27. Ali Çelebi el-Izniki, *Mecmuatü'l-mücerrebat*, MS Hekimoğlu Ali Paşa 541, fol. 2a–b.

28. Ali Çelebi el-Izniki, *Sırr al-Rabbani fi ilm al-mizan*, MS A6247, fol. 1b.

29. Anawati, "Jaldaki."

30. Strohmaier, "Al-Djildaki," 270.

31. Holmyard, "Al-Jildaki," 51.

32. The lack of any specific Turkish term to denote an alchemical workshop is notable. Apparently, no term existed in Arabic or Persian either to denote an alchemist's workplace in medieval Islam. Jabir's and al-Razi's workplaces were only termed "laboratory" by modern scholars without mentioning the original term, if any. In the alchemical manuscript ascribed to the Abd al-Jabbar al-Hamadani (d. 1025), the place where al-Hamadani conducted his alchemical experiments and met fellow alchemists is designated merely as his house (*dar*) or chamber (*bayt*). See Leube, *Die Rezepte*, 104–11. G. Leube also indicated in private correspondence that *dar al-darb* (mint) and *karkhana* (workplace, including a workplace dealing with metals) are frequently encountered in premodern Arabic and Persian texts. I cordially thank him for this information.

33. Ali Çelebi el-Izniki, *Mecmuatü'l-mücerrebat*, MS Ali Nihat Tarlan 197, fol.17a.

34. Ali Çelebi el-Izniki, *Mecmuatü'l-mücerrebat*, MS Hekimoğlu Ali Paşa 541/1, fols. 1b–43a.

35. Ruska, *Al-Razi's Buch*; Taylor, *The Alchemy of Al-Razi*.

36. Ali Çelebi el-Izniki, *Mecmuatü'l-mücerrebat*, MS Ali Nihat Tarlan 197, fols. 22b, 9b, 12a, 14a, 17b, 21a, 18b–19a, 23a.

37. Câbir b. Hayyan, *Tanımlar Kitabı*, 50, 54, 60, 62.

38. For the epilogue, see Mehmet Tahir, *Osmanlı Müellifleri*, 1:48–49.

39. Fazlıoğlu, "Fazıl Ali Bey," 1:442; Artun, "Hearts of Gold," 55; Inalcık, *Economic and Social History*, 122–25; Pamuk, *Monetary History*.

40. Barkan, "Istila Devirlerinin," 290, 292, 297.

41. Wolper, *Cities and Saints*, 26–27, 32, 42.

42. For the saying, see Barkan, "Istila Devirlerinin," 286; for Üftade Efendi, see Azamat, "Üftade," 282–83.

43. Erzi, "Bursa'da Ishaki."

44. Taşğın, *Horasandan Balkanlara.*

45. Clayer, "Confréries," 285.

46. Yediyıldız, "Şer'iye Sicillerine," 31. An akçe was an Ottoman silver coin minted between the fourteenth and the seventeenth centuries. Over time its weight decreased from 6 *kirat* (1.1 g.) to 1 *kirat* (0.2 g.), and its silver content from 90 to 50 percent.

47. Hızlı, *Osmanlı Klasik*, 5–7.

48. Unan, *Kuruluşundan Günümüze*, 273 (for the imam of Fatih Mosque); Kemal Beydilli, *Osmanlıda Imamlar.*

49. Unan, *Kuruluşundan Günümüze*, 273.

50. Gümüşalan, *15. ve 16. Yüzyıllarda*, 22.

51. Kılıç, "XVI. Yüzyılın."

52. Theisen, "Attraction of Alchemy"; Matus, *Franciscans.*

53. Fahreddin, *Risale-i ferahiyye*, fols. 63b–83b.

54. Fahreddin, *Risale-i ferahiyye*, fol. 66b.

55. Fahreddin, *Risale-i ferahiyye*, fols. 81b–83b.

56. Ottoman alchemical manuscripts from the fourteenth through the seventeenth centuries examined so far have a similar format of about ten–sixteen centimeters. They have no traces that would indicate workshop use. The script in all of them is legible (nothing like grimoires), indicating scribal copying within a textual tradition. Although some works of Ali Çelebi el-Izniki have between ten and twenty copies in libraries, the copies of other alchemical works do not exceed three or four. As the place of copying is not recorded on the manuscripts, it is not possible to know how they circulated. The copyist's name and date of copying is not indicated, thus the only way to know that knowledge was transmitted from Egypt to Bursa is by the author's name (al-Jildaki, for example) and that of the translator (Ali el-Izniki, for example), but this does not indicate the precise itinerary the manuscript followed.

57. Erünsal, "Türk Edebiyatı Tarihinin Arşiv Kaynakları 4."

58. Türk, "Tasavvuf Kültüründe," 48–62.

59. Bachour, *Oswaldus Crollius*; Bachour, "Iatrochemistry and Paracelsism," 111.

60. Bachour, *Oswaldus Crollius*, 38.

61. A. Maurocordatos's thesis was a study based on W. Harvey's *De Motu Cordis* (1628). See Maurocordatos, *Pneumaticum Instrumentum.*

62. Valléry-Radot, "Médécin diplomate"; Valléry-Radot, "Un puissant personnage."

63. More than thirty mineral compounds are listed in Günergun, *14.–17. Yüzyıllarda.*

64. Kahl, *The Dispensatory.*

65. Mehmed Tahir, *Osmanlı Müellifleri*, 3:228.

66. The colophon of the author's copy notes that the text was completed in 1695 (H. 1107) in Istanbul (Konstantiniyye) (MS T7083, fol. 359b, Istanbul University Library, Istanbul). The illuminated front page of this copy reads the title *Künuz-i hayat el-insan kanun-ı etibba-i feylesofan.* The same manuscript was recorded as *Şifa el-müminin* (Health for the believers) in Istanbul University Library's catalogue, the database of

manuscripts in Turkey at https://www.yazmalar.gov.tr/index.php?dill=eng/, and in Ihsanoğlu et al., *Osmanlı Tabii ve Tatbiki Bilimler*, 1:329. The title *Şifa el-müminin* is absent in MS T7083, and it is unclear why the manuscript was recorded as such. The MS T7083 seems to be the author's master copy from which other abridged versions were produced, focusing less on Paracelsus and the description of *spagyric art* and more on theoretical medicine, such as Hippocrates's four temperaments. The number of recipes in copies of *Künuz-i hayat* are reduced: the 1695 master copy includes 130 recipes for oils (*dühn*), while the 1737 copy at Süleymaniye Library (MS Şehit Ali Paşa 2085) and the 1794 copy (Istanbul University Library MS T7142) include only 94 recipes.

67. Ömer bin Sinan, *Künuz-i hayat*, MS T7083, fols. 4a–4b.

68. Ömer bin Sinan, *Künuz-i hayat*, MS T7083, fol. 343b.

69. Ömer bin Sinan, *Künuz-i hayat*, MS T7140, fol. 1a.

70. Bachour, *Oswaldus Crollius*, 370.

71. Ömer bin Sinan, *Künuz-i hayat*, MS T7083, fol.1b.

72. His gravestone facing the Mawlawi lodge in Bursa reads "Bursavi Ömer Efendi." In *Mürşid el-muhtar*, his full name is given as Ebu'l-Abbas Derviş Ömer Şifai el-Mevlevi. See Mehmed Tahir, *Osmanlı Müellifleri*, 3:229. For a list of his works, see Ihsanoğlu et al., *Osmanlı Tıbbi Bilimler*, 1:365–74.

73. Uzluk, "Bursalı Tabib Mevlevi Ömer Şifai."

74. Ömer Şifai el-Burusavi, *Mürşid el-muhtar.*

75. The text here (MS Hamidiye 1020, Süleymaniye Library) is dated 1698. The name of the author is given as "Derviş Ömer Şifai" in the preface (fol. 2a) and as "Şeyhzade Ömer Çelebi" in the epilogue (87a). The title *Cevher el-ferid fi tıbb el-cedid* is mentioned on fol. 2b. The date of compilation is 1698 (1110 H), fol. 87b.

76. See preface (fol. 2b) and epilogue (fol. 87a) of the MS Hamidiye 1020. For borrowing recipes, see Bachour, *Oswald Crollius*, 371, 383.

77. Ömer Şifai el-Burusavi, *Cevher el-ferid fi tıbb el-cedid*, fol. 97a–100b.

78. Thompson, *Alchemy and Alchemists*, 114.

79. Forbes, *Short History*, 117, 119 (figure 41).

80. Ihsanoğlu et al., *Osmanlı Tıbbi Bilimler*, 1:117–19. Ali Münşi is buried next to a Bayramiyye lodge at Üsküdar, Istanbul. He stated in *Bizaat el-Mübtedi* that his master in medical sciences was Ömer Şifai. See Uzluk, "Bursalı Hekim Ali Münşi Efendi," 332, 333.

81. Adıvar, *Osmanlı Türklerinde Ilim*, 165–66.

82. Günergun and Etker, "From Quinaquina to 'Quinine Law'."

83. See the chapter on "Osmanlı Medrese Müntesipleri ve Kimya" (Ottoman madrasa adherents and chemistry) in Izgi, *Osmanlı Medreselerinde Ilim*, 2:145–74.

84. Akıncı, "Kitab-ı Düstûr'." For Vesim, see Ünver, "Hekim Vesim Abbas." Dervish Abbas Vesim was a member of the Khalwati, Qadiri, and Naqshi orders. See Mehmed Tahir, *Osmanlı Müellifleri*, 3:237.

85. Akıncı, "Kitab-ı Düstûr'," 140.

86. The translation was titled *Mürşidü'l-elibba fi tercemet-i ispagirya* (Guide of the wise man for the translation of Spagyrik). For a comparison of the Arabic text with the Turkish translation, see Bachour, *Oswaldus Crollius*, 368–69.

87. The 1733 translation was made the physician Ahmed-i Sani. See Zülfikar, "18. Yüzyıla ait," 101–20. The *Tuhfah* was translated into Latin by the French missionary and linguist Ange de Saint Joseph (Joseph de la Brosse), *Pharmacopoeia Persica*. For a translation into modern Turkish, see Sarı et al., *Klasik Dönem*.

88. Kieffer, "Laboratories of Art and Alchemy," 105–27. An early seventeenth-century laboratory was set up by Wolfgang II von Hohenlohe in Schloss Weikersheim in Germany. See Smith, "Laboratories," 290–92. For the pursuit of natural knowledge in domestic spaces in early modern Europe, see Cooper, "Homes and Households."

89. Merçil, *Türkiye Selçuklularında Meslekler*, 172; Evliya Çelebi, *Günümüz Türkçesiyle*, vol. 1, bk. 2, 488, 571, 573, 607, 660.

90. Debus, *English Paracelsians*, 176.

91. Debus, *Chemistry and Medical Debate*, 18–19.

92. An eighteenth-century dictum pointed to the harm caused by unqualified frankish (European) physicians who had abandoned the traditional cures and administered remedies under the name *tıbb-ı cedid* (new medicine) in Istanbul. See Altunay, *Hicri Onikinci*, 37. The dictum is dated 1703, a year before Omar b. Sinan compiled the formulary Şifa el-müminin, 1704. See Baylav, *Eczacılık Tarihi*, 191.

93. For the 1503 inventory of the palace library (in Istanbul) and its alchemical manuscripts, see Erünsal, "Türk Edebiyat Tarihinin Arşiv Kaynakları 6"; Artun, "Hearts of Gold," 19–21.

94. Şen, "Reading the Stars," 598–99.

95. It is related that Murad IV allowed a Maghrebi to perform transmutation in the imperial palace in Istanbul and allocated one thousand florins to a female alchemist in Diyarbakır during his Iraq campaign. See Abdülhak Adnan (Adıvar), *La science*, 96; Artun, "Hearts of Gold," 45. The Maghrebi was given for once the permission to put his claims in action, but this does not indicate Murad IV's patronage of alchemical experiments; rather, it might simply have been curiosity or even an inquisition.

96. Smith, "Alchemy"; Smith, *Business of Alchemy*; Purš and Karpenko, *Alchemy and Rudolf II*.

Chapter 12: A Wooden Skeleton Emerges in the Knowledge Hub of Edo Japan

I would like to express my greatest gratitude to Pamela H. Smith for her extensive assistance and suggestions in language and in the analytical framework. I particularly appreciate her patience, which was an essential factor in the completion of this paper.

1. In 1783 John Hunter acquired the corpse of the Irish Giant, Charles Byrne (1761–1783), and boiled it down to the bones. However, he did not openly display the reassembled skeleton until 1787. See Eric Cubbage, "Tragic Story of Charles Byrne."

2. See Morinaga, *Secrecy in Japanese Art*.

3. Michel, "Looking at Corpses."

4. Uchiie, Ōedo shitai kō, 144; Ishii Ryōsuke and Harafuji Hiroshi, *Bakumatsuo-furegaki shūsei*, 2533.

5. Sugidatsu Yoshikazu, "Edo jidai kaibō."

6. Sugita Genpaku, *Kyōi no gen*.

7. For example, Ōtsuki Gentaku titled his primer of Dutch learning *Rangaku kaitei* 蘭学階梯 (Stairs to Dutch learning), which was finished in 1783 and published in 1788. Also see Numata Jirō, *Western Learning*, 56.

8. This episode has several versions. Here I combined the information from Ōtsuki, "Shinkangi setsu," 268; Egawa Yoshio, *Hiroshimaken ijinden*, 6; and Asada Ritsuen's 浅田栗園 *Kyorin zatsudan* 杏林雑談 [Tittle-tattles of the medical field], quoted in Fujigawa, "Mokkotsu kō," 3.

9. Suzuki Takao, *Hone kara miru Nihonjin*, 39–54.

10. Hattori Toshirō, *Heian jidai igakushi kenkyū*, 229.

11. Nihon gakushiin, *Meijizen*, 3:236–38.

12. Asami Dōsai, *Kinsō hidenhū*, 904:310.

13. Goble, *Confluences*, 89–111.

14. Kaibara, *Nihon seikei*, 71–75.

15. Goble, *Confluences*, 113–19.

16. Nakura, *Edo no honeduki*, 220–22. For a diagram of the family lineage, see 19 and 23.

17. Mayanagi, "Riben Jianghu."

18. Cook, *Matters of Exchange*, 339–77.

19. Hara, *Nihon de hajimete*, 1.

20. Kaibara, *Nihon seikei*, 125–34.

21. Kōshi, *Honedugi ryōji jūhōki* 1:123–32.

22. It is not entirely clear whether Kōshi or another writer authored this essay in the book. *Honedugi ryōji jūhōki*, 1:133–44.

23. Anonymous, *Kahō Namba*, 1:263–65.

24. Anonymous, *Kahō Namba*, 1:266–86.

25. Anonymous, *Kahō Namba*, 1:346–50.

26. Anonymous, *Kahō Namba*, 1:265.

27. The site Hohosoki of the island where Hoshino processed the body is now linked to the Hiroshima mainland, after the island was later reclaimed from the sea.

28. Kataoka, "Shinkangi (Hoshino mokotsu)."

29. Kataoka, Suzaki, and Ajima,"The Hoshino Wooden Skeleton."

30. The date of the essay is the first lunar month—namely, January 24 to February 21—of 1792. Hoshino received the corpses on May 8, 1791. Assuming the carpentry started on the next day and it took exactly three hundred days, it would be finished on February 23, 1792.

31. Sakai, *Shinsōhan Kaitai shinsho*, 206. The page number of the original is 58.

32. This essay survives. Kaibara discovered it and showed its image in Kaibara, *Nihon seikei*, 266–69.

33. Nakayama, "Hoshinno Ryōetsu no kaishi."

34. Thunberg was a Swedish botanist and physician, student of Carl Linnaeus (1707–1778).

35. Sugimoto, "Kindai igaku no genryū."

36. Sakai, "Kaitai shinsho to Jūtei Kaitai shinsho."

37. Thanks to Professor Chen Liwei's advice, I found the original copy of the chart in the exhibition commemorating the centenary of the establishment of applied chemistry in Waseda University, November 8, 2017.

38. See Ōtsuki's preface for *Shinkangi setsu*, 269–71.

39. Kaibara, *Nihon seikei*, 273–75.

40. Koishi's compliments can be found in Torii, *Nihon seikotsujutsu zensho*, 1:1204–7.

41. Ōtsuki, *Jū*tei Kaitai shinsho, 11:9b–10a.

42. Yoshida was a samurai and did not learn the arts from his family.

43. Nihon gakushiin, *Meijizen*, 3:266–67.

44. Kōshi revealed some tips from secret transmission; however, the main part of his book is citing texts. Much of the contents are from the imperial compilation of the medical encyclopedia *Yuzuan yizong jinjian* 御纂醫宗金鑑 [Golden mirror of the orthodox lineage of medicine, 1742], which could not possibly be transmitted by Chen.

45. Koga, *Nagasaki yōgakushi*, 231–32.

46. Kaibara, *Nihon seikei*, 164.

47. Kaibara made a list of the thirteen recorded orders of wooden skeletons, including two by Hoshino himself, but it cannot be certain that all were completed. See *Nihon seikei*, 257.

48. Kagami's preface of his *Seikotsu shinsho*, quoted in Kaibara, *Nihon seikei*, 276–77.

49. Kagami left contributions in various branches in Japanese history of medicine. For his biographical information, see Nihon gakushiin, *Meijizen*, 3:267; Fujigawa, "Mokkotsu kō," 8–16.

50. Kagami, *Mokotsu Teian*, 4a.

51. Nakano, *Osaka meiiden*, 75.

52. Kaibara, "*Ikei kyokotsuki*."

53. Fujigawa, "Mokkotsu kō," 8.

54. Kaibara, *Nihon seikei*, 279–81.

55. Kagami's representative work is *Seikotsu shinsho*.

56. Kagami, *Mokotsu teian*, 7a.

57. Kaibara, *Nihon seikei*, 282.

58. Kaibara, *Nihon seikei*, 303–10.

59. After the downfall of the shogunate in 1867, this skeleton passed to the Meiji government and is currently preserved in the University's Museum. Fujigawa, "Mokkotsu kō," 1.

Bibliography

Archive Collections

Ali Çelebi el-Izniki. *Mecmuatü'l-mücerrebat.* MS Hekimoğlu Ali Paşa 541, Millet Library, Istanbul; MS Ali Nihat Tarlan 197, Süleymaniye Library, Istanbul.

Ali Çelebi el-Izniki. *Sırr al-Rabbani fi ilm al-mizan.* MS A6247, Istanbul University Library, Istanbul.

Ali Münşi. *Risale-i cedvar.* MS T4234/4, Istanbul University Library, Istanbul.

Ibn Sallum, Salih b. Nasrullah. *Gayat al-itkan fi tabdir badan al-insan.* MS Ayasofya 3628, Süleymaniye Library, Istanbul.

Mehmed Fahreddin. *Risale-i ferahiyye.* MS T9324, Istanbul University Library, Istanbul.

Ms. Fr. 640. Bibliothèque nationale de France, Paris.

Ms. IOL Tib J 748. Stein Collection, British Library.

Ms. IOL Tib J 756. Stein Collection, British Library.

Ms. Pelliot tibétain 127. Pelliot Collection, Bibliothèque nationale de France.

Ms. Pelliot tibétain 1044. Pelliot Collection, Bibliothèque nationale de France.

Mullah Arab. *Talhis al-burhan.* MS 631, Zeytinoğlu Library, Kütahya.

Mullah Arab. *Talhis ar-rutbat al-hakim li Majriti.* MS Ali Emiri 2835/2, Millet Library, Istanbul.

Ömer bin Sinan el-Izniki. *Künuz-i hayat el-insan kanun-ı etibba-i feylesofan.* MSS T7083, T7140, T7142, Istanbul University Library, Istanbul; MS Şehit Ali Paşa 2085, Süleymaniye Library, Istanbul.

Ömer Şifai el-Burusavi. *Cevher el-ferid fi tıbb el-cedid.* MS Hamidiye 1020/1, Süleymaniye Library, Istanbul.

Ömer Şifai el-Burusavi. *Mürşid el-muhtar fi ilm el-esrar.* MS T7118, Istanbul University Library, Istanbul.

Safiyüddin Efendi. *Kaside-i Şeyh Safi.* MS Karaçelebizade 359/10, Süleymaniye Library, Istanbul.

Primary and Secondary Sources

Abbou Hershkovitz, Keren. "The Historiography of Science between the 10th and the 14th Centuries." PhD thesis (in Hebrew), Ben-Gurion University, Israel, 2008.

Abulafia, David. *The Great Sea: A Human History of the Mediterranean*. Oxford: Oxford University Press, 2011.

Acker, William R. B., trans. *Some T'ang and Pre-T'ang Texts on Chinese Painting*. Vol. 2. Leiden: Brill, 1974.

Adıvar, Abdülhak Adnan. *Osmanlı Türklerinde Ilim*. 1943. 5th ed. Istanbul: Remzi Kitabevi, 1981.

Adıvar, Abdülhak Adnan. *La science chez les Turcs Ottomans*. Paris: G. P. Maisonneuve, 1939.

Ágoston, Gabor. "'The Most Powerful Empire': Ottoman Flexibility and Military Might." In *Empires and Superpowers: Their Rise and Fall*, edited by Georg Zimmar and David Hicks, 121–71. Washington, DC: Society for the Preservation of Greek Heritage, 2005.

Akasoy, Anna. "The Buddha and the Straight Path: Rashīd al-Dīn's *Life of the Buddha:* Islamic Perspectives." In *Rashīd al-Dīn: Agent and Mediator of Cultural Exchanges in Ilkhanid Iran*, edited by Anna Akasoy, Charles Burnett, and Ronit Yoeli-Tlalim, 173–96. London: The Warburg Institute; Turin: Nino Aragno Editore, 2013.

Akasoy, Anna, Charles Burnett, and Ronit Yoeli-Tlalim, eds. *Islam and Tibet: Interactions along the Musk Routes*. Farnham: Ashgate, 2011.

Akasoy, Anna, Charles Burnett, and Ronit Yoeli-Tlalim, eds. *Rashīd al-Dīn: Agent and Mediator of Cultural Exchanges in Ilkhanid Iran*. London: Warburg Institute, 2013.

Akasoy, Anna, and Ronit Yoeli-Tlalim. "Along the Musk Routes: Exchanges between Tibet and the Islamic World." *Asian Medicine: Tradition and Modernity, Journal of the International Association for the Study of Traditional Asian Medicine* 3, no. 2 (2007): 217–40.

Akıncı, Sırrı. "Kitab-ı Düstûr'ü Vesim fi't Tıbb'il Cedidi ve'l Kadim'in İncelemesinden Ortaya Çıkan Sonuçlar." *Yeni Tıp Alemi* 15 (1964): 137–54.

Akrich, Madeleine. "A Gazogene in Costa Rica: An Experiment in Techno-sociology." In *Technological Choices: Transformation in Material Cultures since the Neolithic*, edited by Pierre Lemonnier, 289–337. London: Routledge, 1993.

Alam, Muzaffar, and Sanjay Subrahmanyam. *Indo-Persian Travels in the Age of Discoveries, 1400–1800*. Cambridge: Cambridge University Press, 2007.

Allsen, Thomas T. *Culture and Conquest in Mongol Eurasia*. Cambridge: Cambridge University Press, 2001.

Allsen, Thomas T. *The Royal Hunt in Eurasian History*. Philadelphia: University of Pennsylvania Press, 2006.

Almeida, Francisco Alonso. *A Middle English Remedy Book Edited from Glasgow University Library MS Hunter 185*. Heidelberg: Universitätsverlag Winter GmbH, 2014.

Altunay, Ahmet Refik. *Hicri Onikinci Asırda Istanbul Hayatı (1100–1200)*. Istanbul: Türk Tarih Encümeni, 1930.

Anawati, Georges C. "Jaldaki." In *Encyclopaedia of the History of Arabic Science*. Vol. 3, *Technology, Alchemy and Life Sciences*, edited by Roshdi Rashed, 874. London: Routledge, 1996.

Andaya, Barbara Watson. *The Flaming Womb: Repositioning Women in Early Modern Southeast Asia*. Honolulu: University of Hawai'i Press, 2006.

Andaya, Barbara Watson. "From Temporary Wife to Prostitute: Sexuality and Economic Change in Early Modern Southeast Asia." *Journal of Women's History* 9 (1998): 11–34.

Andaya, Barbara Watson, and Leonard Andaya. *History of Malaysia*. Basingstoke: Palgrave, 2001.

Anderson, Christy, Anne Dunlop, and Pamela H. Smith, eds. *The Matter of Art: Materials, Practices, Cultural Logics, c. 1250–1750*. Manchester: Manchester University Press, 2014.

Anderson, Webster. *Gems: Their Sources, Descriptions, and Identification*. 4th ed. London: Butterworths, 1983.

Ange de Saint Joseph (Joseph de la Brosse). *Pharmacopoeia Persica, ex Idiomate Persico in Latinum Conversa*. Paris: Typis Stephani Michallet, 1681.

Anonymous. *Buyi Leigong paozhi bianlan* 補遺雷公炮製便覽 (Complete synopsis of Master Thunder's pharmaceutical preparations). Imperial ed., 1591. 3 vols. Facsimile, Shanghai: Shanghai cishu chubanshe, 2005.

Anonymous. *Kahō Namba honedugi hiden* 家法難波骨継秘伝 (Family disciplines of the Namba secret transmission of bone setting, 1770). In Torii, *Nihon seikotsujutsu zensho* 1:258–351.

Anonymous. "*A Text-Book of Tea Planting and Manufacture:* review." *Nature* 1466, no. 57 (December 3, 1897): 96–97.

Antrobus, H. A. *A History of the Assam Company, 1839–1953*. Edinburgh: T. and A. Constable, 1957.

Antrobus, H. A. *A History of the Jorehaut Tea Co. Ltd., 1859–1946*. London: Tea and Rubber Mail, 1948.

Appadurai, Arjun. "How Histories Make Geographies: Circulation and Context in a Global Perspective." *Transcultural Studies* 1 (2010): 4–13.

Appadurai, Arjun, ed. *The Social Life of Things: Commodities in Cultural Perspective.* Cambridge: Cambridge University Press, 1986.

Arber, Agnes Robertson. *Herbals, Their Origin and Evolution: A Chapter in the History of Botany, 1470–1670.* Cambridge: Cambridge University Press, 1938.

Archivio Segreto Vaticano. "Processus in causa Aloysii a Cruce, laici professi Ordinis Fratrum Minorum, martyris Malacen (anni 1622–1627)." *Cong. Riti, Processus, 1647–1649*, vol. 1. 3 vols.

Artun, Tuna. "Bâyezîd-i Kimya'î: Osmanlı Kimya Metinlerinde II. Bayezid." *Türklük Bilgisi Araştırmaları—Journal of Turkish Studies* 39 (2013): 181–86.

Artun, Tuna. "Hearts of Gold and Silver: The Production of Alchemical Knowledge in the Early Modern Ottoman World." PhD dissertation, Princeton University, Princeton, NJ, 2013.

Asami, Dōsai 浅見道斎. *Kinsō hidenshū* 金瘡秘伝集 (Collections of secret transmissions about sword cuts). In *Zoku gunsho ruiju* 続群書類従 (Various books classified after categories, continued), edited by Hanawa Hokiichi 塙保己一. Vol. 904. Tokyo: Zoku gunsho ruiju kanseikai, 1931.

Atasoy, Nurhan, and Julian Raby. *Iznik: The Pottery of Ottoman Turkey.* 1989. Reprint, London: Laurence King, 1994.

Atçıl, Abdurrahman. "Mobility of Scholars and Formation of a Self-Sustaining Scholarly System in the Lands of the Rūm during the Fifteenth Century." In *Islamic Literature and Intellectual Life in Fourteeenth- and Fifteenth-Century Anatolia*, edited by Andrew Charles Spencer Peacock and Sara Nur Yıldız, 328–29. Orient-Institut Istanbul, Istanbuler Texte u. Studien 34. Würzburg: Egon Verlag, 2016.

The Athenæum Journal of Literature, Science, and the Fine Arts for the Year 1844. London: James Holmes, Took's Court, Chancery Lane, 1844. https://books.google.com/books?id=fk9FAQAAMAAJ& (accessed December 5, 2016).

Aubaile-Sallenave, F. "Atr." In *Encyclopædia Iranica*, vol. 3, fasc. 1, 14–16. http://www.iranicaonline.org/articles/atr-perfume (accessed December 3, 2016).

"Australian Art Gallery to Return 2000-Year-Old Buddha Idol to India." https://indianexpress.com/article/india/india-news-india/australian-art-gallery-to-return-2000-year-old-buddha-idol-to-india (accessed June 24, 2018).

Azamat, Nihat. "Üftade." *TDV Islam Ansiklopedisi.* Istanbul: Türkiye Diyanet Vakfı, 2012.

Azevedo, Antonio. "Apontamentos pera Vossa Magestade ver sobre as cosas do Estado da India." In *Documentaçao ultramarina portuguesa*, vol. 2, edited by António da Silva Rego, 161–64. Lisbon: Centro de Estudos Históricos Ultramarinos, 1960.

Bachour, Natalia. "Iatrochemistry and Paracelsism in the Ottoman Empire in the

Sixteenth and Seventeenth Centuries." *Intellectual History of the Islamicate World* 6 (2018): 82–116.

Bachour, Natalia. *Oswaldus Crollius und Daniel Sennert im frühneuzeitlichen Istanbul—Studien zur Rezeption des Paracelsismus im Werk des osmanischen Artzes Salih b. Nasrullah ibn Sallum al-Halabi*. Freiburg: Centaurus, 2012.

Bagchi, P. C. *India and China: A Thousand Years of Cultural Relations*. Rev. ed. Bombay: Hind Kitabs, 1950.

Baig, Amita, and William Henderson. *A Centenary of Planting in the Kanan Devan Hills Concession, 1878–1978*. Coimbatore: Tata-Finlay, 1978.

Bailey, H. W. "The Preface to the Siddhasāra-śāstra." In *A Locust's Leg: Studies in Honour of S. H. Taqizadeh*, edited by Walter B. Henning and Ehsan Yarshater, 31–38. London: Percy Lund, Humphries, 1962.

Bala, Arun. *The Dialogue of Civilizations in the Birth of Modern Science*. New York: Palgrave Macmillan, 2006.

Balfour, John Hutton. *Description of Asafoetida Plants* (*Narthex asafetida*, *Falconer*). Cambridge, MA: Harvard University Herbarium, 1860.

Ball, Samuel. *An Account of the Cultivation and Manufacture of Tea in China: derived from personal observation with remarks on the experiments now making for the introduction of the culture of the tea tree in other parts of the world*. London: Longman, 1848.

Barad, Karen. *Meeting the Universe Halfway: Quantum Physics and the Entanglement of Matter and Meaning*. Durham: Duke University Press, 2007.

Barbosa, Duarte. *The Book of Duarte Barbosa: An Account of the Countries Bordering on the Indian Ocean and Their Inhabitants, written by Duarte Barbosa and Completed about the Year 1518 A.D.* Translated by Mansel Longworth Dames. 2 vols. New Delhi: Asian Education Services, 2002.

Barkan, Ömer Lütfi. "Istila Devirlerinin Kolonizatör Türk Dervişleri ve Zaviyeler / Les fondations pieuses comme méthodes de peuplement et de colonisation." *Vakıflar Dergisi* 2 (1942): 279–304.

Barker, George M. *A Tea Planter's Life in Assam*. Calcutta: Thacker, Spink, 1884.

Bauer, Ralph. "The Blood of the Dragon: Alchemy and Natural History in Monardes's *Historia medicinal*." In *Medical Cultures of the Early Modern Spanish Empire*, edited by John Slater, Maria Luz Lopez, and Jose Pardo Tomas, 67–88. London: Routledge, 2016.

Baxter, Alan, and Patrick de Silva. *A Dictionary of Kristang*. Canberra: Pacific Linguistics, 2004.

Baylav, Naşit. *Eczacılık Tarihi* (History of pharmacy). Istanbul: Yörük Matbaası, 1968.

Bayly, C. A. "'Archaic' and 'Modern' Globalization in the Eurasian and African Arena, ca. 1750–1850." In *Globalization in World History*, edited by A. G. Hopkins, 46–73. New York: Norton, 2002.

Beckwith, Christopher. *Empires of the Silk Road: A History of Central Eurasia from the Bronze Age to the Present*. Princeton, NJ: Princeton University Press, 2011.

Beckwith, Christopher. *The Tibetan Empire in Central Asia: A History of the Struggle for Great Power among Tibetans, Turks, Arabs, and Chinese during the Early Middle Ages*. Princeton, NJ: Princeton University Press, 1993.

Beckwith, Christopher. *Warriors of the Cloisters: The Central Asian Origins of Science in the Medieval World*. Princeton, NJ: Princeton University Press, 2012.

Behal, Rana P. "Coolie Drivers or Benevolent Paternalists? British Tea Planters in Assam and the Indenture Labour System." *Modern Asian Studies* 44, no. 1 (2010): 29–51.

Bellwood, Peter. "Austronesian Prehistory in Southeast Asia: Homeland, Expansion and Transformation." In *The Austronesians: Historical and Comparative Perspective*, edited by Peter Bellwood and James J. Fox, 96–111.Canberra: Australian National University, 1995.

Benedict, Carol. *Golden-Silk Smoke: A History of Tobacco in China, 1550–2010*. Berkeley: University of California Press, 2011.

Bentinck, Sir William, Governor General of India and Chairman of the Tea Commission. *The Measures Adopted for introducing the Cultivation of the Tea Plant within the British Possessions in India*. British Parliamentary Papers, January 24, 1834.

Berg, Maxine, with Felicia Gottmann, Hanna Hodacs, and Chris Nierstrasz, eds. *Goods from the East, 1600–1800: Trading Eurasia*. London: Palgrave Macmillan, 2015.

Besky, Sarah. *The Darjeeling Distinction: Labor and Justice on Fair-Trade Tea Plantations in India*. Berkeley: University of California Press, 2013.

Beydilli, Kemal. *Osmanlıda Imamlar: Bir Imamın Günlüğü*. Istanbul: Yitik Hazine Yayınları, 2013.

Bhabha, Homi K. "Introduction: On Disciplines and Destinations." In *Territories and Trajectories: Cultures in Circulation*, edited by Diana Sorensen, 1–12. Durham: Duke University Press, 2018.

Bian He. "Assembling the Cure: *Materia Medica* and the Culture of Healing in Late Imperial China." PhD dissertation, Harvard University, Cambridge, MA, 2014.

Binbaş, İlker Evrim. *Intellectual Networks in Timurid Iran: Sharaf al-Dīn 'Alī Yazdī and the Islamicate Republic of Letters*. Cambridge: Cambridge University Press, 2016.

Bīrūnī, Muhamad ibn Ahmad. *Al-Bīrūnī, In den Gärten der Wissenschaft: Ausgewählte Texte aus den Werken des muslimischen Universalgelehrten* (Al-Bīrūnī, in the gardens of science: Selected texts from the works of the Muslim universal scholar). Edited and translated by Gotthard Strohmeier. Leipzig: Reclam, 1991.

Bleichmar, Daniela, and Meredith Martin. *Objects in Motion in the Early Modern World*. Special issue of *Art History* 38, no. 4 (2015).

Blench, Roger. *Archeology, Language and the African Past*. Oxford: AltaMira Press 2006.

Blondé, Bruno, and Wouter Ryckbosch. "Arriving to a Set Table: The Integration of Hot Drinks in the Urban Consumer Culture of the Eighteenth-Century Southern Low Countries." In *Goods from the East*, edited by Maxine Berg with Felicia Gottmann, Hanna Hodacs, and Chris Nierstrasz, 309–30. London: Palgrave Macmillan, 2015.

Blondeau, Anne-Marie. *Matériaux pour l'étude de l'hippologie et de l'hippiatrie tibétaines: à partir des manuscrits de Touen-houang* (Materials for the study of Tibetan hippology: based on the Dunhuang manuscripts). Genève: Librairie Droz, 1972.

Boer, Hugo J. de, and Crystle Cotingting. "Medicinal Plants for Women's Healthcare in Southeast Asia: A Meta-Analysis of Their Traditional Use, Chemical Constituents, and Pharmacology." *Journal of Ethnopharmacology* 151, no. 2 (2014): 747–67.

Bondt, Jakob de. *De medicina Indorum libri IV*. Leiden: Hackius, 1642. Translated by Thomas Noteman, as *An Account of the Diseases, Natural History and Medicines of the East Indies*. London: Noteman, 1769.

Boomgaard, Peter. "The Development of Colonial Healthcare in Java: An Explanatory Introduction." *Bijdragen tot de Taal-, Land- en Volkenkunde* 149, no. 1 (1993): 77–93.

Boomgaard, Peter. "Early Globalization: Cowries as Currency, 600 BCE–1900." In *Linking Destinies: Trade, Towns and Kin in Asian History*, edited by Peter Boomgaard, Henk Schulte Nordholt, and Dick Kooima, 13–28. Amsterdam: KITLV Press, 2008.

Boomgaard, Peter. *Frontiers of Fear: Tigers and People in the Malay World, 1600–1950*. New Haven, CT: Yale University Press, 2001.

Boomgaard, Peter. "Sacred Trees and Haunted Forests in Indonesia: Particularly Java, Nineteenth and Twentieth Centuries." In *Asian Perceptions of Nature: A Critical Approach*, edited by Ole Bruun and Arne Kalland, 48–62. Richmond, Surrey: Curzon Press: 1995.

Bosworth, C. E. "Bukhara ii. From the Arab Invasions to the Mongols." *Encyclopaedia Iranica* 513–15. http://www.iranicaonline.org/articles/bukhara-ii (accessed November 9, 2017).

Bosworth, C. E. "Merv." In *Historic Cities of the Islamic World*, edited by C. E. Bosworth, 401–6. Leiden: Brill, 2007.

Bourguet, Marie Noelle, Christian Licoppe, and H. Otto Sibum, eds. *Instruments, Travel and Science: Itineraries of Precision from the Seventeenth to the Twentieth Century*. London: Routledge, 2002.

Boxer, C. R. "A Note on Portuguese Reactions to the Revival of the Red Sea Spice Trade and the Rise of Atjeh, 1540–1600." *Journal of Southeast Asian History* 10, no. 3 (1969): 415–28.

Boxer, C. R. *Women in Iberian Expansion Overseas, 1415–1815: Some Facts, Fancies, and Personalities.* New York: Oxford University Press, 1975.

Bray, Francesca, ed. *Agriculture.* Vol. 6, part 2 of *Science and Civilisation in China*, edited by Joseph Needham. Cambridge: Cambridge University Press, 1984.

Bray, Francesca. "Introduction: The Power of Tu." In *Graphics and Text in the Production of Technical Knowledge in China*, edited by Francesca Bray, Vera Dorofeeva-Lichtmann, and Georges Métailié, 1–78. Leiden: Brill, 2007.

Bray, Francesca. "Science, Technique, Technology: Passages between Matter and Knowledge in Imperial Chinese Agriculture." *British Journal for the History of Science* 41, no. 3 (2008): 319–44.

Bray, Francesca. "Technological Transitions." In *The Cambridge History of the World*, vol. 6, part 1, *The Construction of a Global World, 1400–1800 CE*, edited by Merry Wiesner-Hanks, Jerry H. Bentley, and Sanjay Subrahmanyam, 76–106. Cambridge: Cambridge University Press, 2015.

Bray, Francesca. *Technology and Gender.* Berkeley: University of California Press, 1997.

Bray, Francesca. *Technology, Gender and History in Imperial China: Great Transformations Reconsidered.* London: Routledge, 2013.

Bray, Francesca. "Thinking with Diagrams in the History of Technology." *EASTS (East Asian Science, Technology and Society)* forthcoming, 2019.

Bréard, Andrea. "Problems of Pursuit: Recreational Mathematics or Astronomy?" In *From China to Paris: 2,000 Years of Transmission of Mathematical Ideas*, edited by Yvonne Dold-Samplonius, Joseph W. Dauben, Menso Folkerts, and Benno van Dalen, 57–86. Stuttgart: Franz Steiner Verlag, 2002.

Breman, Jan. *Labour Migration and Rural Transformation in Colonial India.* Amsterdam: Free University Press, 1990.

Brentjes, Sonja. "Narratives of Knowledge in Islamic Societies: What Do They Tell Us about Scholars and Their Contexts?" *Almagest* 4, no. 1 (May 2013): 74–95.

Brockelmann, Carl. *Geschichte der arabischen Literatur.* Zweiter Supplementband. Leiden: E. J. Brill, 1938.

Brook, Timothy. *The Confusions of Pleasure: Commerce and Culture in Ming China.* Berkeley: University of California Press, 1998.

Brown, Cecil H. "Folk Zoological Life-Forms: Their Universality and Growth." *American Anthropologist* n.s. 81, no. 4 (1979): 791–817.

Bruce, C. A., Superintendent of Tea Culture. *Report on the Manufacture of Tea, and on the Extent and Produce of the Tea Plantations in Assam: Transactions of the Agricultur-*

al and Horticultural Society of India, vol. 7 (1839). Reprinted as Appendix 3 in Antrobus, *A History* (1957): 463–76. This paper was presented by the Tea Committee and read at the meeting of the society on Wednesday, August 14, 1839.

Buell, Paul, Eugene Anderson, and Charles Perry, trans. and eds. *A Soup for the Qan: Chinese Dietary Medicine of the Mongol Era as Seen in Hu Sihui's Yinshan Zhengyao—Introduction, Translation, Commentary and Chinese Text*. Leiden: Brill, 2010.

Câbir b. Hayyan. *Tanımlar Kitabı—Kitab al-Hudud*. Arabic text and translation to Turkish by Aygün Akyol and Iclal Arslan. Ankara: Elis Yayınları, 2015.

Cahen, Claude. "Futuwwa." *The Encyclopaedia of Islam*. 2nd ed. Leiden: E. J. Brill, 1983.

Cai Tao. *Tie wei shan con tan* (Miscellaneous records at Mount Tiewei). Repr., Hangzhou: zhi buzu zhai, 1819.

Callataÿ, Godefroid de. "Magia en al-Andalus: Rasa'il Ijwan al-Safa', Rutbat al-Hakim y Gayat al-Hakim (Picatrix)." *Al-Qantara: Revista de Estudios Árabes* 34 (2013): 297–344.

Callataÿ, Godefroid de, and Sébastien Moureau. "Towards the Critical Edition of the Rutbat al-Hakim: A Few Preliminary Observations." *Arabica* 62 (2015): 385–94.

Cammann, Schuyler. "Magical and Medicinal Woods in Old Chinese Carvings." *Journal of American Folklore* 74, no. 292 (1961): 116–25.

Canby, Sheila R. "Depictions of Buddha Sakyamuni in the *Jami 'al-Tavarikh* and the *Majmà 'al-Tavarikh*." *Muqarnas* 10 (1993): 299–310.

Cardoso, George [*sic*]. *Agiologio Lusitano dos sanctos e varoens illustres em virtude do reino de Portugal, e svas Conqvistas*. Vol. 1. Lisbon: Na Officina Craesbeekiana, 1652.

Carrillo, Ana María. "Nacimiento y muerta de una profesión: Las parteras tituladas en México" (Birth and death of a profession: midwives in Mexico). *Dynamis* 19 (1999): 167–90.

Carrubba, Robert. "The First Report of the Harvesting of Asafetida in Iran." *Agricultural History* 53, no. 2 (1979): 451–56.

Casale, Giancarlo. "The Ottoman Administration of the Spice Trade in the Sixteenth-Century Red Sea and Persian Gulf." *Journal of the Economic and Social History of the Orient* 49, no. 2 (2006): 170–98.

Casale, Giancarlo. *The Ottoman Age of Exploration*. Oxford: Oxford University Press, 2010.

Caudmont, Jean. "Présence de la langue portugaise dans le sud-est de l'Asie" (Presence of the Portuguese language in Southeast Asia). *La Linguistique* 30, no. 1 (1994): 9–28.

Chakravarti, Ranabir. "Early Medieval Bengal and the Trade in Horses: A Note." *Journal of the Economic and Social History of the Orient* 42, no. 2 (1999): 194–211.

Chang Bei 長北. "*Xiushi lu* banben jiaokan ji《髹飾錄》版本校勘記 (On the different editions of *Xiushi lu*)." *Gugong bowuyuan yuankan* 1 (2006): 108–22.

Chemla, Karine, ed. *History of Science, History of Text*. Dordrecht: Springer, 2004.

Chen Kaijun. "The Rise of Technocratic Culture in High-Qing China: A Case Study of Bondservant (Booi) Tang Ying (1682–1756)." PhD dissertation, Columbia University, New York, 2014.

Chen Paul C. Y. "Traditional and Modern Medicine in Malaysia." *Social Science and Medicine* 15, no. 2 (1981): 127–36.

Chen Cheng 陳誠. *Xiyu fan guo zhi*西域番國志 (A chronicle of barbaric states in the Western Region). Beijing: Zhonghua shuju, 2000.

Chen Jiamo 陳嘉謨. *Bencao mengquan*本草蒙筌 (Tools for beginners in materia medica). 1565. Beijing: Renmin weisheng chubanshe, 1988.

Chen Ming. "La médecine dans la région de Turfan à l'époque médiévale: interactions entre chinois et hu" (Medicine in the Turfan region in medieval times: Interactions between Chinese and hu). In Despeux, Médecine, religion et société, 2:641–709.

Chen Ming. "The Transmission of Indian Ayurvedic Doctrines in Medieval China: A Case Study of *Aṣṭāṅga* and *Tridoṣa* Fragments from the Silk Road." In *Annual Report of the International Research Institute for Advanced Buddhology at Soka University for the Academic Year 2005* (2006): 201–30.

Chen Ming 陳明. *Yindu fanwen yidian "Yili jinghua" yanjiu*印度梵文醫典"醫理精華"研究 (A study on the Sanskrit medical text *Siddhasāra*). Rev. ed. Beijing: Shangwu yinshuguan, 2014.

Chen Shiduo 陳士鐸. *Bencao milu*本草秘錄 (Secret records on materia medica). Taiyuan: Shanxi kexue jiaoyu chubanshe, 1982.

Chen Yu. *Duanyan minsu kao*. Beijing: Wenwu chubanshe, 2010.

Chen Zhugui and Zhu Zizhen, comps. *Zhongguo chaye lishi ziliao xuanji* (Selected historiographical materials on tea in China). Beijing: Agriculture Press, 1981.

Chin, Tamara. "The Invention of the Silk Road, 1877." *Critical Inquiry* 40, no. 1 (2013): 194–219.

Christian, David. "Silk Roads or Steppe Roads? The Silk Roads in World History." *Journal of World History* 11, no. 1 (2000): 1–26.

Churchill, Wendy D. *Female Patients in Early Modern Britain: Gender, Diagnosis and Treatment*. London: Routledge, 2016.

Çizakça, Murat. "A Short History of the Bursa Silk Industry (1500–1900)." *Journal of the Economic and Social History of the Orient* 23 (1980): 142–52.

Clayer, Nathalie. "Confréries." In *Dictionnaire de l'Empire Ottoman*, edited by

François Georgeon, Nicolas Vatin, and Gilles Veinstein, 285–86. Paris: Fayard, 2015.

Clunas, Craig. *Superfluous Things: Material Culture and Social Status in Early Modern China*. Urbana and Chicago: University of Illinois Press and Polity Press, 1991.

Cook, Harold J. "Global Economies and Local Knowledge in the East Indies: Jacobus Bontius Learns the Facts of Nature." In *Colonial Botany: Science, Commerce, and Politics in the Early Modern World*, edited by Londa L. Schiebinger and Claudia Swan, 100–118. Philadelphia: University of Pennsylvania Press, 2005.

Cook, Harold J. *Matters of Exchange: Commerce, Medicine, and Science in the Dutch Golden Age*. New Haven, CT: Yale University Press, 2007.

Cook, Harold J. "Trading in Medical Simples and Developing New Science: De Orta and His Contemporaries." In *Medicine, Trade and Empire: Garcia de Orta's Colloquies on the Simples and Drugs of India (1563) in Context*, edited by Palmira Fontes da Costa, 129–46. Farnham: Ashgate, 2015.

Cooper, Alix. "Homes and Households." In *The Cambridge History of Science*, vol. 3 (Early Modern Science), edited by Katherine Park and Lorraine Daston, 224–33. Cambridge: Cambridge University Press, 2008.

Cotgrave, Randle. *A Dictionarie of the French and English Tongues*. London: Adam Islip, 1611. http://www.pbm.com/~lindahl/cotgrave/search/493r.html (accessed December 16, 2016).

Coupaye, Ludovic. "Ways of Enchanting: *Chaînes opératoires* and Yam Cultivation in Nyamikum Village, Maprik, Papua New Guinea." *Journal of Material Culture* 14, no. 4 (2009): 433–58.

Crole, D. *Tea: A text book of tea planting and manufacture. With some account of the laws affecting labour in tea gardens in Assam and elsewhere*. London, Crosby Lockwood & Son, 1879.

Crosby, Alfred W. *The Columbian Exchange: Biological and Cultural Consequences of 1492*. 1972. Reprint, Westport, CT: Praeger Publishers, 2003.

Crossley, Pamela. *A Translucent Mirror: History and Identity in Qing Imperial Ideology*. Berkeley: University of California Press, 2002.

Cubbage, Eric. "The Tragic Story of Charles Byrne 'The Irish Giant'." http://www.thetallestman.com/pdf/charlesbyrne.pdf (accessed May 1, 2011).

Culpeper, Nicholas. *Culpeper's last legacy left and bequeathed to his dearest wife, for the publicke good, being the choicest and most profitable of those secrets*. London, 1655.

Culpeper, Nicholas. *The English physician. Or, a treatise of practical physic, on the diseases of women*. London, 1652. Text Creation Partnership digital edition.

Cummins, Alexander. "Textual Evidence for the Material History of Amulets in

Seventeenth-Century England." In *Physical Evidence for Ritual Acts, Sorcery, and Witchcraft in Christian Britain: A Feeling for Magic*, edited by Ronald Hutton, 164–87. New York, Palgrave MacMillan, 2015.

Curry, John J. *The Transformation of Muslim Mystical Thought in the Ottoman Empire: The Rise of the Halveti Order, 1350–1650*. Edinburgh: Edinburgh University Press, 2010.

D'Andrea, Jeanne. *Ancient Herbs in the J. Paul Getty Museum Gardens*. Malibu: J. Paul Getty Museum, 1982.

Dakhlia, Jocelyne. *Lingua Franca*. Arles: Actes Sud, 2008.

Dalby, Andrew. *Dangerous Tastes: The Story of Spices*. Berkeley: University of California Press, 2000.

Daston, Lorraine, and Katherine Park. *Wonders and the Order of Nature, 1150–1750*. New York: Zone Books, 1998.

Debus, Allen G. *Chemistry and Medical Debate*. Nantucket, MA: Science History Publications, 2001.

Debus, Allen G. *The English Paracelsians*. London: Oldbourne Press, 1965.

DeCaroli, Robert. *Image Problems: The Origin and Development of the Buddha's Image in Early South Asia*. Seattle: University of Washington Press, 2015.

Deeg, Max. "Wailing for Identity: Topical and Poetic Expressions of Cultural Belonging in Chinese Buddhist Literature." In *Buddhist Encounters and Identities across East Asia*, edited by Ann Heirman, Carmen Meinert, and Christoph Anderl, 227–52. Leiden: Brill, 2018.

de Groot, Jakob Jan Maria. *The religious system of China, its ancient forms, evolution, history and present aspect, manners, customs and social institutions connected therewith. Vol. 5, The Soul, and Ancestral Worship (465–928)*. Leyden: E. J. Brill, 1892.

Dennis, Joe. *Writing, Publishing, and Reading Local Gazetteers in Imperial China, 1100–1700*. Cambridge, MA: Harvard University Asia Center, 2015.

Denny, Walter B. *Iznik: The Artistry of Ottoman Ceramics*. 2004. Reprint, London: Thames and Hudson, 2015.

Desouza, Shaila. "Health Practices and People's Identity: Re-Locating Identity." *Lusotopie* (2000): 455–68.

Desouza, Shaila. "Tradition, Colonialism and Modernity: Women's Health in Goa." *Gender, Technology and Development* 7, no. 2 (2003): 189–208.

Despeux, Catherine, ed. *Médecine, religion et société dans la Chine médiévale: Étude de manuscrits chinois de Dunhuang et de Turfan* (Medicine, religion, and society in medieval China: A study of the Chinese manuscripts from Dunhuang and Turfan). 3 vols. Paris: Collège de France, 2010.

Dey, Arnab. "Of Planters, Ecology, and Labor: Plantation Worlds, Human History and Nonhuman Actors in Eastern India (Assam)." PhD dissertation, University of Chicago, 2012.

Deyell, John S. "Cowries and Coins: The Dual Monetary System of the Bengal Sultanate." *Indian Economic and Social History* 47, no. 1 (2010): 63–106.

Diamond, Norma. "The Miao and Poison: Interactions on China's Southwest Frontier." *Ethnology* 27, no. 1 (1988): 1–25.

Diaz-Mas, Paloma. *Sephardism: The Jews from Spain*. Translated by George K. Zucker. Chicago: University of Chicago Press, 1992.

Dickson, Sarah Augusta. *Panacea or Precious Bane: Tobacco in Sixteenth Century Literature*. New York: New York Public Library, 1954.

Digby, Simon. "The Maritime Trade of India." In *The Cambridge Economic History of India, Volume 1: c. 1200–c. 1750*, edited by Tapan Raychaudhuri and Irfan Habib, 125–59. Cambridge: Cambridge University Press, 1982.

Dodoens, Rembert. *A nievve herball; or, Historie of plantes. Translated by Henry Lyte.* London: G. Dewes, 1578. http://www.biodiversitylibrary.org/item/189507#page/292/mode/2up (accessed December 10, 2016).

Dorje, Gyurme, comm. and trans. *Tibetan Elemental Divination Paintings: Illuminated Manuscripts from The White Beryl of Sangs-rgyas rGya-mtsho: With the Moonbeams Treatise of Lo-chen Dharmaśri.* London: John Eskenazi, 2001.

Dorofeeva-Lichtmann, Vera. "Spatial Organization of Ancient Chinese Texts (Preliminary Remarks)." In *History of Science, History of Text*, edited by Karine Chemla, 3–53. Dordrecht: Springer, 2004.

Dressler, Markus. "How to Conceptualize Inner-Islamic Plurality/Difference: 'Hetereodoxy' and 'Syncretism' in the Writings of Mehmet F. Köprülü (1890–1966)." *British Journal of Middle Eastern Studies* 37 (2010): 241–60.

Duan Chengshi 段成式. *Youyang zazu*酉陽雜俎 (Miscellaneous jottings from south of Mount *You*). Taibei: Shangwu yinshuguan, 1965. Reprint of the Sibu Congkan edition of the 1608 edition.

Duan Wenjie 段文傑. "Shiliu guo, Beichao shiqi de Dunhuang shiku yishu" 十六國、北朝時期的敦煌石窟藝術 (The Art of Dunhuang Grottoes from the Sixteen Kingdoms and Northern Dynasties Period). In *Duan Wenjie Dunhuang shiku yishu lunwen ji* 段文傑敦煌石窟藝術論文集 (Collection of Essays by Duan Wenjie on the Art of Dunhuang Grottoes), edited by Duan Wenjie, 1–41. Lanzhou: Gansu renmin chubanshe, 1994.

Duan Yuming 段玉明. *Zhikong: Zuihou yiwei lai Hua de Yindu gaoseng* 指空—最後一位來華的印度高僧 (Zhikong: The last eminent Indian monk to come to China). Chengdu: Bashu shushe, 2007.

Duara, Prasenjit. *The Crisis of Global Modernity: Asian Traditions and a Sustainable Future.* Cambridge: Cambridge University Press, 2015.

Dugan, Holly. *The Ephemeral History of Perfume: Scent and Sense in Early Modern England.* Baltimore, MD: Johns Hopkins University Press, 2011.

Du Yongtao. *The Order of Places: Translocal Practices of the Huizhou Merchants in Late Imperial China.* Leiden: Brill, 2015.

Dymock, William, Warden, C.J.H., Hooper, David. *Pharmacographia Indica. A History of the Principal Drugs of Vegetable Origin Met with in British India.* 3 vols. London: Kegan Paul, Trench, and Trubner, 1890-93. Reproduced in *Hamdard*, the organ of the Institute of Health and Tibbi Research, Pakistan 1972.

Edwards, R. "The Cave Relief at Ma Hao." *Artibus Asiae* 17 (1954): 5–28, 103–29.

Egawa Yoshio 江川義雄. *Hiroshimaken ijinden* 廣島縣醫人傳 (Biographies of the physicians of Hiroshima prefecture). Hiroshima: Egawa Obstetrics Clinic, 1986.

Ellis, Markman, Richard Coulton, and Matthew Mauger. *Empire of Tea: The Asian Leaf that Conquered the World.* London: Reaktion Books, 2015.

Elman, Benjamin A. *From Philosophy to Philology: Intellectual and Social Aspects of Change in Late Imperial China.* Cambridge, MA: Council on East Asian Studies, Harvard University, 1990.

Elman, Benjamin A. *On Their Own Terms: Science in China, 1550–1900.* Cambridge, MA: Harvard University Press, 2005.

Emmerick, Ronald E. "New Light on the 'Siddhasāra.'" *Bulletin of the School of Oriental and African Studies, University of London* 37, no. 3 (1974): 628–54.

Emmerick, Ronald E. "On Ravigupta's 'Gaṇas.'" *Bulletin of the School of Oriental and African Studies, University of London* 34, no. 2 (1971): 363–75.

Emmerick, Ronald E. "Ravigupta's Place in Indian Medical Tradition." *Indologica Taurinensia* 3–4 (1975–1976): 209–21.

Emmerick, Ronald E. "Ravigupta's *Siddhasāra* in Arabic." In *Studien zur Geschichte und Kultur des vorderen Oriens, Festschrift für Bertold Spuler zum siebzigsten Geburtstag*, edited by Hans Robert Roemer and Albrecht Noth, 28–31. Leiden: Brill, 1981.

Emmerick, Ronald E. "The Sanskrit Text of the Siddhasāra." *Bulletin of the School of Oriental and African Studies, University of London* 34, no. 1 (1971): 91–112.

Emmerick, Ronald E. *The Siddhasāra of Ravigupta.* Vol 1. Wiesbaden: Franz Steiner Verlag, 1980.

Emmerick, Ronald E. "Some Remarks on Translation Techniques of the Khotanese." In *Sprachen des Buddhismus in Zentralasien*, edited by K. Röhrborn and W. Veenker, 17–26. Wiesbaden: Otto Harrassowitz Verlag, 1983.

Emmerick, Ronald E. *Tibetan Texts concerning Khotan*. London: Oxford University Press, 1967.

Erédia, Manoel Godinho de. "Description of Malaca, Meridional India and Cathay." Edited and translated by J. V. Mills. *Journal of the Malayan Branch of the Royal Asiatic Society* 8, part 1 (1930): 1–288.

Erünsal, Ismail E. "Türk Edebiyatı Tarihinin Arşiv Kaynakları 4: Lami'i Çelebi'nin Terekesi." *Journal of Turkish Studies (Fahir Iz Armağanı I)* 14 (1990): 179–93.

Erünsal, Ismail E. "Türk Edebiyat Tarihinin Arşiv Kaynakları 6: 909/1503 Tarihli Defter-i Kütüb." *Türklük Bilgisi Araştırmaları* 32 (2008): 203–19.

Erzi, H. Adnan. "Bursa'da Ishaki Dervişlerine Mahsus Zaviyenin Vakfiyesi." *Vakıflar Dergisi* 2 (1942): 423–29.

Escauriaza-Lopez, Valéria. "Aurel Stein's Methods and Aims in Archaeology on the Silk Road." In *Sir Aurel Stein, Colleagues and Collections*, edited by Helen Wang. London: British Museum Publication, 2012. https://www.britishmuseum.org/pdf/2_Lopez.pdf (accessed October 25, 2018).

Esherick, Joseph W., and Mary Backus Rankin, eds. *Chinese Local Elites and Patterns of Dominance*. Berkeley: University of California Press, 1990.

European Commission, Directorate-General for Translation. *Studies on Translation and Multilingualism: Lingua Franca: Chimera or Reality?* Luxembourg: EUR-OP, 2011.

Evans, Jennifer. "Fumigating for Health." Early Modern Medicine (blog). http://earlymodernmedicine.com/fumigating-for-health (accessed December 4, 2016).

Evliya Çelebi. *Günümüz Türkçesiyle Evliya Çelebi Seyahatnamesi*. Vol. 1, bk. 2, 5th ed., translated by Seyit Ali Kahraman and Yücel Dağlı. Istanbul: Yapı Kredi Yayınları, 2008.

Evliya Çelebi. *Günümüz Türkçesiyle Evliya Çelebi Seyahatnamesi*. Vol. 2, bk. 1, 2nd ed., translated by Yücel Dağlı and Seyit Ali Kahraman. Istanbul: Yapı Kredi Yayınları, 2008.

Fan, Fa-ti. *British Naturalists in Qing China*. Cambridge, MA: Harvard University Press, 2004.

Faroqhi, Suraiya. "At the Ottoman Empire's Industrious Core: The Story of Bursa." In *City in the Islamic World*, edited by Salma K. Jayyusi, Renate Holod, Attilio Petruccioli, and André Raymond, 357–81. Leiden: Brill, 2008.

Faroqhi, Suraiya. "Bursa at the Crossroads, Iranian Silk European Competition and the Local Economy, 1470–1700." In *Making a Living in the Ottoman Lands, 1480–1820*, edited by Suraiya Faroqhi, 113–48. Istanbul: Isis Press, 1995.

Fayuan zhulin 法苑珠林 (Pearl-grove of the Garden of the Law). By Daoshi 道世 (d. 668?). T. (53) 2122.

Fazlıoğlu, Ihsan. "Fazıl Ali Bey." *Yaşamları ve Yapıtlarıyla Osmanlılar Ansiklopedisi*. Vol. 1. Istanbul: Yapı Kredi Yayınları, 1999.

Feng, H. Y., and J. K. Shryock. "The Black Magic in China Known as Ku." *Journal of the American Oriental Society* 55, no. 1 (1935): 1–30.

Ferguson, A. M. *"Iňgē Vā," or the Sinnai Durai's Pocket Tamil Guide*. 8th ed. Colombo: The "Ceylon Observer" Press, 1924.

Fèvre, Francine. "Drôles de bestioles: qu'est-ce qu'un chong?" *Anthropozoologica* 18 (1993): 57–65.

Few, Martha. "Medical *Mestizaje* and the Politics of Pregnancy in Colonial Guatemala, 1660–1730." In *Science in the Spanish and Portuguese Empires, 1500–1800*, edited by Daniela Bleichmar, Paula De Vos, Kristin Huffine, and Kevin Sheehan, 132–46. Redwood City, CA: Stanford University Press, 2008.

Fierro, Maribel. "Bāṭinism in Al-Andalus: Maslama b. Qasim al-Qurtubī (d. 353/964), Author of the 'Rutbat al-Hakim' and the 'Ghayat al-Hakim (Picatrix).'" *Studia Islamica* 84 (1996): 87–112.

Finlay, Robert. *The Pilgrim Art: Cultures of Porcelain in World History*. Berkeley: University of California Press, 2010.

Finlay, Robert. "The Pilgrim Art: The Culture of Porcelain in World History." *Journal of World History* 9, no. 2 (1998): 141–87.

Finlay, Robert. "Portuguese and Chinese Maritime Imperialism: Camões's Lusiads and Luo Maodeng's Voyage of the San Bao Eunuch." *Comparative Studies in Society and History* 34, no. 2 (1992): 225–41.

Flecker, Michael. "A Ninth-Century A.D. Arab or Indian Shipwreck in Indonesia: First Evidence for Direct Trade with China." *World Archaeology* 32, no. 3 (2001): 335–54.

Folsom, Kenneth. *Friends, Guests, and Colleagues: The Mu-Fu System in the Late Ch'ing Period*. Berkeley: University of California Press, 1968.

Forbes, Robert J. *The Short History of the Art of Distillation*. Leiden: E. J. Brill, 1970.

Forêt, Philippe. *Mapping Chengde: The Qing Landscape Enterprise*. Honolulu: University of Hawai'i Press, 2000.

Forêt, Philippe, and Andreas Kaplony, eds. *The Journey of Maps and Images of the Silk Road*. Leiden: Brill, 2008.

Forrest, D. M. *A Hundred Years of Ceylon Tea, 1867–1967*. London: Chatto and Windus, 1967.

Fortune, Robert. *A Journey to the Tea Countries of China and India*. London: John Murray, 1852. Facsimile reprint, Cambridge: Cambridge University Press, 2012.

Fortune, Robert. *Three Years' Wanderings in the Northern Provinces of China, Including a Visit to the Tea, Silk and Cotton Countries*. London: John Murray, 1847.

Fox, Ariel. "Precious Bodies: Money Transformation Stories from Medieval to Late Imperial China." *Harvard Journal of Asiatic Studies* 76, nos. 1–2 (2016): 43–85.

Fox, Kate. *The Smell Report*. Oxford: Social Issues Research Centre, 2009. http://www.sirc.org/publik/smell.pdf (accessed December 4, 2016).

Franck, Irene M., and David M. Brownstone, *The Silk Road: A History*. New York: Facts on File Publications, 1986.

Frank, Andre Gunder. "The Centrality of Central Asia." *Studies in History* 8, no. 1 (1992): 43–97.

Fromer, Julie E. "'Deeply Indebted to the Tea-Plant': Representations of English National Identity in Victorian Histories of Tea." *Victorian Literature and Culture* 36, no. 2 (2008): 531–47.

Fryer, John. *A New Account of East India and Persia: Being Nine Years' Travels, 1672–1681*. Vol. 2. London: Hakluyt Society, 1912.

Fujigawa Yū 富士川游. "Mokkotsu kō" 木骨考 (Studies of wooden skeletons). In *Fujigawa chosakushū* 富士川游著作集 (Works of Fujigawa Yū), edited by Fujigawa Eirō 富士川英郎, 10:3–25. Kyoto: Shibungaku, 1972.

Fu Liming 付黎明 and Xie Yunxia 謝云霞, "Lun Li Yu zaowu sheji sixiang zhong de pingmin yishi 論李漁造物設計思想中的平民意識 (On the Perspective of Ordinary Folks in Li Yu's Theory of Creation)." *Hunan shehui kexue* 3 (2013): 251–53.

Gardella, Robert. *Harvesting Mountains: Fujian and the China Tea Trade, 1757–1937*. Berkeley: University of California Press, 1994.

Ghazanfar, Shahina A., and Martin Fisher. *Vegetation of the Arabian Peninsula*. Dordrecht: Springer Netherlands, 1998.

Ghosh, Amitav. *In an Antique Land*. New York: Knopf, 1993.

Gianno, Rosemary. "Women Are Not Brave Enough; Semelai Male Midwives in the Context of Southeast Asian Cultures." *Bijdragen tot de Taal-, Land- en Volkenkunde* 160, no. 1 (2004): 31–71.

Giddens, Anthony. *A Contemporary Critique of Historical Materialism*. Vol. 1, *Power, Property and the State*. Berkeley: University of California Press, 1981.

Gilman, Sander L., and Zhou Xun. "Introduction." In *Smoke: A Global History of Smoking*, edited by Sander L. Gilman and Zhou Xun, 9–28. London: Reaktion Books, 2004.

Glasgow University Archives. *Guide to James Finlay & Co Managers' and Assistants' Letterbooks*. 2015. http://www.gla.ac.uk/services/archives/collections/business/features/finlay (accessed April 15, 2016).

Glover, Ian C., and Bérénice Bellina. "Ban Don Ta Phet and Khao Sam Kaeo: The Earliest Indian Contacts Re-assessed." In *Early Interactions between South and*

Southeast Asia: Reflections on Cross-Cultural Exchange, edited by Pierre-Yves Manguin, A. Mani, and Geoff Wade, 17–45. Singapore: Institute of Southeast Asian Studies, 2011.

Goble, Andrew E. *Confluences of Medicine in Medieval Japan: Buddhist Knowledge, Islamic Formulae, and Wounds of War.* Honolulu: University of Hawai'i Press, 2011.

Gohlman, William E. *The Life of Ibn Sina: A Critical Edition and Annotated Translation*. New York: State University of New York Press, 1974.

Golas, Peter. "'Like Obtaining a Great Treasure': The Illustrations in Song Yingxing's The Exploitation of the Works of Nature." In *Graphics and Text in the Production of Technical Knowledge in China: The Warp and the Weft*, edited by Francesca Bray, Vera Dorofeeva-Lichtmann, and Georges Métailié, 569–614. Leiden: Brill, 2007.

Golden, Peter B. *Central Asia in World History*. New York: Oxford University Press, 2011.

Golinski, Jan. *Making Natural Knowledge: Constructivism and the History of Science*. Chicago: University of Chicago Press, 2005.

Göttler, Christine, and Wolfgang Neuber, eds. *Spirits Unseen: The Representation of Subtle Bodies in Early Modern European Culture*. Leiden: Brill, 2008.

Gracias, Fatima da Silva de. *Kaleidoscope of Women in Goa, 1510–1961*. New Delhi: New Concept, 1996.

Granet, Marcel. *La pensée chinoise*. Paris: La Renaissance du Livre, 1934.

Grasskamp, Anna, and Monica Juneja, eds. *EurAsian Matters: China, Europe, and the Transcultural Object, 1600–1800*. Berlin: Springer, 2018.

Grehan, James. "Smoking and 'Early Modern' Sociability: The Great Tobacco Debate in the Ottoman Middle East." *American Historical Review* 111 (2006): 1352–77.

Guha-Thakurta, Tapati. *Monuments, Objects, Histories: Institutions of Art in Colonial and Postcolonial India*. New York: Columbia University Press, 2013.

Guibourt, Nicolas J.-B. *Histoire naturelle des drogues simples. Tome 3e*. Paris: Chez J.-B. Baillière, 1850.

Gümüşalan, Nermin. *15. ve 16. Yüzyıllarda Bursa'da Sağlık Hayatı*. Bursa: Bursa Büyükşehir Belediyesi, 2015.

Guṇabhadra求那跋陀羅, translated. *Yang que moluo jing* 央掘魔羅經 (Aṅgulimālika-sūtra). Chinese Buddhist Electronic Text Association, T02, no. 120.

Günergun, Feza. *14–17. Yüzyıllarda Osmanlı Imparatorluğunda Kullanılan Anorganik Ilaçlar* (Inorganic drugs used in the Ottoman Empire during the fourteenth–seventeenth centuries). Istanbul: n.p., 1986.

Günergun, Feza. "Ottoman Encounters with European Science: Sixteenth- and Seventeenth-century Translations into Turkish." In *Cultural Translation in Early*

Modern Europe, edited by Peter Burke and R. Po-chia Hsia, 192–211. Cambridge: Cambridge University Press, 2007.

Günergun, Feza. "Science in the Ottoman World." In *Imperialism and Science* by George N. Vlahakis, Isabel Maria Malaquias, Nathan M. Brooks, François Regourd, Feza Günergun, and David Wright, 71–118. Santa Barbara, CA: ABC Clio, 2006.

Günergun, Feza. "Travellers in the Field: Conveying Botanical Material from Ottoman Turkey to Western Europe (16th and 17th c.)." Paper presented at the conference "Knowledge, Exchange, Encounter: Europe and the Ottoman Empire, 1453–1718," July 8–9, 2013, at CRASSH, University of Cambridge, England.

Günergun, Feza, and Şeref Etker. "From Quinaquina to 'Quinine Law': A Bitter Chapter in the Westernisation of Turkish Medicine." *Osmanlı Bilimi Araştırmaları / Studies in Ottoman Science* 14, no. 2 (2013): 41–68.

Günergun, Feza, and Dhruv Raina, ed. *Science between Europe and Asia: Historical Studies on the Transmission, Adoption, and Adaptation of Knowledge*. Dordrecht: Springer, 2011.

Gutas, Dimitri. *Greek Thought, Arabic Culture: The Graeco-Arabic Translation Movement in Baghdad and Early 'Abbāsid Society (2nd–4th/8th–10th centuries)*. London: Routledge, 1998.

Gyatso, Thinley, and Chris Hakim. *Essentials of Tibetan Traditional Medicine*. Berkeley, CA: North Atlantic Books, 2010.

Haberland, Detlef, ed. *Engelbert Kaempfer (1651–1716): Ein Gelehrtenleben zwischen Tradition und Innovation*. Wiesbaden: Harrassowitz Verlag, 2004.

Habib, Irfan. "Non-agricultural Production." In *History of Science, Philosophy and Culture in Indian Civilization, Volume VIII, Part I: Economic History of Medieval India, 1200–1500*, edited by Irfan Habib, 85–103. New Delhi: Centre for Studies in Civilizations, 2011.

Halleux, Robert. "The Reception of Arabic Alchemy in the West." In *Encyclopedia of the History of Arabic Science*, edited by Roshdi Rashed, 3:886–903. London: Routledge, 1996.

Handler, Sarah. *Austere Luminosity of Chinese Classical Furniture*. Berkeley: University of California Press, 2001.

Han E, comp. *Sishi zuanyao* (Compendium of essentials for the four seasons). Possibly late ninth century. Modern ed., *Sishi zuanyao jiaoshi* (Annotated modern translation of the *Sishi zuanyao*), by Miao Qiyu. Beijing: Agriculture Press, 1981.

Hansen, Valerie. *The Silk Road: A New History*. Oxford: Oxford University Press, 2012.

Hansen, Valerie, ed. *The Silk Road: Key Papers*. Leiden: Brill, 2012.

Hanser, Jessica. "Teatime in the North Country: Consumption of Chinese Imports in North-East England." *Northern History* 49, no. 1 (2012): 51–74.

Hara Sanshin 原三信. *Nihon de hajimete honyakushita kaibōsho* 日本で初めて翻訳した解剖書 (The earliest translated anatomy book in Japan). Fukuoka: Rokudai Hara Sanshin Ranhōi sanpyakunen kinen shōgakkai, 1995.

Hardie, Alison. *The Craft of Gardens*. Shanghai: Shanghai Books, 2012.

Hardie, Alison. "Ji Cheng's Yuan Ye in Its Social Setting." In *The Authentic Garden*, edited by Erik de Jong and Leslie Tjon Sie Fat, 207–14. Amsterdam: Clusius Foundation, 1991.

Harriot, Thomas. *A brief and true report of the new found land of Virginia*. 1588 quarto. Facsimile, Ann Arbor: Clements Library Associates, 1951.

Hartwell, Robert. "Foreign Trade, Monetary Policy and Chinese 'Mercantilism.'" In *Collected Studies on Sung History Dedicated to James T. C. Liu in Celebration of His Seventieth Birthday*, edited by Kinugawa Tsuyoshi, 453–88. Kyoto: Dohosha, 1989.

Harvey, Susan Ashbrook. *Scenting Salvation: Ancient Christianity and the Olfactory Imagination*. Berkeley: University of California Press, 2006.

Hattori Toshirō 服部敏良. *Heian jidai igakushi kenkyū* 平安時代医学史研究 (Studies of medical history of the Heian period). Tokyo: Yoshikawa kōbunkan, 1975.

Hauser, I. L. *Tea: Its Origin, Cultivation, Manufacture and Use*. Chicago, Rand, McNally, 1890.

Hayami, Yujiro, and A. Damodaran. "Towards an Alternative Agrarian Reform: Tea Plantations in South India." *Economic and Political Weekly* 39, no. 36 (2004): 3992–97.

He Chuanyao. *Baoyantang yanbian*. N.p., 1837.

Heirman, Ann, Carmen Meinert, and Christoph Anderl, eds. *Buddhist Encounters and Identities across East Asia*. Leiden: Brill, 2018.

Heller, Amy. "Preliminary Remarks on Painted Coffin Panels from Tibetan Tombs." In *Scribes, Texts and Rituals in Early Tibet and Dunhuang*, edited by Brandon Dotson, Kazushi Iwao, and Tsuguhito Takeuchi, 11–23. Wiesbaden: Reichert Verlag, 2013.

Henderson, John B. *The Development and Decline of Chinese Cosmology*. New York: Columbia University Press, 1984.

An Herbal (1525). Edited and translated by Sanford Larkey and Thomas Pyles. Brattleboro: E. L. Hildreth, 1941.

Herman, Kiriama. "Mutual Exchange: Chinese and East African Trade in the Late

First/Early Second Millennium BCE." In *Ancient Silk Trade Routes: Selected Works from Symposium on Cross Cultural Exchanges and Their Legacies in Asia*, edited by Dashu Qin and Jian Yuan, 137–48. Singapore, SGP: World Scientific Publishing, 2015.

Herren, Madeleine, Martin Rüesch, and Christiane Sibille. *Transcultural History: Theories, Methods, Sources*. Berlin: Springer, 2012.

Hilaire-Pérez, Liliane. *La pièce et le geste: artisans, marchands et savoir technique à Londres au XVIIIe siècle*. Paris: Albin Michel, 2013.

Hinrichs, T. J., and Linda L. Barnes, eds. *Chinese Medicine and Healing: An Illustrated History*. Cambridge, MA: Belknap Press of Harvard University Press, 2013.

Hinsch, Bret. *The Rise of Tea Culture in China: The Invention of the Individual*. London: Rowman and Littlefield, 2015.

Hippocrates. *Ancient Medicine: Airs, Waters, Places. Epidemics 1 and 3. The Oath. Precepts. Nutriment*. Translated by W. H. S. Jones, Loeb Classical Library 147. Cambridge, MA: Harvard University Press, 1923.

Hitzel, Frédéric. "Iznik." In *Dictionnaire de l'Empire Ottoman*, edited by François Georgeon, Nicolas Vatin, and Gilles Veinstein, 630. Paris: Fayard, 2015.

Hızlı, Mefail, *Osmanlı Klasik Döneminde Bursa Medreseleri*. Istanbul: Iz Yayınları, 1998.

Hodacs, Hanna, and Leos Müller. "Chests, Tubs and Lots of Tea: The European Market for Chinese Tea and the Swedish East India Company, c. 1730–1760." In *Goods from the East*, edited by Maxine Berg with Felicia Gottmann, Hanna Hodacs, and Chris Nierstrasz, 277–93. London: Palgrave Macmillan, 2015.

Hodder, Ian. *Studies in Human-Thing Entanglement*. Creative Commons, 2016. http://www.ian-hodder.com/books/studies-human-thing-entanglement (accessed May 5, 2016).

Holmyard, Eric John. "Al-Jildaki." *Iraq* 4 (1937): 47–53.

Holmyard, Eric John. "Maslama al-Majriti and the Rutbatu'l-Hakim." *Isis* 6 (1924): 293–305.

Horden, Peregrine, and Nicholas Purcell. *The Corrupting Sea: A Study of Mediterranean History*. Oxford: Blackwell, 2000.

Horden, Peregrine, and Nicholas Purcell. "The Mediterranean and the New Thalassology." *American Historical Review* 111, no. 3 (2006): 722–40.

Howard, Angela F. "Buddhist Cave Sculpture of the Northern Qi Dynasty: Shaping a New Style, Formulating New Iconographies." *Archives of Asian Art* 49 (1996): 6–25.

Howes, David, and Constance Classen. *Ways of Sensing: Understanding the Senses in Society*. London: Routledge, 2014.

Huang, H. T., ed. *Fermentations and Food Science*. Vol. 6, part 5 of *Science and Civilisation in China*, edited by Joseph Needham. Cambridge: Cambridge University Press, 2000.

Huang Cheng 黃成. *Xiushi lu tushuo* 髹飾錄圖說 (Illustrations on Xiushi lu). Jinan: Shangdong huabao chubanshe, 2007.

Huang Cheng 黃成. *Xiushi lu* 髹飾錄 (Record on lacquer decoration). Edited by Wang Shixiang 王世襄. Beijing: Zhongguo renmin daxue chubanshe, 2004.

Huang Nengfu 黃能馥. *Zhongguo fuzhuang shi* 中國服裝史(A history of Chinese clothing). Beijing: Zhongguo lüyou chubanshe, 1995.

Huang Shu 黃庶. *Fa tan ji*伐檀集 (Cutting down sandal trees: Collection of essays). Vol. 1092. Siku Quanshu series, 1985. Facsimile printing, Taibei: Shangwu Yinshuguan.

Huijiao 慧皎. *Gaoseng zhuan*高僧傳 (Biographies of prominent monks, ca. 919–1002). 5th–6th centuries. Vol. 50. Tripitaka ed. Taibei: Xinwenfeng chuban gongsi, 1983.

Huilin 慧琳. *Yiqie jing yin yi*一切經音義 (The sounds and meanings of the Tripitaka). 8th–9th centuries. T54, no. 2128, CBETA.

Hume, Ivor Noël. "Hunting for a Little Ladle: Tobacco Pipes." *CW Journal* (Winter 2003–2004). http://www.history.org/foundation/journal/winter03–04/pipes.cfm (accessed December 11, 2016).

Hu Sihui 忽思慧. *Yinshan zhengyao*飲膳正要 (Essentials of food and drink). 1330. Beijing: Renmin weisheng chubanshe, 1986.

Ibn Shahriyār, Buzurg. *Kitab ʿajāyib al-Hind* (Livre des merveilles de l'Inde). Arab text with French translation by L. Marcel Devic. Leiden: Brill, 1883–1886.

Ihsanoğlu, Ekmeleddin, and Ramazan Şeşen, M. Serdar Bekar, Gülcan Gündüz, and Veysel Bulut. *Osmanlı Tabii ve Tatbiki Bilimler Literatürü Tarihi* (History of the literature of natural and applied sciences during the Ottoman period), vol. 1. Istanbul: IRCICA, 2006.

Ihsanoğlu, Ekmeleddin, and Ramazan Şeşen, M. Serdar Bekar, Gülcan Gündüz, and Veysel Bulut. *Osmanlı Tıbbi Bilimler Literatürü Tarihi* (History of the literature of medical sciences during the Ottoman period), vol. 1. Istanbul: IRCICA, 2008.

Inalcık, Halil. *An Economic and Social History of the Ottoman Empire*. Edited by Halil Inalcık and Donald Quataert. Cambridge: Cambridge University Press, 2000.

Inalcık, Halil. *The Ottoman Empire: Classical Age (1300–1600)*. London: Phoenix, 1994.

"Indian Rupee to Euro Currency Converter." http://www.likeforex.com/currency-converter/indian-rupee-inr_eur-euro.htm/1890 (accessed April 12, 2014).

Ingold, Tim. *Lines: A Brief History*. London: Routledge, 2007.

Ingold, Tim. *The Perception of the Environment: Essays on Livelihood, Dwelling and Skill*. London: Routledge, 2000.

International Dunhuang Project, The British Library. Last revised September 2016. http://www.idp.bl.uk.

Ishii Ryōsuke 石井良助 and Harafuji Hiroshi 服藤弘司, eds. *Bakumatsuofuregaki shusei* 幕末御触書集成 (Collections of the public announcements in the late [Edo] Shogunate period). Tokyo: Iwanami Shoten, 1992–1995.

Isnard, Christophe. *Memoires et instructions pour le plant des meuriers blancs, nourriture des vers a soye: et l'art de filer, mouliner, & aprester les soyes dans Paris & lieux cirvonvoisins*. Paris, 1665.

Izgi, Cevat. *Osmanlı Medreselerinde Ilim* (Science in Ottoman medreses). Vol. 2. Istanbul: Iz Yayıncılık, 1997.

Jacobson, J. J. L. L., Inspector of Tea Cultivation. *Handboek v.d. Kultuur en fabrikatie von Thee*. Batavia: ter Landsdrukkerij, 1843.

Jahn, Karl. *RASHĪD AL-DĪN's History of India: Collected Essays with Facsimiles and Indices*. The Hague: Mouton, 1965.

Janelle, Donald. "Global Interdependence and Its Consequences." In *Collapsing Space and Time: Geographic Aspects of Communications and Information*, edited by Stanley D. Brunn and Thomas Leinbach, 49–81. London: HarperCollins, 1991.

Jeffery, Roger. "Merchant Capital and the End of Empire: James Finlay, Merchant Adventurers." *Economic and Political Weekly* 17, no. 7 (1982): 241–48.

Johnson, Trevor. "Blood, Tears and Xavier-Water: Jesuit Missionaries and Popular Religion in the Eighteenth-Century Upper Palatinate." In *Popular Religion in Germany and Central Europe, 1400–1800*, edited by Bob Scribner and Trevor Johnson, 183–202. Basingstoke: Macmillan Education UK, 1996.

Joshi, Sanjay. "Review of *Empire's Garden: Assam and the Making of Modern India* by Javeeta Sharma." *Social History* 38, no. 3 (2013): 397–98.

Jung, Dinah. "The Cultural Biography of Agarwood: Perfumery in Eastern Asia and the Asian Neighbourhood." *Journal of the Royal Asiatic Society* 23, no. 1 (2013): 103–25.

Kaempfer, Engelbert. *Amoemitatum Exoticarum*. Lemgovia (Lippe, Germany): Tyous & Impensis Henrici Wilhelmi Meyeri Aulae Lippiacae Typographi, 1712.

Kagami Bunken 各務文献. *Mokotsu teian* 摸骨呈案 (Petition surrounding the model skeleton). Unpublished original manuscript, 1819, preserved at the Kyou shoya Library, Takeda Science Foundation.

Kagami Bunken 各務文献. *Seikotsu shinsho* 整骨新書 (New book for bone setting). Osaka: Shinsaibashidori denmacho, 1810.

Kahl, Oliver. *The Dispensatory of Ibn at-Tilmīd, Arabic Text, English Translation, Study and Glosseries*. Leiden: Brill, 2007.

Kahl, Oliver. *Sābūr ibn Sahl's Dispensatory in the Recension of the 'Adudī Hospital*. Leiden: Brill, 2009.

Kahl, Oliver. *The Sanskrit, Syriac and Persian Sources in the Comprehensive Book of Rhazes*. Leiden: Brill, 2015.

Kaibara Hiroshi 蒲原宏. "Ikei kyokotsuki to Kagami Bunken: Seikei gekashi ni okeru sono igi"異形頸骨記と各務文献：整形外科史に於けるその意義 ("Abnormal neck" and Kagami Bunken: Its meaning in the history of plastic surgery). *Journal of the Japanese Society for the History of Medicine* 16 (1970): 88–92.

Kaibara Hiroshi. *Nihon seikei geka zenshi* 日本整形外科前史 (Early history of plastic surgery in Japan). Osaka: Orientoru Shuppansha, 1984.

Kalinowski, Marc. "Mantic Texts in Their Cultural Context." In *Medieval Chinese Medicine: The Dunhuang Medical Manuscripts*, edited by Vivienne Lo and Christopher Cullen, 109–33. London: RoutledgeCurzon, 2005.

Kara, Mustafa. "Tekke." *TDV Islam Ansiklopedisi*. Istanbul: Türkiye Diyanet Vakfı, 2012.

Karamustafa, Ahmet T. *God's Unruly Friends: Dervish Groups in the Islamic Later Middle Period, 1200–1550*. Salt Lake City: University of Utah Press, 1994.

Karataş, Hasan. "The Ottomanization of the Halvetiye Sufi Order: A Political Story Revisited." *Journal of the Ottoman and Turkish Studies Association* 1 (2014): 71–89.

Karetzky, Patricia E. "Wu Zetian and Buddhist Art of the Tang Dynasty." *T'ang Studies* 20–21 (2002–2003): 113–50.

Kataoka Katsuko 片岡勝子. "Shinkangi (Hoshino mokotsu) no seisaku kattei ni kansuru kenkyū" 身幹儀（星野木骨）の制作過程に関する研究 (On the procedures for the manufacture of Hoshino's wooden skeleton). *Journal of the Japanese Society for the History of Medicine* 51 (2005): 188–89.

Kataoka Katsuko, Suzaki Etsuko, and Ajima Noriaki. "The Hoshino Wooden Skeleton: The First Wooden Model of a Human Skeleton, Made during the Edo Era in Japan." *Anatomical Science International* 82 (2007): 38–45.

Kauz, Ralph. "The Maritime Trade of Kish during the Mongol Period." In *Beyond the Legacy of Genghis Khan*, edited by Linda Komaroff, 51–67. Leiden: Brill, 2006.

Kazutaka, Unno. "Cartography in Japan." In *The History of Cartography*, edited by J. B. Harley and David Woodward, vol. 2, bk. 2. Chicago: University of Chicago Press, 1994.

Kieffer, Fanny. "The Laboratories of Art and Alchemy at the Uffizi Gallery in Renaissance Florence: Some Material Aspects." In *Laboratories of Art: Alchemy and*

Art Technology from Antiquity to the 18th Century, edited by Sven Dupré, 105–27. Archimedes, New Studies in the History and Philosophy of Science and Technology, 37. Cham, Switzerland: Springer, 2014.

Kieschnick, John. *The Impact of Buddhism on Chinese Material Culture*. Princeton, NJ: University of Princeton Press, 2003.

Kılıç, Orhan. "XVI. Yüzyılın İkinci Yarısında Osmanlı Devleti'nde Kalpazanlık Faaliyetleri." In *Osmanlı*, vol. 3, edited by Hasan Celal Güzel, Kemal Çiçek and Güler Eren, 180–87. Ankara: Yeni Türkiye Yayınları, 1999.

Kim Young-Oak. "The Philosophy of Wang Fu-chih (1619–1692)." PhD dissertation, Harvard University, Cambridge, MA, 1982.

King, Victor Terry. "Cursing, Special Death and Spirits in Embaloh Society." *Bijdragen tot de Taal-, Land- en Volkenkunde* 132 (1976): 124–45.

Knowles, L. C. A. *The Economic Development of the British Overseas Empire*. Vol. 1, *The British Tropics*. London: Routledge, 1924.

Koga Jūjirō 古賀十二郎. *Nagasaki Yōgakushi* 長崎洋学史 (History of Western learning in Nagasaki). Nagasaki: Nagasaki bunkensha, 1967.

Köhbah, Markus. "Vom Asketen zum Glaubenskämpfer: Geyiklü Baba, ein Betrag zur Osmanischen Hagiographie." *Journal of Ottoman Studies* 3 (1982): 45–51.

Kolta, Kemal Sabri. "Paracelsus im Urteil des osmanischen Oberhofartzes Salih b. Nasrullah b. Sallum." In *Die medizinischen Beziehungen zwischen Deutschland und der Türkei*. Verhandlungen des am 18. und 19. Oktober 1976 in Istanbul abgehaltenen Symposiums, edited by Heinz Goerke and Arslan Terzioğlu, 102–11. Munich: Münchener Vereinigung für Geschichte der Medizin e.V., 1978.

Kōshi Hōyoku 高志鳳翼. *Honedugi ryōji jūhōki* 骨継療治重宝記 (Treasured notes for healing bones, 1746). In Torii, *Nihon seikotsujutsu zensho*, 1:80–257.

Kou Zongshi 寇宗奭. *Bencao yanyi* 本草衍義 (Augmented materia medica). Annotated by Zhang Lijun and Ding Kan. Beijing: Zhongguo yiyao keji chubanshe, 2012.

Kuhn, Dieter, ed. *Textile Technology: Spinning and Reeling*. Vol. 5, part 9 of *Science and Civilisation in* China, edited by Joseph Needham. Cambridge: Cambridge University Press, 1988.

Kumamoto, Hiroshi. "Khotanese Official Documents in the Tenth Century." PhD dissertation, University of Pennsylvania, Philadelphia, 1982.

Kuzmina, Elena Efimovna. *The Prehistory of the Silk Road*. Philadelphia: University of Pennsylvania Press, 2007.

Larkin, Deirdre. "The Virtues of Rosemary." *The Medieval Garden Enclosed* (blog), The Metropolitan Museum of Art (February 10, 2012). http://blog.metmuseum.org/cloistersgardens/2012/02/10/the-virtues-of-rosemary (accessed December 7, 2016).

Latour, Bruno. *Reassembling the Social: An Introduction to Actor-Network Theory*. London: Oxford University Press, 2005.

Laufer, Berthold. "Asbestos and Salamander." *T'oung Pao*, 2nd ser., 16, no. 3 (1915): 299–373.

Laufer, Berthold. *The Diamond: A Study in Chinese and Hellenistic Folk-Lore*. Publication 184, Field Museum of Natural History, Anthropological Series. Vol. 15, no. 1 (1915).

Laufer, Berthold. "Loan-Words in Tibetan." *T'oung Pao* 17 (1916): 404–552.

Laufer, Berthold. *Sino-Iranica: Chinese Contribution to the History of Civilization in Ancient Iran*, vol. 15, no. 3. Chicago: Field Museum of Natural History Publication no. 201, 1919.

Laufer, Berthold. "The Story of the Pinna and the Syrian Lamb." *Journal of American Folk-Lore* 28 (1915): 103–28.

Laufer, Berthold. "Three Tokharian Bagatelles." *T'oung Pao* 16, no. 1 (1915): 272–88.

Leach, Maria, and Jerome Fried, eds. *The Standard Dictionary of Folklore, Mythology, and Legend*. New York: Funk and Wagnalls, 1950.

Leclerc, Lucien. *Traité des simples*. Translated from Al Baytar's pharmacopeia. Paris: Institut du monde arabe, 1912.

Le Gall, Dina. *A Culture of Sufism: Naqshbandis in the Ottoman World, 1450–1700*. Albany: State University of New York Press, 2005.

Lemonnier, Pierre, ed. *Technological Choices: Transformation in Material Cultures since the Neolithic*. London: Routledge, 1993.

Leong, Elaine. "Making Medicines in the Early Modern Household." *Bulletin of the History of Medicine* 82 (2008): 145–68.

Leube, Georg. *Die Rezepte der Freiburger alchemistischen Handschrift des Abd al-Gabbar al-Hamadani, Edition, Übersetzung und Commentar*. Islamkundliche Untersuchungen Band 315. Berlin: Klaus Schwarz Verlag, 2013.

Leung, Angela K. C. "The Evolution of the Idea of Chuanran Contagion." In *Health and Hygiene in Chinese East Asia*, edited by Angela K. C. Leung and Charlotte Furth, 25–50. Durham: Duke University Press, 2010.

Leung, Angela K. C. *Leprosy in China: A History*. New York: Columbia University Press, 2009.

Lev, Efraim, and Leigh Chipman, eds. *Medical Prescriptions in the Cambridge Genizah Collections: Practical Medicine and Pharmacology in Medieval Egypt*. Leiden: Brill, 2012.

Lev, Efraim, and Zohar Amar. *Practical Materia Medica of the Medieval Eastern Mediterranean according to the Cairo Genizah*. Leiden: Brill, 2008.

Li, Jian-min. "They Shall Expel Demons: Etiology, the Medical Canon and the Transformation of Medical Techniques before the Tang." In *Early Chinese Religion. Part 1: Shang through Han (1250BC–220AD)*, edited by John Lagerwey and Marc Kalinowski, 1103–50. Leiden: Brill, 2009.

Lidai minghua ji 歷代名畫記 (Records of famous painters through the ages). Compiled by Zhang Yanyuan 張彥遠 (815–?). Congshu jicheng ed. Taibei: Shangwu yinshuguan, 1986.

Lifchez, Raymond, ed. *The Dervish Lodge: Architecture, Art, and Sufism in Ottoman Turkey*. Berkeley: University of California Press, 1992.

Ligeti, Louis. "À propos du 'Rapport sur les rois demeurant dans le nord'" (About the "Report on kings living in the North"). In Études tibétaines dédiées à la mémoire de Marcelle Lalou (Tibetan studies dedicated to the memory of Marcelle Lalou), edited by Ariane Macdonald, 166–89. Paris: Adrien Maisonneuve, 1971.

Lin, Shen-Yu. "The Tibetan Image of Confucius." *Revue d'Études Tibétaines* 12 (2007): 105–29.

Lin Tianwei 林天蔚. *Songdai xiang yao maoyi shigao* 宋代香藥貿易史稿 (Preliminary history of the spice trade in the Song). Hong Kong: Zhongguo xueshe, 1960.

Li Shizhen. *Bencao gangmu* (Systematic materia medica). Beijing: Renmin weisheng chubanshe, 1975.

Li Shizhen 李時珍. *Bencao gangmu* 本草綱目 (Systematic materia medica). Taibei: Wenguang tushu, 1982. Based on the 1885 edition.

Liu, Andrew B. "The Birth of a Noble Tea Country: On the Geography of Colonial Capital and the Origins of Indian Tea." *Journal of Historical Sociology* 23, no. 1 (2010): 73–100.

Liu Wentai 劉文泰 et al. *Bencao pinhui jingyao*本草品彙精要 (Collection of categorized essential materia medica). 1505.

Liu Xinru. *Ancient India and Ancient China: Trade and Religious Exchanges, a.d. 1–600.* New Delhi: Oxford University Press, 1988.

Liu Xinru. *The Silk Road: Overland Trade and Cultural Interactions in Eurasia*. Washington, DC: American Historical Association, 1998.

Liu Xinru. *The Silk Road in World History*. New York: Oxford University Press, 2010.

Liu Xinru. "Silks and Religions in Eurasia, c. AD 600–1200." *Journal of World History* 6, no. 1 (Spring 1995): 25–48.

Li Yingcun 李應存, Li Jintian 李金田, and Shi Zhenggang 史正剛. "E cang Dunhuang wenxianДх. 02822 'Mengxue zi shu' zhong zhi yiyao zhishi" 俄藏敦煌文獻 Дх. 02822 <蒙學字書>中之醫藥知識 (Medical knowledge as shown by the primer Дх. 02822 of the Dunhuang collection in Russia). *Gansu zhongyi xueyuan xuebao* 4 (2006): 38–42.

Li Yukun 李玉昆. "Song-Yuan shiqi Quanzhou di xiangliao maoyi" 宋元時期泉州的香料貿易 (Spice trade in Quanzhou of the Song and Yuan periods). *Haijiao shi yanjiu* 1 (1998): 58–67.

Lloyd, Geoffrey, and Nathan Sivin. *The Way and the Word: Science and Medicine in Early China and Greece*. New Haven, CT: Yale University Press, 2002.

Lo, Vivienne. "Manuscrits de Dunhuang et de Khotan sur la moxibustion." In Despeux, *Médecine, religion et société*, 1:239–84.

Lo, Vivienne, and Christopher Cullen, eds. *Medieval Chinese Medicine: The Dunhuang Medical Manuscripts*. London: RoutledgeCurzon, 2005.

Lo, Vivienne, and Li Jianmin. "Manuscripts, Received Texts and the Healing Arts." In *China's Early Empires: A Re-appraisal*, edited by Michael Nylan and Michael Loewe, 367–97. Cambridge: Cambridge University Press, 2010.

Lo, Vivienne, and Ronit Yoeli-Tlalim. "Travelling Light: Sino-Tibetan Moxa-Cautery from Dunhuang." In *Imagining Chinese Medicine*, edited by Vivienne Lo and Penelope Barrett, 271–90. Leiden: Brill, 2018.

Lockwood, Howard, ed. "Perfumes and the Perfumer's Art." *American Stationer* 23 (1888): 859. https://babel.hathitrust.org/cgi/pt?id=nyp.33433090917224;view=1up;seq=857 (accessed December 12, 2016).

Lourdusamy, John Bosco. "Tea and Technology: The Case of Colonial India." Paper presented at the Center for the Advanced Study of India, University of Pennsylvania, Philadelphia, October 30, 2014. https://casi.sas.upenn.edu/events/lourdusamy (accessed May 17, 2016).

Lu Gwei-Djen and Nathan Sivin, ed. *Medicine*. Vol. 6, part 6 of *Science and Civilisation in China,* edited by Joseph Needham. Cambridge: Cambridge University Press, 2000.

Luo Bingfen, et al., eds. *Tun hong nas thon pa'i bod kyi gso rig yig cha gces bsdus* (A selection of Tibetan medical texts from Dunhuang). Beijing: Mi rigs dpe skrun khang, 2002.

Lutgendorf, Philip. "Making Tea in India: Chai, Capitalism, Culture." *Thesis Eleven* 113, no. 1 (2012): 11–31.

Lu Weijing. "Beyond the Paradigm: Tea-Picking Women in Imperial China." *Journal of Women's History* 15, no. 4 (2004): 19–46.

Lu Yinglong. *Xian chuan kuo yi zhi* (Leisure collection of marvels). Repr., Beijing: Zhonghua shuju, 1985.

Lu Yitian 陸以湉. *Lenglu yihua*冷廬醫話 (Medical writings in the cold study). 1858. Taiyuan: Shanxi kexue jishu chubanshe, 1993.

Ma, Debin. "The Great Silk Exchange: How the World Was Connected and Developed." In *Textiles in the Pacific, 1500–1900*, edited by Debin Ma, 1–32. New York: Routledge, 2005.

Macdonald, Ariane. "Une lecture des Pelliot tibétain 1286, 1287, 1038, 1047, et 1290: essai sur la formation et l'emploi des mythes politiques dans la religion royale de Sroṅ-bcan Sgam-Po" (A reading of Pelliot tibétain 1286, 1287, 1038, 1047, and 1290: An essay on the formation and use of political myths in the royal religion of Sroṅ-bcan Sgam-Po). In Études tibétaines dédiées à la mémoire de Marcelle Lalou (Tibetan studies dedicated to the memory of Marcelle Lalou), edited by Ariane Macdonald, 190–391. Paris: Adrien Maisonneuve, 1971.

Macfarlane, Alan, and Iris Macfarlane. *Green Gold: The Empire of Tea*. London: Ebury Press, 2004.

Mackillop, Andrew. "A North Europe World of Tea: Scotland and the Tea Trade, c. 1690–c. 1790." In *Goods from the East*, edited by Maxine Berg with Felicia Gottmann, Hanna Hodacs, and Chris Nierstrasz, 294–308. London: Palgrave Macmillan, 2015.

Maggi, Mauro. "Khotanese Literature." In *The Literature of Pre-Islamic Iran: Companion Volume 1*, edited by Ronald E. Emmerick and Maria Macuch (vol. 17 of *A History of Persian Literature*, edited by Ehsan Yarshater), 330–417. London: I. B. Tauris, 2009.

Mahdi, Waruno. *Malay Words and Malay Things: Lexical Souvenirs from an Exotic Archipelago in German Publications before 1700*. Wiesbaden: Harrassowitz Verlag, 2007.

Mair, Victor. "Kinesis versus Stasis, Interaction versus Independent Invention." In *Contact and Exchange in the Ancient World*, edited by Victor Mair, 1–16. Honolulu: University of Hawai'i Press, 2006.

Mäkinen, Martti. "Henry Daniel's Rosemary in MS X.90 of the Royal Library, Stockholm." *Neuphilologische Mitteilungen* 103 (2002): 305–27.

Manderson, Lenore. "Roasting, Smoking and Dieting in Response to Birth: Malay Confinement in Cross-Cultural Perspective." *Social Science and Medicine. Part B, Medical Anthropology* 15, no. 4 (1981): 509–20.

Manderson, Lenore. "Shaping Reproduction: Maternity in Early Twentieth-Century Malaya." In *Maternities and Modernities: Colonial and Postcolonial Experiences in Asia and the Pacific*, edited by Kalpana Ram and Margaret Jolly, 26–49. Cambridge: Cambridge University Press, 1998.

Manderson, Lenore. "Traditional Food Classifications and Medical Theory in Peninsular Malaysia." *Ecology of Food and Nutrition* 11, no. 2 (1981): 81–92.

"Maqam." *Merriam-Webster's Encyclopedia of World Religions*. Springfield, MA: Merriam-Webster, 1999.

Martin, Dan. "An Early Tibetan History of Indian Medicine." In *Soundings in Tibetan Medicine: Anthropological and Historical Perspectives*, edited by Mona Scrempf, 307–25. Leiden: Brill 2007.

Martin, Dan. "Greek and Islamic Medicines' Historical Contact with Tibet: A Reassessment in View of Recently Available but Relatively Early Sources on Tibetan Medical Eclecticism." In *Islam and Tibet: Interactions along the Musk Routes*, edited by Anna Akasoy, Charles Burnett, and Ronit Yoeli-Tlalim, 117–43. Farnham: Ashgate, 2011.

Matthee, Rudi. "From Coffee to Tea: Shifting Patterns of Consumption in Qajar Iran." *Journal of World History* 7, no. 2 (1996): 199–230.

Mattioli, Pietro. *Commentaires de M. Pierre André Matthiole medecin senois, sur les six livres de Pedacius Dioscoride Anazarbeen*. Lyon, 1579.

Matus, Zachary A. *Franciscans and the Elixir of Life: Religion and Science in the Later Middle Ages*. Philadelphia: University of Pennsylvania Press, 2017.

Mau Chuan-Hui. "Les progrès de la sériciculture sous les Yuan (13e–14e siècles) d'après le *Nongsang jiyao* 農桑輯要." *Revue de Synthèse* 131 (2010): 193–217.

Maurocordatos, Alexandro. *Pneumaticum Instrumentum Circulandi Sanguinis sive De Motu et Usu Pulmonum* (Bononiæ (Bologna), 1664). Edited by M. P. Vretos. Lipsiae: F. A. Brockhaus, 1870.

Mayanagi Makoto 真柳誠. "Riben Jianghu shiqi chuanru di Zhongguo yishu ji qi Heke" 日本江户时期传入的中国医书及其和刻 (The Chinese medical books imported to Japan in the Edo period, and their Japanese editions). *China Historical Materials Science and Technology* 23, no. 3 (2002): 232–54.

McClive, Cathy. *Menstruation and Procreation in Early Modern France*. London: Routledge, 2016.

McCormick, Melissa. *Tosa Mitsunobu and the Small Scroll in Medieval Japan*. Seattle: University of Washington Press, 2009.

McDermott, Joseph P. "'Noncommercial' Private Publishing in Late Imperial China." In *The Book Worlds of East Asia and Europe*, edited by Joseph McDermott and Peter Burke, 105–45. Hong Kong: Hong Kong University Press, 2015.

McEwan, John, FGRS, FR Met. Soc. (J. McE). "Tea." In *Encyclopaedia Britannica*, 476–83. 11th ed. New York: Encyclopaedia Britannica Company, 1911.

McGowan, Alexander Thorburn. *Tea Planting in the Outer Himalayah*. London: Smith, Elder, 1861.

McKeown, Arthur Philip. "From Bodhgayā to Lhasa to Beijing: The Life and Times of Śāriputra (c. 1335–1426), Last Abbot of Bodhgayā." Unpublished PhD dissertation, Harvard University, Cambridge, MA, 2010.

McNeill, William. "Foreword." In *The World System: Five Hundred Years or Five Thou-*

sand Years? edited by A. Gunder Frank and B. K. Gills, vii–xiii. London: Routledge, 1993.

Mehmed Tahir, Bursalı. *Osmanlı Müellifleri*. Vol. 1, translated by A. Fikri Yavuz and Ismail Özen. Istanbul: Meral Yayınevi, 1972.

Mehmed Tahir, Bursalı. *Osmanlı Müellifleri*. Vol. 3, translated by Ismail Özen. Istanbul: Meral Yayınevi, 1975.

Melikian-Chirvani, Assadullah Souren. "Iran to Tibet." In *Islam and Tibet: Interactions along the Musk Routes*, edited by Anna Akasoy, Charles Burnett, and Ronit Yoeli-Tlalim, 80–115. Farnham: Ashgate, 2011.

Merçil, Erdoğan. *Türkiye Selçuklularında Meslekler*. Ankara: Türk Tarih Kurumu, 2000.

Meyer, Frederick G., Emily Emmart Trueblood, and John L. Heller, eds. *The Great Herbal of Leonhart Fuchs: De historia stirpium commentarii insignes, 1542*. Stanford, CA: Stanford University Press, 1999.

Meyerhof, Max. "'Ali aṭ-Ṭabarī's 'Paradise of Wisdom,' One of the Oldest Arabic Compendiums of Medicine." *Isis* 16 (1931): 6–54. Reprinted in *'Alī Ibn Rabban al-Ṭabarī (d.c. 250/864): Texts and Studies*, collected and reprinted by Fuat Segzin, *Islamic Medicine*, 30: 90–138. Frankfurt am Main: Institute for the History of Arabic-Islamic Science at the Johann Wolfgang Goethe University, 1996.

Miao Xiyong 繆希雍. *Shennong bencao jing shu*神農本草經疏 (Exegesis of the Divine Farmer's classic of materia medica). In *Yixue quanji* (Complete works), 1–420. Beijing: Xueyuan chubanshe, 2000.

Miao Xiyong 繆希雍. *Xianxing zhai yixue guang biji*先醒齋醫學廣筆記 (Medical notes in the Xianxing study). In *Yixue quanji* (Complete works), 687–787. 1642. Bejing: Xueyuan chubanshe, 2000.

Michel, Wolfgang. "Looking at Corpses: Negoro Toshoku's 'True Shape of Human Bones' and Its Place in Japanese Medical History." In *Source Materials and Personalities*, vol. 4, edited by Wolfgang Michel, Yōichi Yoshida, and Akihide Yōshima, 42–89. Nakatsu: Board of Education, 2012.

Miksic, John. "Introduction." In *Ancient Silk Trade Routes: Selected Works from Symposium on Cross Cultural Exchanges and Their Legacies in Asia*, edited by Dashu Qin and Jian Yuan, 1–17. Singapore, SGP: World Scientific Publishing Company, 2015.

Mills, J. V. G. *Hsing-ch'a sheng-lan: "The Overall Survey of the Star of the Raft by Fei Hsin."* Translated and revised by Roderich Ptak. Wiesbaden: Harrassowitz, 1996.

Millward, James. *Eurasian Crossroads: A History of Xinjiang*. New York: Columbia University Press, 2007.

Millward, James. *The Silk Road: A Very Short Introduction*. New York: Oxford University Press, 2013.

Millward, James, Ruth Dunnell, Mark Elliott, and Philippe Forêt, eds. *New Qing Imperial History: The Making of Inner Asian Empire at Qing Chengde*. London: RoutledgeCurzon, 2004.

Millward, James, Shinmen Yasushi, and Sugawara Jun, eds. *Studies on Xinjiang Historical Sources in the 17th–20th Centuries*. Tokyo: Toyo Bunko Research Library, 2010.

Money, E. *Tea—Cultivation and Manufacture. Fourth edition, revised and supplemented by additional chapters*. London, W. B. Whittingham, 1883.

Morinaga, Maki Isaka. *Secrecy in Japanese Art: "Secret Transmission" as a Mode of Knowledge*. New York: Palgrave Macmillan, 2005.

Morris, Melissa. "Cultivating Colonies: Tobacco and the Upstart Empires, 1580–1640." PhD dissertation, Columbia University, New York, 2017.

Mote, Frederick. "Yuan and Ming." In *Food in Chinese Culture*, edited by Kwang-chih Chang, 193–258. New Haven, CT: Yale University Press, 1977.

Motherby, G. *A new medical dictionary, or, General repository of physic: containing an explanation of the terms and a description of various particulars relating to anatomy, physiology, physic, surgery, materia medica, pharmacy . . . : each article, according to its importance, being considered in every relation to which its usefulness extends in the healing art*. London: Printed for J. Johnson, 1775.

Mukerji, Chandra. *Impossible Engineering: Technology and Territoriality on the Canal du Midi*. Princeton, NJ: Princeton University Press, 2009.

Mukerji, Chandra. "Tacit Knowledge and Classical Technique in Seventeenth-Century France: Hydraulic Cement as a Living Practice among Masons and Military Engineers." *Technology and Culture* 47, no. 4 (2006): 713–33.

Nakano Misao 中野操. *Osaka meiiden* 大阪名醫傳 (Biographies of famous physicians in Osaka). Kyoto: Shibunkaku, 1983.

Nakayama Yoku 中山沃. "Hoshinno Ryōetsu no kaishi" 星野良悦解屍 (Hoshino Ryōetsu's dissections). *Journal of the Japanese Society for the History of Medicine* 18 (1972): 180–81.

Nakura Yumio 名倉弓雄. *Edo no honeduki* 江戸の骨つぎ (Bone setting in Edo). Tokyo: Mainichi shimbunsha, 1974.

Nappi, Carla. *The Monkey and the Inkpot: Natural History and Its Transformations in Early Modern China*. Cambridge MA: Harvard University Press, 2009.

Naquin, Susan. "The Material Manifestations of Regional Culture." Unpublished paper, 2016.

Nasr, Seyyed Hossein, and Mehdi Mohaghegh, eds. *Al-As'ilah wa'l-Ajwibah (Questions and Answers)*. Kuala Lumpur: International Institute of Islamic Thought and Civilisation, 1995.

Neelis, Jason. *Early Buddhist Transmission and Trade Networks: Mobility and Exchange within and beyond the Northwestern Borderlands of South Asia*. Leiden: Brill, 2010.

Nguyễn, Đông Phong, and Tấn Hưng. *Monts et merveilles au Pays du Bois d'Aigle: Cultures migratoires de la province vietnamienne de Khán Hòa* (Mountains and wonders in the Land of Eagle Wood: Migratory cultures of the Vietnamese province of Khán Hòa). Paris: Éditions Publibook, 2009.

Niehof, A. "Traditional Medication at Pregnancy and Childbirth in Madura, Indonesia." In *The Context of Medicines in Developing Countries: Studies in Pharmaceutica Anthropology*, edited by Sjaak van der Geest and Susan Reynolds Whyte, 235–52. Amsterdam: Het Spinhuis, 1991.

Nie Hongyin 聶鴻音. "Xixia 'Tiansheng lü ling' li di Zhong yao ming" 西夏"天盛律令"裡的中藥名 (Chinese medicinal names in the "Tiansheng law code" of the Tangut Kingdom). *Zhonghua wenshi luncong* 4 (2009): 291–312.

Nierstraz, Chris. "The Popularization of Tea: East India Companies, Private Traders, Smugglers and the Consumption of Tea in Western Europe, 1700–1760." In *Goods from the East*, edited by Maxine Berg et al., 263–77. London: Palgrave Macmillan, 2015.

Nihon gakushiin 日本学士院 (The Japan Academy), ed. *Meijizen Nihon igakushi* 明治前日本医学史 (The Pre-Meiji medical history of Japan). Tokyo: Nihon koigaku shiryō senta, 1978.

Ning Qiang. *Art, Religion and Politics in Medieval China: The Dunhuang Cave of the Zhai Family*. Honolulu: University of Hawai'i Press, 2004.

Nitin Varma, M. A. "Producing Tea Coolies? Work, Life and Protest in the Colonial Tea Plantations of Assam, 1830s–1920s." PhD dissertation, Philosophy Faculty III, Humboldt University, Berlin, 2011.

Noncelli, Carmen. "The Erotics of Mercantile Imperialism: Cross-Cultural Requitedness in the Early Modern Period." *Journal for Early Modern Cultural Studies* 8, no. 1 (2008): 134–52.

Norton, Marcia. *Sacred Gifts, Profane Pleasures: A History of Tobacco and Chocolate in the Atlantic World*. Ithaca: Cornell University Press, 2008.

Norton, Marcia. "Tasting Empire: Chocolate and the European Internalization of Mesoamerican Aesthetics." *American Historical Review* 111 (June 2006): 660–91.

Numata Jirō. *Western Learning*. Tokyo: Japan-Netherlands Institute, 1992.

Okajima Akihiro. "Kanji bunka ken to wa" (Concerning the "cultural zone of Chinese characters"). Unpublished paper, 2004. http://uwazura.up.seesaa.net/image/89AA938791E498p.pdf (accessed January 27, 2014).

Okanishi Tameto 岡西為人. *Chongji xinxiu bencao* 重輯新修本草 (Re-edition of the new compilation of materia medica by Su Jing 蘇敬). Taibei: Guoli Zhongguo yiyao yanjiusuo, 1959.

"The Old Campaigner." *History of the High Range Planting District, North Travancore.* Bangalore: Scripture Literature Press, 1930.

Ong, Aihwa, and Stephen J. Collier, eds. *Global Assemblages: Technology, Politics, and Ethics as Anthropological Problems.* Oxford: Blackwell, 2005.

Öngören, Reşat. *Tarihte Bir Aydın Tarikatı: Zeyniler.* Istanbul: Insan Yayınları, 2003.

Orta, Garcia da. *Colóquios dos simples e drogas da India* (Colloquies on the simples and drugs of India). Edited by F. M. C. de Mello Ficalho, 2 vols. Lisbon: Imprensa Nacional, 1891–1895.

Orta, Garcia da. *Colóquios dos simples e drogas e cousas medicinais da India* (*Colloquies on the Simples and Drugs of India* [1563]). Translated by Clements Markham. London: Henry Sotheran, 1913.

Ortelius, Abraham. *Theatrum Orbis Terrarum.* Antwerp, 1570.

Ortiz Gómez, Teresa, and Montserrat Cabré I Pairet. "Mujeres y salud: práticas y saberes. Presentación." *Dynamis* 19 (1999): 17–24.

Osbeck, Peter. *A Voyage to China and the East Indies.* Translated by John Forster. Vol. 1. London: Benjamin White, 1771.

Oswald, Adrian. *Clay Pipes for the Archaeologist.* Oxford: British Archaeological Reports, 1975.

Otsuki Bansui 大規磐水 and Otsuki Banri 大規磐里. *Ran'en tekihō* 蘭畹摘芳 (Extracts of Dutch botanical learning). Edo (Tokyo): Suharaya Mohee, 1815.

Ōtsuki Gentaku 大槻玄沢. *Jūtei Kaitai shinsho* 重訂解体新書 (Revised new book of anatomy). Edo (Tokyo): Senshōbō, 1826.

Ōtsuki Gentaku. "Shinkangi setsu" 身幹儀說 (On the apparatus of the body's mainstay). In *Hiroshimaken shi* 廣島縣史 (Histories of the Hiroshima Prefecture), edited by Hiroshima Prefecture. Vols. 6–12. Hiroshima: Hiroshima Prefecture, 1976.

Özbaran, Salih. "The Ottoman Turks and the Portuguese in the Persian Gulf, 1534–1581." *Journal of Asian History* 6, no. 1 (1972): 45–87.

Özcan, Tahsin. "Molla Arap." *TDV Islam Ansiklopedisi.* Istanbul: Türkiye Diyanet Vakfı, 2005.

Palmer, Richard. "In Bad Odour: Smell and Its Significance in Medicine from Antiquity to the Seventeenth Century." In *Medicine and the Five Senses*, edited by W. F. Bynum and Roy Porter, 61–68. Cambridge: Cambridge University Press, 1993.

Pamuk, Şevket. *A Monetary History of the Ottoman Empire.* Cambridge: Cambridge University Press, 2000.

Parker, G. *The Making of Roman India*. Cambridge: Cambridge University Press, 2008.

Peacock, Andrew Charles Spencer, and Sara Nur Yıldız, eds. *Islamic Literature and Intellectual Life in Fourteenth- and Fifteenth-Century Anatolia*. Istanbuler Texte und Studien Herausgegeben vom Orient-Institut Istanbul, Band 34. Würzburg: Ergon Verlag 2016.

Peletz, Michael G. "Poisoning, Sorcery and Healing Rituals in Negeri Sembilan." *Bijdragen Tot De Taal-, Land- En volkenkunde* 144 (1988): 132–64.

Pelliot, Paul. "Artistes des Six Dynasties et des T'ang." *T'oung Pao* 22 (1923): 215–91.

Perdue, Peter. *China Marches West: The Qing Conquest of Central Eurasia*. Cambridge, MA: Belknap Press, 2005.

Pereira, Jonathan. *Elements of Materia Medica and Therapeutics*. Vol. 2. London: Longman, Brown, Green, and Longmans, 1842.

Pettegree, Andrew, Malcolm Walsby, and Alexander Wilkinson, eds. *French Vernacular Books: Books Published in the French Language before 1601*. Leiden: Brill, 2007.

Pinch, Trevor, and Wiebe Bijker. "The Social Construction of Facts and Artefacts: Or How the Sociology of Science and the Sociology of Technology Might Benefit Each Other." *Social Studies of Science* 14, no. 3 (1984): 399–441.

Piontek-Ma, Elke. *Der Bericht von Sun Pei über die kaiserliche Seidenmanufactur von Suzhou im 17. Jahrhundert*. Heidelberg: Edition Forum, 1999.

Piper, Alan. "The Milk of the Goat Heidrun: An Investigation into the Sacramental Use of Psychoactive Milk and Meat." In *Entheogens and the Development of Culture: The Anthropology and Neurobiology of Ecstatic Experience*, edited by John Rush, 211–78. Berkeley, CA: North Atlantic Press, 2013.

Pliny the Elder (Secundus, Gaius Plinius). *The Natural History of Pliny*. Vol. 3. Translated and edited by John Bostock and H. T. Riley. London: Henry G. Bohn, 1855.

Poliquin, Rachel. "Vegetal Prejudice and Healing Territories." In *Textual Healing: Essays on Medieval and Early Modern Medicine*, edited by Elizabeth Lane Furdell, 169–93. Leiden: Brill, 2005.

Pollock, Sheldon. *The Language of the Gods in the World of Men: Sanskrit, Culture, and Power in Premodern India*. Berkeley: University of California Press, 2006.

Porter, David. *Ideographia: The Chinese Cipher in Early Modern Europe*. Stanford, CA: Stanford University Press, 2001.

Porter, Roy. *The Greatest Benefit to Mankind: A Medical History of Humanity*. New York: W. W. Norton, 1997.

Potts, Daniel T. "Technological Transfer and Innovation in Ancient Eurasia." In *The Globalization of Knowledge in History*, edited by Jürgen Renn, 105–20. Berlin: Edition Open Access, 2012.

Prazniak, Roxann. "Siena on the Silk Roads: Ambrogio Lorenzetti and the Mongol Global Century, 1250–1350." *Journal of World History*, 21, no. 2 (June 2010): 177–217.

Purš, Ivo, and Vladimír Karpenko, eds. *Alchemy and Rudolf II: Exploring the Secrets of Nature in Central Europe in the 16th and 17th Centuries*. Prague: Artefactum, 2016.

Qian Chaoding. *Shuikeng shiji*. In *Duanxi yankao jicheng*, compiled by Guangling guji keyinshe. Nanjing: Jiangsu guji, 1999.

Qin Dashu. "Archaeological Investigations of Chinese Ceramics Excavated from Kenya." In *Ancient Silk Trade Routes: Selected Works from Symposium on Cross Cultural Exchanges and Their Legacies in Asia*, edited by Dashu Qin and Jian Yuan, 87–109. Singapore, SGP: World Scientific Publishing, 2015.

Raby, Julian. "A Seventeenth-Century Description of Iznik-Nicea." *Istanbuler Mitteilungen* 26 (1976): 149–88.

Raj, Kapil. *Relocating Modern Science: Circulation and the Construction of Knowledge in South Asia and Europe, 1650–1900*. Basingstoke: Palgrave Macmillan, 2007.

Rambelli, Fabio. *Buddhist Materiality: A Cultural History of Objects in Japanese Buddhism*. Stanford: Stanford University Press, 2007.

Rambelli, Fabio. "The Idea of India (*Tenjiku*) in Pre-modern Japan: Issues of Signification and Representation in the Buddhist Translation of Cultures." In Sen, *Buddhism across Asia*, 259–90.

Ray, Himanshu Prabha. *Monastery and Guild: Commerce under the Sātavāhanas*. New Delhi: Oxford University Press, 1986.

Ray, Himanshu Prabha. *The Return of the Buddha: Ancient Symbols for a New Nation*. New Delhi: Routledge, 2014.

Ray, Himanshu Prabha. *The Winds of Change: Buddhism and the Maritime Links of Early South Asia*. New Delhi: Oxford University Press, 1994.

Reed, Annette Yoshiko. "'Ancient Jewish Sciences' and the Historiography of Judaism." In *Ancient Jewish Sciences*, edited by Jonathan Ben Dov and Seth Sanders, 195–254. New York: New York University Press, 2014.

Rees, Abraham. "Galls." *The Cyclopædia; or, Universal Dictionary of Arts, Sciences, and Literature*, vol. 15. London: n.p., 1819.

Rego, António da Silva, ed. "Do Principio do Reyno de Ormus." In *Documentaçao Ultramarina Portuguesa*, edited by António da Silva Rego, 2:79–95. Lisbon: Centro de Estudos Históricos Ultramarinos, 1960.

Reid, Anthony. "Elephants and Water in the Feasting of Seventeenth-Century Aceh." *Journal of the Malaysian Branch of the Royal Asiatic Society* 62, no. 2 (1989): 25–44.

Reyes, Raquel A. G. "Glimpsing Southeast Asian Naturalia in Global Trade, c. 300 BCE–1600 CE." In *Environment, Trade and Society in Southeast Asia: A Longue*

Durée Perspective, edited by David Henley and Henk Schulte Nordholt, 96–119. Leiden, Netherlands: Brill, 2015.

Rezakhani, Khodadad. "The Road That Never Was: The Silk Road and Trans-Eurasian Exchange." *Comparative Studies of South Asia, Africa and the Middle East* 30, no. 3 (2010): 420–33.

Rhi, Juhyung. "Reading Coomaraswamy on the Origin of the Buddha Image." *Artibus Asiae* 70, no. 1 (2010): 151–72.

Rhie, Marylin M. *Early Buddhist Art of China and Central Asia, Volume 1: Later Han, Three Kingdoms, and Western Chin in China and Bactria to Shan-shan in Central Asia*. Leiden: E. J. Brill, 1999.

Richthofen, Ferdinand von. "Über die zentralasiatischen Seidenstrassen bis zum 2.Jh n. Chr" (On the Central Asian silk roads until the 2nd century A.D.). *Verhandlungen der Gesellschaft für Erdkunde zu Berlin* 4 (1877): 96–122.

Riello, Giorgio, and Tirthankar Roy, eds. *How India Clothed the World: The World of South Asian Textiles, 1500–1850*. Leiden: Brill, 2009.

Robertson, A. S. *A History of Tea Planting in South India*. Unpublished MS with illustrations in author's family collection. 2 volumes. Compiled 1976.

Rodgers, Daniel T. "Cultures in Motion: An Introduction." In *Cultures in Motion*, edited by Daniel T. Rodgers, Bhavani Raman, and Helmut Reimitz. Princeton, NJ: Princeton University Press, 2014.

Rong Xinjiang. "Dunhuang Studies and the History of Science and Technology." In *Eighteen Lectures on Dunhuang*, translated by Imre Galambos, 413–26. Leiden: Brill, 2013.

Rong Xinjiang. "New Evidence on the History of Sino-Arabic Relations: A Study of Yang Liangyao's Embassy to the Abbasid Caliphate." In *Imperial China and Its Southern Neighbors*, edited by Victor H. Mair and Liam Kelly, 239–67. Singapore: Institute of Southeast Asian Studies, 2015.

Rong Xinjiang. "Official Life at Dunhuang in the Tenth Century: The Case of Cao Yuanzhong." In *The Silk Road: Trade, Travel, War and Faith*, edited by Susan Whitfield, 57–62. London: British Library, 2004.

Rossabi, Morris. "A Translation of Ch'en Ch'eng's Hsi-Yü Fan-Kuo Chih." *Ming Studies* 17 (1983): 49–59.

Rothschild, N. Harry. *Emperor Wu Zhao and Her Pantheon of Devis, Divinities, and Dynastic Mothers*. New York: Columbia University Press, 2015.

Rowe, William T. *Saving the World: Chen Hongmou and Elite Consciousness in Eighteenth-Century China*. Stanford, CA: Stanford University Press, 2001.

Royle, John Forbes. *A Manual of Materia Medica and Therapeutics*. London: John Churchill, 1876.

Royle, John Forbes. *Xiyao dacheng* 西藥大成 (Compendium of Western pharmacopeia) (*Xiyao dacheng yaopin Zhong Xi mingmu biao* 西藥大成藥品中西名目表 (A manual of materia medica and therapeutics)). 1876. Translated by Fu Lanya 傅蘭雅 (John Fryer) and Zhao Yuanyi 趙元益. Shanghai: Jiangnan Arsenal Translation Department, 1887.

Ruska, Julius. *Al-Razi's Buch Geheimnis der Geheimnisse mit Einleitung und Erläuterungen in deutscher Übersetzung*. Quellen und Studien zur Geschichte der Naturwissenschaften und der Medizin, Band 6. Berlin: Verlag von Julius Springer, 1937.

Sachau, Edward C. *Alberuni's India*. 2 vols. London: Trübner, 1888. Facsimile reprint, Elibron Classics, Adamant Media, 2005.

Sakai Shizu 酒井シヅ. *Kaitai shinsho* to *Jūtei Kaitai shinsho* 『解体新書』と『重訂解体新書 (New book of anatomy and revised new book of anatomy). In Ōtsuki Gentaku no kenkyū 大槻玄沢の研究 (Studies on Ōtsuki Gentaku), edited by Yōgakushi Kenkyūkai 洋学史研究会, 97–158. Kyoto: Shibunkaku, 1990.

Sakai Shizu, trans. *Shinsōhan Kaitai shinsho: zen gendaigo goyaku* 新装版解体新書：全現代語訳 (New book of anatomy in a new binding: Completely translated into modern language). Tokyo: Kōdansha, 1998.

Sanceau, Elaine, and Aureliano Sampaio. *Mulheres portuguesas no ultramar*. Porto: Livraria Civilização Editoria, 1979.

Sandhu, Kernial Singh, and Paul Wheatley, eds. *Melaka: The Transformation of a Malay Capital c. 1400–1980*. 2 vols. Oxford: Oxford University Press, 1983.

Santos, Diego. "A Note on the Syriac and Persian Sources of the Pharmacological Section of the *Youyang Zazu*." *Collectanea Christiana Orientalia* 7 (2010): 217–29.

Santo-Tomás, Enrique García. "'Offspring of the Mind': Childbirth and Its Perils in Early Modern Spanish Literature." In *Medical Cultures of the Early Modern Spanish Empire*, edited by John Slater, Maríaluz López-Terrada, José Pardo-Tomás, 149–66. London: Routledge, 2014.

Sarabia, Angélica Morales. "The Culture of Peyote: Between Divination and Disease in Early Modern New Spain." In *Medical Cultures of the Early Modern Spanish Empire*, edited by John Slater, Maríaluz López-Terrada, and José Pardo-Tomás, 21–40. London: Routledge, 2014.

Sargent, Matthew. "Global Trade and Local Knowledge: Gathering Natural Knowledge in Seventeenth-Century Indonesia." In *Intercultural Exchange in Southeast Asia: History and Society in the Early Modern World*, edited by Tara Alberts and D. R. M Irving, 144–60. London: IB Tauris, 2013.

Sarı, Nil, Ramazan Tuğ, Ahmet Zeki Izgöer, and Mehmet Yahya Okutan. *Klasik Dönem Ilaç Hazırlama Yöntemleri ve Terkipleri*. Istanbul: Novartis, 2003.

Schäfer, Dagmar. *The Crafting of the 10,000 Things: Knowledge and Technology in Seventeenth-Century China*. Chicago: University of Chicago Press, 2011.

Schäfer, Dagmar, ed. *Cultures of Knowledge: Technology in Chinese History*. Leiden: Brill, 2010.

Schafer, Edward H. *The Golden Peaches of Samarkand: A Study of T'ang Exotics*. Berkeley: University of California Press, 1963.

Schiebinger, Londa, and Claudia Swan, eds. *Colonial Botany: Science, Commerce, and Politics in the Early Modern World*. Philadelphia: University of Pennsylvania Press: 2005.

Schlanger, Nathan. "Mindful Technology: Unleashing the Chaîne Opératoire for an Archaeology of Mind." In *The Ancient Mind: Elements of Cognitive Archeology*, edited by Colin Renfrew and Ezra B. W. Zubrow, 143–51. Cambridge: Cambridge University Press, 1994.

Schopen, Gregory. "On Monks, Nuns and 'Vulgar' Practices: The Introduction of the Image Cult into Indian Buddhism." In *Bones, Stones, and Buddhist Monks: Collected Papers on the Archaeology, Epigraphy, and Texts of Monastic Buddhism in India*, edited by Gregory Schopen, 238–57. Honolulu: University of Hawai'i Press, 1997.

Schottenhammer, Angela. "Transfer of *Xiangyao* 香藥 from Iran and Arabia to China: A Reinvestigation of Entries in the *Youyang zazu* 酉陽雜俎 (863)." In *Aspects of the Maritime Silk Road: From the Persian Gulf to the East China Sea*, edited by Ralph Kauz, 117–49. Wiesbaden: Harrassowitz Verlag, 2010.

Schottenhammer, Angela. "Yang Liangyao's Mission of 785 to the Caliph of Baghdād: Evidence of an Early Sino-Arabic Power Alliance?" *Bulletin de l'École française d'Extrême-Orient* 101 (2015): 177–242.

Seidel, Anna. "Imperial Treasures and Taoist Sacraments: Taoist Roots in the Apocrypha." In *Tantric and Taoist Studies in Honour of R. A. Stein*, vol. 2, edited by Michel Strickman, 297–302. Bruxelles: Institut Belge des hautes études chinoises, 1983.

Seidemann, Johannes. *World Spice Plants: Economic Usage, Botany, Taxonomy*. Berlin: Springer-Verlag, 2005.

Selvakumar, V. "Contacts between India and Southeast Asia in Ceramic and Boat Building Traditions." In *Early Interactions between South and Southeast Asia: Reflections on Cross-Cultural Exchange*, edited by Pierre-Yves Manguin, A. Mani, and Geoff Wade, 197–220. Singapore: Institute of Southeast Asian Studies, 2011.

Şen, Ahmet Tunç. "Reading the Stars at the Ottoman Court: Bāyezīd II (r. 886/1481–918/1512) and His Celestial Interests." *Arabica* 64 (2017): 557–608.

Sen, Samita. "Commercial Recruiting and Informal Intermediation: Debate over the Sardari System in Assam Tea Plantations, 1860–1900." *Modern Asian Studies* 44, no. 1 (2010): 3–28.

Sen, Tansen. "Astronomical Tomb Paintings from Xuanhua: Maṇḍalas?" *Ars Orientalis* 29 (1999): 29–54.

Sen, Tansen. *Buddhism across Asia: Networks of Material, Intellectual and Cultural Exchange*. Vol. 1. Singapore: Institute of Southeast Asian Studies and Manohar Publishers, 2014.

Sen, Tansen. *Buddhism, Diplomacy, and Trade: The Realignment of Sino-Indian Relations, 600–1400*. Honolulu: University of Hawai'i Press, 2003.

Sen, Tansen. "The Formation of Chinese Maritime Networks to Southern Asia, 1200–1450." *Journal of the Economic and Social History of the Orient* 49, no. 4 (2006): 421–53.

Sen, Tansen. "The Impact of Zheng He's Expeditions on Indian Ocean Interactions." *Bulletin of the School of Oriental and African Studies* 79, no. 3 (2016): 609–36.

Sen, Tansen. "The Military Campaigns of Rajendra Chola and the Chola-Srivijaya-China Triangle." In *Nagapattinam to Suvarnadwipa: Reflections on the Chola Naval Expeditions to Southeast Asia*, edited by Hermann Kulke, K. Kesavapany, and Vijay Sakhuja, 61–75. Singapore: Institute of Southeast Asian Studies, 2009.

Sen, Tansen. "Temple Heritage of a Chinese Migrant Community: Movement, Connectivity, and Identity in the Maritime World." In *Travelling Pasts: The Politics of Cultural Heritage in the Indian Ocean World*, edited by Burkhard Schnepel and Tansen Sen. Leiden: Brill, forthcoming.

Sen, Tansen. "Yijing and the Buddhist Cosmopolis of the Seventh Century." In *Texts and Transformations: Essays in Honor of the 75th Birthday of Victor H. Mair*, edited by Haun Saussy, 345–68. Amherst: Cambria Press, 2018.

Sevea, Terenjit Singh. "Pawangs on the Malay Frontier: Miraculous Intermediaries of Rice, Ore, Beasts and Guns." Unpublished PhD thesis, University of California, Los Angeles, 2013.

Shafri, Mohd. "Makkah and Medicine in the Late Ottoman Period from a Malay Medical Treatise." *Journal of the International Society for the History of Islamic Medicine* 14–15 (2015–2016): 146–53.

Shah, N., and Amir Zare. "Asafoetida (heeng): The Well-Known Medicinal-Condiment of India and Iran." *Scitech Journal* 1, no. 4 (2014): 30–36.

Shaked, Shaul. "Judaeo-Persian Letter from Dandan-Uiliq." In *The Silk Road: Trade, Travel, War and Faith*, edited by Susan Whitfield, 221–22. London: The British Library, 2004.

Sharma, Jayeeta. "British Science, Chinese Skill and Assam Tea: Making Empire's Garden." *Indian Economic Social History Review* 43 (2006): 429–55.

Sharma, Jayeeta. "'Lazy' Natives, Coolie Labour and the Assam Tea Industry." *Modern Asian Studies* 43, no. 6 (2009): 1287–324.

Shaw, Stanford J. *History of the Ottoman Empire and Modern Turkey*. Vol. 1. Cambridge: Cambridge University Press, 1976.

Shefer-Mossensohn, Miri, and Keren Abou-Hershkovitz. "Early Muslim Medicine and the Indian Context: A Reinterpretation." *Medieval Encounters* 19, no. 3 (2013): 274–99.

Shigeo, Kamata. "The Transmission of Paekche Buddhism to Japan." In *Introduction of Buddhism to Korea: New Cultural Patterns*, edited by Lewis R. Lancaster and C. S. Yu, 143–60. Berkeley: Asian Humanities Press, 1989.

Shi Sengyou 釋僧佑. *Chu Sanzang ji ji* 出三藏記集 (Compilation of notes on the translation of the Tripitaka). 6th century. T55, no. 2145, CBETA.

Shi song lü 十誦律. Translated from Sarvāstivāda-Vinaya, *Ten Recitations Vinaya*. T23, no. 1435, CBETA.

Sibu yidian 四部醫典. Translated from *rGyud-bzhi, The Four Tantras*. Translated by Li Yongnian. Beijing: Renmin weisheng chubanshe, 1983.

Siddiqi, M. Zubair. "An Early Arabian Author, on the Indian System of Medicine." *Calcutta Review*, 3rd ser., 41 (1931): 277–83. Reprinted in *'Alī Ibn Rabban al-Ṭabarī (d.c. 250/864): Texts and Studies*, collected and reprinted by Fuat Segzin, *Islamic Medicine*, 30:145–51. Frankfurt am Main: Institute for the History of Arabic-Islamic Science at the Johann Wolfgang Goethe University, 1996.

Sima Qian. *Records of the Grand Historian of China: Translated from the Shih chi of Ssu-ma Ch'ie by Burton Watson*. 2 vols. New York: Columbia University Press, 1971.

Sims-Williams, Ursula. "Hoernle, Augustus Frederic Rudolf." *Encyclopædia Iranica* 12, no. 4 (2004): 418–20. http://iranicaonline.org/articles/hoernle-augustus-frederic-rudolf (accessed November 9, 2017).

Skaff, Jonathan Karam. "Sasanian and Arab-Sasanian Silver Coins from Turfan: Their Relationship to International Trade and the Local Economy." *Asia Major*, 3rd ser., 11, no. 2 (1998): 67–115.

Smith, Pamela H. "Alchemy as a Language of Mediation at the Habsburg Court." *Isis* 85 (1994): 1–25.

Smith, Pamela H. *The Body of the Artisan: Art and Experience in the Scientific Revolution*. Chicago: University of Chicago Press, 2004.

Smith, Pamela H. *The Business of Alchemy: Science and Culture in the Holy Roman Empire*. Princeton, NJ: Princeton University Press, 1994.

Smith, Pamela H. "Itineraries of Materials and Knowledge in the Early Modern World." In *The Global Lives of Things*, edited by Anne Gerritsen and Giorgio Riello, 31–61. New York: Routledge, 2015.

Smith, Pamela H. "Knowledge in Motion: Following Itineraries of Matter in the Early Modern World." In *Cultures in Motion*, edited by Daniel T. Rodgers, Bhavani Raman, and Helmut Reimitz, 109–33. Princeton, NJ: Princeton University Press, 2014.

Smith, Pamela H. "Laboratories." In *The Cambridge History of Science*. Vol. 3 (Early Modern Science), edited by Katherine Park and Lorraine Daston, 290–304. Cambridge: Cambridge University Press, 2008.

Smith, Pamela H. "Science." In *A Concise Companion to History*, edited by Ulinka Rublack, 268–97. Oxford: Oxford University Press, 2011.

Smith, Pamela H. "Vermilion, Mercury, Blood, and Lizards: Matter and Meaning in Metalworking." In *Materials and Expertise in Early Modern Europe: Between Market and Laboratory*, edited by Ursula Klein and E. C. Spary, 29–49. Chicago: Chicago University Press, 2010.

Smith, Paul J. *Taxing Heaven's Storehouse: Horses, Bureaucrats, and the Destruction of the Sichuan Tea Industry, 1074–1224*. Cambridge, MA: Harvard University Press, 1991.

Smith, W. L. "'Chinese Sugar': On the Origin of Hindi Cīnī, 'Sugar.'" *Indologica* 12 (1984): 225–32.

Soledade, Fr. Fernando da. *Historia Serafica Chronologica da Ordem de S. Francisco na Provincia de Portugal*. Vol. 5: *Refere os seus progressos em tempo de cento & quarenta & seis annos, do de 1569. atè o de 1715. aos quaes ajuntou as memorias dos tres feguintes* (He reports his progress in a time of one hundred and forty-six years, from 1569, until the year 1715, to which he added the memoirs of the following three). Lisbon: Officina de Antonio Pedrozo Galram, 1721.

Song shi 宋史 (Official history of the Song dynasty). 4 vols. Taibei: Dingwen Publishers, 1980.

Song Xian 宋峴. *Huihui yaofang kaoshi* 回回藥方考釋 (A study on Islamic medicinal recipes). 2 vols. Beijing: Zhonghua shuju, 2000.

Song Yingxing 宋應星. *Tiangong kaiwu* 天工開物 (The works of heaven and the inception of things). Yangzhou: Jiangsu guanglin guji keyinshe, 1997.

Song Yingxing 宋應星. *Tiangong kaiwu* 天工開物 (The works of heaven and the inception of things). Edited by Guan Qiaoling 管巧灵 and Tan Shuchun 谭属春. Changsha: Yuelu shushe, 2002.

Souza, Marie Reynolds de, Teresa Joaquim, Dina Canço, Pedro Pires, and Isabel de Castro, eds. *O rosto feminino da expansão portuguesa, Congresso internacional realizado em Lisboa, Portugal 21–25 novembro de 1994, Actas*. Vol. 2. Lisbon: Comissão para a Igualidade e para os Direitos das Mulheres, 1995.

Starr, S. Frederick. *Ferghana Valley: The Heart of Central Asia*. Hoboken: Taylor and Francis, 2014.

Starr, S. Frederick. *Lost Enlightenment: Central Asia's Golden Age from the Arab Conquest to Tamerlane*. Princeton, NJ: Princeton University Press, 2013.

Starr, S. Frederick, ed. *The New Silk Roads: Transport and Trade in Greater Central Asia*. Washington, DC: Central Asia-Caucasus Institute and Silk Road Studies Program, 2007.

Starr, S. Frederick, Baktybek Beshimov, Inomjon I. Bobokulov, and Pulat Shozimov, eds. *Ferghana Valley: The Heart of Central Asia*. Armonk, NY: M. E. Sharpe, 2011.

Stein, Aurel. *On Central-Asian Tracks: Brief Narrative of Three Expeditions in Innermost Asia and North-Western China*. London: Macmillan, 1933.

Stein, Rolf. *Rolf Stein's Tibetica Antiqua: With Additional Materials*. Edited and translated by Arthur Mckeown. Leiden: Brill, 2010.

Stein, Rolf. "Saint et divin, un titre tibétain et chinois des rois tibétains." *Journal asiatique* 269 (1981): 231–75.

Stein, Sarah Abrevaya. "Creating a Taste for News: Historicizing Judeo-Spanish Periodicals of the Ottoman Empire." *Jewish History* 14, no. 1 (2000): 9–28.

Stern, E. Marianne. "Early Roman Export Glass in India." In *Rome and India: The Ancient Sea Trade*, edited by Vimala Begley and Richard D. De Puma, 23–36. Madison: University of Wisconsin Press, 1991.

Stewart, Rory. *The Places in Between*. Orlando, FL: Harcourt, 2006.

Stolberg, Michael. "'You Have No Good Blood in Your Body': Oral Communication in Sixteenth-Century Physicians' Medical Practice." *Medical History* 59 (2015): 63–82.

Stoler, Ann Laura. *Carnal Knowledge and Imperial Power: Race and the Intimate in Colonial Rule*. 2nd ed. Berkeley: University of California Press, 2010.

Stratmann, F. H., ed. *The Tragicall History of Hamlet, Prince of Denmarke by William Shakespeare*. London: N. Trübner, 1869.

Strohmaier, Gotthard. "Al-Djildaki." *The Encyclopaedia of Islam*. New ed. Vol. 12 supplement. London: E. J. Brill, 2004.

Subrahmanyam, Sanjay, ed. *Explorations in Connected History: From the Tagus to the Ganges*. New York: Oxford University Press, 2005.

Subrahmnayam, Sanjay. "A Note on the Rise of Surat in the Sixteenth Century." *Journal of the Economic and Social History of the Orient* 43, no. 1 (2000): 23–33.

Su E 蘇鶚. *Duyang za bian* 杜陽雜編 (Compiling miscellanies at Duyang). Repr., Beijing: Zhonghua shuju, 1985.

Sugidatsu Yoshikazu 杉立義一. "Edo jidai kaibō no jiseki to sono hankyō" 江戸時代解剖の事跡とその反響 (The acts of dissections in the Edo period and their responses). In *Rekishi chū no byō to igaku* 歴史中の病と医学 (Illnesses and medicine in history), edited by Yamada Keiji 山田慶児 and Kuriyama Shigehisa 栗山茂久, 503–43. Kyoto: Shibungaku, 1997.

Sugimoto Tsutomu 杉本つとむ. "Kindai igaku no genryū 2: Oranda igaku no so, Yoshio Kōgyū" 近代医学の源流 2：オランダ医学の祖・吉雄耕牛 (The origins of modern medicine 2: The founder of Dutch-style medicine, Yoshio Kōgyū). *Nihon ishikai zasshi* 109, no. 2 (1992): 1670–72.

Sugita Genpaku 杉田玄白. *Kyōi no gen* 狂医之言 (A wild physician's words). In *Nihon shisō taikei* 日本思想大系 (Great series of Japanese thoughts), edited by Ienaga Saburō 家永三郎, 64:227–43. Tokyo: Iwanami shoten, 1982.

Suh Soyoung. "Herbs of Our Own Kingdom: Layers of the Local in the Materia Medica of Early Chosŏn Korea." *Asian Medicine: Tradition and Modernity* 4 (2009): 395–422.

Sui shu 隋書 (Official history of the Sui dynasty). Beijing: Zhonghua shuju, 1973.

Sung, Ying-hsing [Song Yinxing]. *T'ien kung k'ai-wu: Chinese Technology in the Seventeenth Century*. Translated from the Chinese and annotated by E-Tu Zen Sun and Shiou-chuan Sun. Mineaola, NY: Dover, 1966/1997.

Sun Simiao 孫思邈. *Beiji qianjin yaofang*備急千金要方 (Essential recipes for urgent use worthy of a thousand gold pieces). Early 7th century. Beijing: Beijing renmin weisheng chubanshe, 1982. Reprint of the Edo igaku facsimile of the Northern Song edition.

Sun Simiao 孫思邈. *Qianjin yifang*千金翼方 (Supplementary recipes worthy of a thousand gold pieces). Late 7th century. Beijing: Renmin weisheng chubanshe, 1955. Based on the 1307 edition.

Sun Yirang 孫詒讓. *Kaogong ji* 考工記. Chuxue she 楚學社 facsimile edition. In *Zhongguo kexue jishu dianji tonghui* 中國科學技術典籍通彙. Zhengzhou: Henan jiaoyu chubanshe, 1993.

Suo Yuming 索予明. "Tihong kao 剔红考." *Gugong jikan* 3 (1972): 6.

Su Song 蘇頌. *Bencao tujing* 本草圖經 (Illustrated classic of materia medica, 1061). Hefei: Anhui kexue chubanshe, 1994.

Suzuki Takao 鈴木隆雄. *Hone kara miru Nihonjin* 骨から見る日本人 (The Japanese people seen from the aspect of bones). Tokyo: Kodansha, 1998.

Tagliacozzo, Eric. *The Longest Journey: Southeast Asians and the Pilgrimage to Mecca*. New York: Oxford University Press, 2013.

Takata Tokio. "Multilingualism in Tun-huang." *Acta Asiatica: Bulletin of the Institute of Eastern Culture: Tun-huang and Turfan Studies* 78 (2000): 49–70.

Takeda Kazuaki 武田和昭. *Hoshi mandara no kenkyū*星曼荼羅の研究 (Research on the Star Maṇḍalas). Kyoto: Hōzōkan, 1995.

Takeuchi Tsuguhito. "Sociolinguistic Implications of the Use of Tibetan in East Turkestan from the End of Tibetan Domination through the Tangut Period (9th–12th c.)." In *Turfan Revisited: The First Century of Research into the Arts and*

Cultures of the Silk Road, edited by D. Durkin-Meisterernst et al., 341–48. Berlin: Dietrich Reimer, 2004.

Takeuchi Tsuguhito and Ai Nishida. "The Present Stage of Deciphering Old Zhang-zhung." In *Issues in Tibeto-Burman Historical Linguistics*, edited by Yasuhiko Nagano, 151–65. Senri Ethnological Studies 75. Osaka: National Museum of Ethnology, 2009.

Tanabe, George J. Jr., and Willa Jane Tanabe, eds. *The Lotus Sutra in Japanese Culture*. Honolulu: University of Hawai'i Press, 1989.

Tanba Yasuyori 丹波康賴. *Ishinpò* 醫心方 (Essential medical recipes, 984). Shenyang: Liaoning kexue jishu chubanshe, 1996.

Tang Kaijian 湯開建 and Tian Yu 田渝. "Wanli sishiwu nian Tian Shengjin 'Bao Qianluo guo jingong shu yanjiu" 萬曆四十五年田生金 "報暹羅國進貢疏" 研究 (A study on the document by Tian Shengjin, "Memorial on tributes by the Siam state" dated 1617). *Ji'nan daxue xuebao* 4 (2007): 123–34.

Tang Shenwei 唐慎微. *Chongxiu zhenghe jingshi zhenglei beiyong bencao* 重修政和經史證類備用本草 (Materia medica for emergency uses, with evidences from classics and historical records, revised during Zhenghe Reign). Beijing: Renmin weisheng chubanshe, 1982.

Tang Shenwei 唐慎微. *Chongxiu zhenghe jingshi zhenglei beiyong bencao* 重修政和經史證類備用本草 (Materia medica for emergency uses, with evidences from classics and historical records, revised during Zhenghe Reign). Edited by Shang Zhijun. 1082. Beijing: Huaxia chubanshe, 1993.

Taşğın, Ahmet. *Horasandan Balkanlara Ulaşan Eren Koyun Baba ve Menakıbı*. Istanbul: Çizgi Yayınları, 2017.

Taylor, Gail Marlow. *The Alchemy of Al-Razi: A Translation of the "Book of Secrets."* s.l.: CreateSpace Independent Publishing Platform, 2014.

Taylor, Jean Gelman. *The Social World of Batavia: Europeans and Eurasians in Colonial Indonesia*. 2nd ed. Madison: University of Wisconsin Press, 2009.

Tea Gazette. *Tea Planter's Vade Mecum*. Calcutta: Office of the Tea Gazette, 1885. Reprint, London: W. B. Whittingham, 1886.

Tek, Abdurrezzak. "Mevleviliğin Bursa'daki Izleri." *Uludağ Üniversitesi Ilahiyat Fakültesi Dergisi* 15 (2006): 1–27.

Tek, Abdurrezzak. "Osmanlı Tasavvuf Tarihi Kaynağı Olarak Bursa Vefeyâtnâmelerinden *Ricâl-i Zeyniyye*." *Uludağ Üniversitesi Ilahiyat Fakültesi Dergisi* 13 (2004): 215–30.

Theisen, Wilfrid R. "The Attraction of Alchemy for Monks and Friars in the 13th–14th Centuries." *American Benedictine Review* 46 (1995): 239–53.

Thompson, Charles John Samuel. *Alchemy and Alchemists*. Mineola, NY: Dover Publications, 2002. Published previously as *The Lure and Romance of Alchemy*. London: G. G. Harrap, 1932.

"Tianbao er-nian Jiaohe jun shigu an" 天寶二年交河郡市估案 (Price list of Jiaohe county, year 743). *Turfan Fragments of Price Lists*. Otani 3043.

Torii Yoshio 鳥居良夫 ed. *Nihon seikotsujutsu zensho* 日本整骨術全書 (Complete books of bone setting in Japan). Tokyo: Azusagawa shobō, 1981.

Trainor, Kevin. *Relics, Ritual, and Representation in Buddhism: Rematerializing the Sri Lankan Theravada Tradition*. Cambridge: Cambridge University Press, 1997.

Trinidade, Paulo da. *Conquista espiritual do Oriente* (Spiritual conquest of the Orient). Edited by Félix Lopes. 3 vols. Lisbon: Centro de Estudos Históricos Ultramarinos, 1962.

Trivellato, Francesca. *The Familiarity of Strangers: The Sephardic Diaspora, Livorno, and Cross-Cultural Trade in the Early Modern Period*. New Haven: Yale University Press, 2009.

Türk, Gönül Gülşen. "Tasavvuf Kültüründe Derviş-Kitap Münasebeti ve Tekke Kütüphaneleri." Masters dissertation, Uludağ University, Bursa, Turkey, 1995.

Twitchett, Denis C., and Frederick W. Mote. *The Cambridge History of China*. Vol. 8. Cambridge: Cambridge University Press, 1998.

Tythacott, Louise. *The Lives of Chinese Objects: Buddhism, Imperialism and Display*. New York: Berghahn Books, 2011.

Uchiie Mikiito 氏家幹人. Ōedo shitai kō 大江戸死体考 (A study of corpses in Great Edo). Tokyo: Heibonsha, 1994.

Unan, Fahri. *Kuruluşundan Günümüze Fâtih Külliyesi*. Ankara: Türk Tarih Kurumu, 2003.

Unschuld, Ulrike. "Traditional Chinese Pharmacology: An Analysis of Its Development in the 13th Century." *ISIS* 68, no. 242 (1977): 224–48.

Ünver, A. Süheyl. "Hekim Vesim Abbas Efendiyi Ruhan Yetiştiren Kimdir?" *Dirim* 22 (1947): 76–77.

Uray, Géza. "The Earliest Evidence of the Use of the Chinese Sexagenary Cycle in Tibetan." In *Tibetan and Buddhist Studies Commemorating the 200th Anniversary of the Birth of Alexander Csoma de Kőrös*, edited by Louis Ligeti, 2:341–60. 2 vols. Budapest: Akademiai Kiado, 1984.

Uray, Géza. "L'emploi du tibétain dans les chancelleries des états du Kan-sou et de Khotan postérieurs à la domination tibétaine" (The use of Tibetan in the chancelleries of the states of Gansu and Khotan after the Tibetan Domination). *Journal Asiatique* 269 (1981): 81–90.

Uray, Géza. "The Old Tibetan Sources of the History of Central Asia up to 751 A.D.:

A Survey." In *Prolegomena to the Sources on the History of Pre-Islamic Central Asia*, edited by Harmatta János, 275–304. Budapest: Akadémiai Kiadó, 1979.

Uzluk, Feridun Nafiz. "Bursalı Hekim Ali Münşi Efendi ?–1146 (?–1733/34)." *Ankara Universitesi Dil ve Tarih Coğrafya Fakültesi Dergisi* 8 (1950): 329–37.

Uzluk, Feridun Nafiz. "Bursalı Tabib Mevlevi Ömer Şifai." *Dirim* 25 (1950): 170–76.

Vaissière, Étienne de la. *Sogdian Traders: A History*. Leiden: Brill, 2005.

Vallery-Radot, Pierre. "Un médécin diplomate: Le docteur Alexandre Mavrocordato (1641–1709)." *La Presse Médicale* 68 (1960): 2349–51.

Vallery-Radot, Pierre. "Un puissant personnage, Alexandre Maurocordato, pacificateur et médecin (1641–1709), grand interprète de la porte sous le règne de Louis XIV (d'après l'ouvrage de la Princesse Bibesco 'Mes vies antérieures')." *Histoire de la Médecine* 10 (1960–1961): 32–37.

Van Bladel, Kevin. "The Arabic History of Science of Abū Sahl ibn Nawbaḥt (fl. ca. 770–809) and Its Middle Persian Sources." In *Islamic Philosophy, Science, Culture, and Religion: Studies in Honor of Dimitri Gutas*, edited by Felicitas Opwis and David Reisman, 41–62. Leiden: Brill, 2012.

Van Bladel, Kevin. "The Bactrian Background of the Barmakids." In *Islam and Tibet: Interactions along the Musk Route*, edited by Anna Akasoy, Charles Burnett, and Ronit Yoeli-Tlalim, 43–88. Farnham: Ashgate, 2011.

Van Gulik, R. H., trans. *Mi Fu on Ink-Stones*. Peking: Henry Vetch, 1936.

Van Schaik, Sam. "Secrets of the Cave." *Early Tibet* (blog). http://earlytibet.com (accessed November 9, 2017).

Van Schaik, Sam, and Imre Galambos. *Manuscripts and Travellers: The Sino-Tibetan Documents of a Tenth-Century Buddhist Pilgrim*. Berlin: De Gruyter, 2012.

Vida, Marco Girolamo. *The silkworm: a poem. In two books. Written by Marcus Hieronymus Vida, and translated into English verse by the Reverend Samuel Pullein, of Trinity College, Dublin, A. M.* Translated by Samuel Pullein. Dublin: S. Powell, 1750.

Villiers, John. "Aceh, Melaka and the 'Hystoria dos cercos de Malaca' of Jorge de Lemos." *Portuguese Studies* 17 (2001): 75–85.

Vogel, Hans Ulrich, and Sabine Hieronymus. "Cowry Trade and Its Role in the Economy of Yunnan: From the Ninth to Mid-Seventh Century (Part I)." *Journal of the Economic and Social History of the Orient* 36, no. 3 (1993): 211–52.

Vogel, Hans Ulrich, and Sabine Hieronymus. "Cowry Trade and Its Role in the Economy of Yunnan: From the Ninth to the Mid-Seventh Century (Part II)." *Journal of the Economic and Social History of the Orient* 36, no. 3 (1993): 309–53.

Wade, Geoff. "An Early Age of Commerce in Southeast Asia, 900–1300 CE." *Journal of Southeast Asian Studies* 40, no. 2 (2009): 221–65.

Wade, Geoff. "The Zheng He Voyages: A Reassessment." *Malaysian Branch of the Royal Asiatic Society* 78, no. 1 (2005): 37–58.

Waley, Arthur. *The Analects of Confucius*. London: George Allen and Unwin, 1938.

Walker, John. *Memorandum*. In *The Measures Adopted*, compiled by Sir William Bentinck, 10–11. 1834.

Walter, Mariko Namba. "Sogdians and Buddhism." *Sino-Platonic Papers* 174 (November 2006): 1–66.

Wang, Michelle C. "Early Chinese Buddhist Sculptures as Animate Bodies and Living Presences." *Ars Orientalis* 46 (2016): 13–38.

Wang Chong 王充. *Lunheng* 論衡. *Sibu beiyao* edition. Taipei: Taiwan zhonghua shuju, 1976.

Wang Fuzhi 王夫之. *Zhouyi waizhuan* 周易外傳. Beijing: Beijing chubanshe, 1999.

Wang Huaiyin 王懷隱, ed. *Taiping shenghui fang*太平聖惠方 (Recipes of sacred benevolence under the great peace). 10th century. Beijing: Renmin weisheng chubanshe, 1958.

Wang Jingfen 王靜芬 (Dorothy Wong). "Zuowei shishenjizhe Wenshu pusa ji qi tuxiang kaoshi: Zhongguo yu Yindu de zaoqi pusa tuxiang" 作為使神跡者文殊菩薩及其圖像考釋: 中國與印度的早期菩薩圖像 (Iconography of the wonder-working Mañjuśrī: Early representations of the bodhisattva in China and India)." In *Wutai shan xinyang duo wenhua, kua zongjiao di xingge yiji guoji xing yingxiang li: Di er ci Wutai shan yantaohui lunwenji* 五臺山信仰多文化、跨宗教的性格以及國際性影響力:第 二次五臺山研討會論文集 (The Wutai shan cult, its multiculturality, cross-religious characteristics and international impact: Proceedings of the Second International Conference on Wutai shan), edited by Venerable Miaojiang 釋妙江 et al., 530–50. Taipei: Xinwenfeng chubanshe, 2018.

Wang Qi 王圻. *Sancai tuhui* 三才圖會 (Collected illustrations of the three realms). *Xuxiu siku quanshu* edition. Shanghai: Shanghai guji chubanshe, 1995–2002.

Wang Shizhen 王使臻. "E cang wenxian Дх. 02822 'zi shu' di laiyuan ji xiangguan wenti" 俄藏文獻Дх. 02822 "字書" 的來源及相關問題 (The origin and related questions on the primer Дх. 02822 kept in Russia). *Xixia xue* 5 (2010): 117–25.

Wang Tao 王燾. *Waitai miyao*外臺秘要 (Secret essentials from the imperial collections). Ca. 752. Osaka: Toyo igaku, 1981.

Wang Zhen. *Nongshu* (Agricultural treatise), 1313. Modern ed. Wang Yuhu, *Wang Zhen Nongshu* (The agricultural treatise of Wang Zhen). Beijing: Agriculture Press, 1981.

Watson, J. F. W. *Essay on the Cultivation and Manufacture of Tea*. Calcutta: T. Black, 1872.

Watt, George. *Dictionary of the Economic Products of India*. Vol. 3. Calcutta, 1889–1896. Reprint, Delhi: Periodical Expert, 1972.

Waugh, Daniel C. "Richthofen's 'Silk Roads': Towards the Archaeology of a Concept." *The Silk Road* 5, no. 1 (2007): 1–10. http://silkroadfoundation.org/newsletter/vol5num1 (accessed November 9, 2017).

Whyte, Susan Reynolds, Sjaak van der Geest, and Anita Hardon, eds. *Social Lives of Medicines*. Cambridge: Cambridge University Press, 2002.

Widrlechner, Mark P. "History and Utilization of *Rosa damascene*." *Economic Botany* 35, no. 1 (1981): 42–58.

Wigen, Kären. "Oceans of History: Introduction." *American Historical Review* 111, no. 3 (2006): 717–21.

Willis, Oliver Rivington. *A Practical Flora for Schools and Colleges*. New York: American Book Company, 1894.

Wolper, Ethel Sara. *Cities and Saints: Sufism and the Transformation of Urban Space in Medieval Anatolia*. Pennsylvania: Pennsylvania State University Press, 2003.

Wong, Dorothy C. *Buddhist Pilgrim-Monks as Agents of Cultural and Artistic Transmission: The International Buddhist Art Style in East Asia, ca. 645–770*. Singapore: NUS Press, 2018.

Woodward, Hiram W. "Bianhong, Mastermind of Borobudur?" *Pacific World: Journal of the Institute of Buddhist Studies* 11(Fall 2009): 25–60.

Woodward, Hiram W. "Tantric Buddhism at Angkor Thom." *Ars Orientalis* 12 (1981): 57–67.

Wu Hung. "Buddhist Elements in Early Chinese Art (2nd and 3rd Centuries a.d.)." *Artibus Asiae* 47, nos. 3–4 (1986): 263–303, 305–52.

Wu Chongyao 伍崇曜. "Afterword." In Wu Lanxiu 吳蘭修, *Duanxi yanshi* 端溪硯史, *Congshu jicheng chubian*, 1. Beijing: Zhonghua shuju, 1991.

Wujastyk, D. "Ravigupta and Vāgbhaṭa." *Bulletin of the School of Oriental and African Studies, University of London* 48, no. 1 (1985): 74–78.

Xie Zhaozhi 謝肇淛. *Wu za zu* 五雜俎 (Five miscellaneous groups (of things)). 1616. Hefei: Huangshan shushe, 2008.

Xiong Zhaoming. "The Hepu Han Tombs and the Maritime Silk Road of the Han Dynasty." *Antiquity* 88 (2014): 1229–43.

Xuan Zang 玄奘. *Da Tang xiyu ji*大唐西域記 (Great Tang record on the Western Regions). T51, no. 2087, CBETA.

Xydopolous, Ioannis. "Alexander's Historians and the Alexander Romance: A Comparative Study of the Representation of India and Indians." In *Memory as History: The Legacy of Alexander in Asia*, edited by Himanshu Prabha Ray and Daniel T. Potts, 19–27. New Delhi: Aryan Books, 2007.

Yang Bin. *Cowrie Shells and Cowrie Money: A Global History*. Oxon: Routledge, Taylor & Francis Group, 2018.

Yang Bin. "Horses, Silver, and Cowries: Yunnan in Global Perspective." *Journal of World History* 15, no. 3 (2004): 281–322.

Yan Guirong 閆貴榮. "Qian yi Songdai Long you shangye maoyi" 淺議宋代隴右商業貿易 (Brief discussion on trade in the Western Long region of the Song period). *Yan'an daxue xuebao* 6 (2007): 103–7.

Yazberdiev, A. B. "The Ancient Merv and Its Libraries." *Journal of Asian Civilizations* 23, no. 2 (2000): 137–57.

Yediyıldız, Mehmet Asım. "Şer'iye Sicillerine Göre Bursa'nın Sosyo Ekonomik Yapısı, 1655–1658." Masters dissertation, Hacettepe University, Ankara, Turkey, 1988.

Ye Kangning 叶康寧. "Cong 'qi yi zai dao' dao 'qi ti dao yong': Mingdai gongyi zaowu sixiang zhuanbian zhengyi" 從 "器以載道" 到 "器體道用": 明代工藝造物思想轉變爭議 (From "qi yi zai dao" to "qi ti dao yong": The controversy over the transformation of the craftsmanship in the Ming dynasty). *Yishu yu shenghuo* 6 (2006): 79–80.

Yijing 義淨. *Nanhai jigui neifa zhuan* 南海寄歸內法傳 (A record of the Buddhist religion as practiced in India and the Malay archipelago). Ca. 671–695. Translated by J. Takakusu. New Delhi: Munshiram Manoharlal Publishers Pvt., 1998. T54, no. 2125, CBETA.

Yıldırım, Rıza. "Sunni Orthodox vs Shi'ite Heterodox? A Reappraisal of Islamic Piety in Medieval Anatolia." In *Islam and Christianity in Medieval Anatolia*, edited by Andrew Charles Spencer Peacock, Bruno de Nicola, and Sara Nur Yıldız, 287–307. Farnham: Ashgate, 2015.

Yoeli-Tlalim, Ronit. "Between Medicine and Ritual: Tibetan 'Medical Rituals' from Dunhuang." In *Tibetan and Himalayan Healing: An Anthology for Anthony Aris*, edited by Charles Ramble and Ulrike Roesler, 739–46. Kathmandu: Vajra Publications, 2015.

Yoeli-Tlalim, Ronit. "Central Asian Mélange: Early Tibetan Medicine from Dunhuang." In *Scribes, Texts, and Rituals in Early Tibet and Dunhuang*, edited by Brandon Dotson, Kazushi Iwao, and Tsuguhito Takeuchi, 53–60. Wiesbaden: Reichert Verlag, 2013.

Yoeli-Tlalim, Ronit. "On Urine Analysis and Tibetan Medicine's Connections with the West." In *Studies of Medical Pluralism in Tibetan History and Society*, edited by Sienna Craig, Mingji Cuomu, Frances Garrett, and Mona Schrempf, 195–211. Andiast, Switzerland: International Institute for Tibetan and Buddhist Studies GmbH, 2010.

Yoeli-Tlalim, Ronit. "Rashīd al-Dīn's *Life of the Buddha:* Some Tibetan Perspectives."

In *Rashīd al-Dīn: Agent and Mediator of Cultural Exchanges in Ilkhanid Iran*, edited by Anna Akasoy, Charles Burnett, and Ronit Yoeli-Tlalim, 197–211. London: The Warburg Institute; Turin: Nino Aragno Editore, 2013.

Yoeli-Tlalim, Ronit. *Reorienting Histories of Medicine: Encounters along the Silk Roads.* London: Bloomsbury, in press.

Yoeli-Tlalim, Ronit. "Review of Arun Bala, *The Dialogue of Civilizations in the Birth of Modern Science.*" *Annals of Science* 70, no. 4 (2013): 557–59.

Yoeli-Tlalim, Ronit. "Re-visiting 'Galen in Tibet.'" *Medical History* 56, no. 3 (July 2012): 355–65.

Yokkaichi Yasuhiro. "Chinese and Muslim Diasporas and the Indian Ocean Trade Network under Mongol Hegemony." In *The East Asian "Mediterranean": Maritime Crossroads of Culture, Commerce and Human Migration*, edited by Angela Schottenhammer, 73–102. Wiesbaden: Harrassowitz Verlag, 2008.

Yu Gengzhe. "Xu gu zhi di: yi xiang wenhua qishi fuhao de qianzhuan liuyi" (Land of sorcery: The shift of a symbol of cultural discrimination). In *Zhongguo shehui kexue* (Social sciences in China) 2 (2006): 191–204.

Yule, Henry, et al. *Hobson-Jobson: A Glossary of Colloquial Anglo-Indian Words and Phrases, and of Kindred Terms, Etymological, Historical, Geographical and Discursive.* London: J. Murray, 1903.

"Zaji shi yao yong zi" 雜集時要用字 (Glossary of today's miscellaneous common words). Dunhuang Document P 3391.

Zanning 贊寧. *Song gao seng zhuan*宋高僧傳 (Biographies of venerable monks compiled in the Song period). 10th century. T50, no. 2061, CBETA.

Zarcone, Thierry. "Qadiriyya." In *Dictionnaire de l'Empire Ottoman*, edited by François Georgeon, Nicolas Vatin, and Gilles Veinstein, 1003–4. Paris: Fayard, 2015.

Zeng Xiongsheng 曾雄生. *Zhongguo nongxue shi* 中國農學史 (History of agriculture in China). Fuzhou: Fujian renmin chubanshe, 2008.

Zhang Binlun, "A Preliminary Analysis of Scientific Development and Its Causes in Anhui Province during the Ming and Qing Dynasties." In *Chinese Studies in the History and Philosophy of Science and Technology*, edited by Fan Dainian and Robert S. Cohen, 327–44. Dordrecht: Springer, 1996.

Zhang Guangda and Rong Xinjiang. "Les noms du royaume de Khotan: Les noms d'ère et la lignée royale de la fin des Tang au début des Song" (The names of the kingdom of Khotan: The names of the era and the royal lineage from the end of the Tang to the beginning of the Song). In Vol. 3 of *Contributions aux études de Touen-Houang* (Contributions to the studies of Touen-Houang), edited by Michel Soymié, 23–46. Paris: École Française d'Extrême Orient, 1984.

Zhang Xing. "Buddhist Practices and Institutions of the Chinese Community in Kolkata." In *Buddhism across Asia: Networks of Material, Intellectual and Cultural Exchange*, edited by Tansen Sen, 429–57. Singapore: Institute of Southeast Asian Studies, 2015.

Zhao Ji 趙佶. *Sheng ji jing* 聖濟經 (Classic of imperial charity). Beijing: Renmin weisheng chubanshe, 1990. Based on the 1887 edition.

Zhao Rukua 趙汝适. *Zhu fan zhi jiaoshi*諸蕃志校釋 (Annotation of the chronicles of foreign states). Annotated by Yang Bowen. Beijing: Zhonghua shuju, 1996.

Zhouyi zhengyi 周易正義. *Shisanjing zhushu* edition. 1998. Reprint, Beijing: Zhonghua shuju, 2003.

Zhu Xi 朱熹. *Huian xiansheng zhuwengong wenji* 晦庵先生朱文公文集 (Collected Works of Master Hui'an, Zhu Wengong). Taipei: Taiwan shangwu yinshuguan, 1979.

Zieme, Peter. "Notes on Uighur Medicine, Especially on the Uighur Siddhasāra Tradition." *Asian Medicine: Tradition and Modernity* 3, no. 2 (2007): 308–22.

Zülfikar, Bedizel. "18. Yüzyıla ait Türkçe 'Müfredat' Kitapları ve Türk Tıp Tarihindeki Yeri." *Tıp Tarihi Araştırmaları* 7 (1998): 101–20.

Zürcher, Erik. *The Buddhist Conquest of China: The Spread and Adaptation of Buddhism in Early Medieval China*. 3rd ed. Leiden: Brill, 2007.

Contributors

Tara Alberts is a senior lecturer in early modern history at the University of York, United Kingdom. She previously held a research fellowship at Jesus College, Cambridge, and was a Max Weber Fellow at the European University Institute, Italy. Her research explores encounters and exchanges between Europe and Asia between c. 1500 and c. 1700. Her first book, *Conflict and Conversion* (Oxford University Press, 2012) explored conversion to Catholicism in sixteenth- and seventeenth-century Southeast Asia. The focus of her current project is the circulation of medicines and the exchange of ideas about health and healing between Europe and Asia.

Francesca Bray is a historian and anthropologist of science, technology, and medicine. Her works include *The Rice Economies* (1986); *Technology and Gender: Fabrics of Power in Late Imperial China* (1997); *Graphics and Text in the Production of Technical Knowledge in China* (2007); *Technology, Gender and History in Imperial China: Great Transformations Reconsidered* (2013); *Rice: Global Networks and New Histories* (2015); and *Moving Crops and the Scales of History* (in preparation). Professor Emerita of Social Anthropology at the University of Edinburgh, Bray is also Past President of the Society for the History of Technology.

Chang Che-chia is an associate research fellow at the Institute of Modern History, Academia Sinica. His major field is medical history in East Asia. He won the Zhu Kezhen Award in 2008 (from the International Society for History of Science, Technology, and Medicine) for his studies of the rhubarb trade between China, and the Islamic and European worlds. He continues his studies of pharmacy, is involved in projects on forensic medicine and ideas of toxicology in early modern China, and focuses now on the transmission of anatomy into China and Japan (from the sixteenth through the twentieth centuries).

Ming Chen is currently professor and head of the department of South Asian studies at Peking University. Since he was awarded a doctoral degree by Peking University in 1999, with a dissertation on Indian medical science, he has focused on the history of cultural communication between China and Central and South Asia in the medieval period, mainly but not exclusively on medicine. His academic interests have also extended to the transmission of materia medica in global history. He has published nine books in Chinese, including *Foreign Medicine and Culture in Medieval China* (2013).

Joslyn DeVinney is a PhD student in the Department of History at Columbia University. Her research is centered on histories of science and the materiality of knowledge making across early modern Eurasia, using Persian, French, and English sources. She is currently writing about British colonial historians of India and their translations of precolonial Persian scientific texts.

Sasha Grafit is a second-year student in the Master of Fine Arts (Creative Writing) program at Columbia University, where he also serves as the translations editor for the *Columbia Journal*. He is the author of *Ways to Upset Your Mother*, an illustrated, self-published book for children and mature adults, and he is currently writing a novel. He thanks Pamela Smith for introducing him to the study of history through the reconstruction of craft techniques.

Feza Günergun (born Baytop) is professor and head of the Department of History of Science, Istanbul University. Trained as a chemical engineer, her research includes the introduction of modern sciences to Turkey, and the histories of educational institutions, chemistry, scientific instruments, and metrology. She is the editor of the journal *Osmanlı Bilimi Araştırmaları* (*Studies in Ottoman Science*, Istanbul). Among the recent books she coedited are *Science between Europe and Asia* (2011, D. Raina) and *Entre Trois Mers – Cartographie Ottomane et française des Dardanelles et du Bosphore* (2016, J.-F. Pérouse). Gunergun curated the exhibition *Pursuing Knowledge* (2016) on scientific instruments and authored its catalogue.

A native of Hong Kong, **Dorothy Ko** is professor and chair of the Department of History at Barnard College, Columbia University. Her recent book, *The Social Life of Insktones: Artisans and Scholars in Early Qing China* (Washington, 2017), is a finalist for the Morey Prize of the College Art Association. She is also the author of *Cinderella's Sisters: A Revisionist History of Footbinding*, among others. A cultural historian of early modern China, she works at the intersection of ecology, art history, gender, and the history of technology.

Angela Ki Che Leung is director and chair professor of history of the Hong Kong Institute for the Humanities and Social Sciences at the University of Hong Kong. She works on medicine, food, and health in late imperial and modern China. Her recent books include *Leprosy in China: A History* (2009); *Health and Hygiene in Chinese East Asia: Policies and Publics in the Long 20th Century* (coedited with C. Furth, 2010); and *Gender, Health and History in Modern East Asia* (coedited with I. Nakayama, 2017). She is currently coordinating a collaborative project on everyday technologies and the making of modern East Asia.

Xiaomeng Liu is a PhD candidate in the Hong Kong Institute for the Humanities and Social Sciences at the University of Hong Kong. His research focuses on the history of medicine and material culture in early modern and modern China. He is currently working on a dissertation on the materia medica and pharmaceutical practices in nineteenth-century China.

Dagmar Schäfer is honorary professor of history of technology, at the Technical University, Berlin; adjunct professor at the Institute of Sinology, Freie Universität, Berlin; and guest professor, at both the School of History and Culture of Science, Shanghai Jiao Tong University, and Tianjin University, China. Her monograph *The Crafting of the 10,000 Things* (University of Chicago Press, 2011) won the History of Science Society Pfizer Award in 2012 and the Association for Asian Studies Joseph Levenson Prize (Pre-1900) in 2013. Her current interest is in the history and sociology of technology of China, focusing on the paradigms configuring the discourse on technological development, past and present.

Tansen Sen is director of the Center for Global Asia and professor of history, NYU Shanghai; Global Network Professor, NYU; and the author of *Buddhism, Diplomacy, and Trade: The Realignment of Sino-Indian Relations, 600–1400* (2003, 2016) and *India, China, and the World: A Connected History* (2017). He has coauthored (with Victor H. Mair) *Traditional China in Asian and World History* (2012) and edited *Buddhism across Asia: Networks of Material, Cultural and Intellectual Exchange* (2014). He is currently working on a book about Zheng He's maritime expeditions in the early fifteenth century and coediting (with Engseng Ho) the *Cambridge History of the Indian Ocean*, volume 1.

Pamela H. Smith is Seth Low Professor of History at Columbia University and founding director of the Center for Science and Society and of its cluster project The Making and Knowing Project (www.makingandknowing.org). Her *Body of the Artisan* (2004); *From Lived Experience to the Written Word: Recovering Skill and Art* (forthcoming); *Ways of Making and Knowing* (ed. with Amy R. W. Meyers and Harold Cook, paperback 2017); and *The Matter of Art* (ed. with Christy Anderson and Anne Dunlop, paperback 2016) all examine craft and practice. The Making and Knowing Project investigates the intersection of craft making and scientific knowing by text-, object-, and laboratory-based research on historical sources.

Ronit Yoeli-Tlalim is a senior lecturer in the History Department at Goldsmiths, University of London. She has coedited three volumes with Anna Akasoy and Charles Burnett: *Rashīd al-Dīn as an Agent and Mediator of Cultural Exchanges in Ilkhanid Iran* (2013); *Islam and Tibet: Interactions along the Musk Routes* (2011); and *Astro-Medicine: Astrology and Medicine, East and West* (2008). She has also coedited (with Vivienne Lo) the Silk Roads special issue of *Asian Medicine: Tradition and Modernity* (2007). Her book *ReOrienting Histories of Medicine: Encounters along the Silk Roads* is forthcoming with Bloomsbury.

Index

Note: Page numbers in italics indicate illustrative material.